50.00

Doyle

Integrative Approaches to Molecular Biology

Integrative Approaches to Molecular Biology

edited by Julio Collado-Vides, Boris Magasanik, and Temple F. Smith

The MIT Press
Cambridge, Massachusetts
London, England

This book was set in Palatino by Asco Trade Typesetting Ltd., Hong Kong and was printed and bound in the United States of America.

Library of Congress Cataloging-in-Publication Data

Integrative approaches to molecular biology / edited by Julio Collado-Vides,
  Boris Magasanik, and Temple F. Smith.
     p.   cm.
Consequence of a meeting held at the Center for Nitrogen Fixation,
National Autonomous University of Mexico, Cuernavaca, in Feb. 1994.
Includes bibliographical references and index.
ISBN 0-262-03239-2 (hc : alk. paper)
  1. Molecular biology—Congresses. I. Collado-Vides, Julio.
II. Magasanik, Boris. III. Smith, Temple F.
QH506.I483   1996
574.8′8—dc20   95-46156
                                                                    CIP

Contents

Preface

There are several quite distinct levels of integration essential to modern molecular biology. First, the field of molecular biology itself developed from the integration of methods and approaches drawn from traditional biochemistry, genetics, and physics. Today the methodologies employed encompass those drawn from the additional disciplines of mathematics, computer science, engineering, and even linguistics. There is little doubt that a discipline that employs such a range of methodologies, to say nothing of the range and complexity of the biological systems under study, will require new means of integration and synthesis. The need for synthesis is particularly true if the wealth of data being generated by modern molecular biology is to be exploited fully.

Few, if any, would claim that a complete description of all the individual molecular components of a living cell will allow one to understand the complete life cycle and environmental interactions of the organism containing that cell. On the other hand, without the near-complete molecular level description, is there any chance of understanding those higher-level behaviors at more than a purely descriptive level? Probably not, if molecular biology's mentor discipline of (classical) physics is a reasonable analog. Note that our current views of cosmology are built on a detailed knowledge of fundamental particle physics, including the details of the interaction of the smallest constituents of matter with light. The latter provided insight into the origin of the 2.7-degree background cosmological microwave radiation, which in turn provides one of the major supporting arguments for the so-called big bang theory. Similarly, our current views of the major biological theories of the evolutionary relatedness of all terrestrial life and its origins have been greatly enhanced by the growing wealth of molecular data. One has only to recall the introduction of the notion of neutral mutation or the RNA enzyme, the first notion leading to the concept of neutral evolution and, indirectly, to that of "punctuated equilibrium," and the second to that of the "RNA-world" origin. Even such classic biological areas as taxonomy have been greatly enhanced by our new molecular data. It is difficult to envision the future integration of our new wealth of information but, in the best sense of the word *reduce*, biology may well become a simpler, reduced set of ideas and concepts, integrating many of the emergent properties of living systems into understandable and analyzable units.

Undoubtedly, integration of biology must ultimately involve an evolutionary perspective. However, evolutionary theory does not yet provide an explicit conceptual system strong enough to explain the biological organization at the level of detail common in molecular biology—protein and DNA structure, gene regulation and organization, metabolism, cellular structure—briefly, the structure of organisms at the molecular level. This reveals another area from which the required integration clearly is missing—that of molecular biology as a discipline wherein theories have been rather limited in their effectiveness. What makes biological systems so impenetrable to theories?

Nobody doubts that organisms obey the laws of physics. This truism is, nonetheless, a source of misunderstanding: A rather naive attitude will demand that biology become a science similar to physics—that is, a science wherein theory plays an important role in shaping experiments, wherein predictions at one level have been shown to fit adequately with a large body of observations at a higher level—*the* successful natural science. But surely biology is limited in what it can achieve in this direction when faced with the task of providing an understanding of complex and historical systems such as cells, organs, organisms, or ecological communities.

To begin, physics itself struggles to explain physical phenomena that occur embedded within biological organisms. Such physical phenomena encompass a vast array of problems—for example, turbulence inside small elastic tubules (blood vessels); temperature transference; laws of motion of middle-size objects at the surface of the planet; chemical behavior of highly heterogeneous low-ionic-strength solutions; evaluation of activity coefficients of molecules in a heterogeneous mixture; interactions involving molecules with atoms numbering in the thousands and at concentrations of some few molecules per cell; and finding the energy landscape for protein folding. These phenomena are difficult problems for classical physics and have contributed to the development of new descriptive tools such as fractal geometries and chaos theory. In fact, they have little to do with the boundaries within which classical physics works best: ensembles with few interacting objects and very large numbers of identical objects.

Biological processes occur within physical systems with which simple predictive physics has trouble, processes that no doubt obey the laws of quantum and statistical physics but wherein prediction is not yet as simple as predicting the clockwise movement of stars. It is within such highly heterogeneous, rather compartmentalized, semiliquid systems that processes of interest to the biologist occur. As complex as they are, however, they are not at the core of what constitute the biological properties that help better to identify an organism as a biological entity—reproduction and differentiation, transference of information through generations; informational molecule processing, editing, and proofreading; and ordered chemical reactions in the form of metabolic pathways and regulatory networks.

This underlying physical complexity makes the analysis of biological organisms difficult. Perhaps a mathematical system to name and reveal plausible universal biological principles has not yet been invented. It also is plausible that a single formal method will not be devised to describe biological organisms, as is the aim of much of modern physics. The various contributions to this book illustrate how rich and diverse are the methods currently being developed and tested in pursuit of a better understanding and integration of biology at the molecular level.

Important difficulties also exist at the level of the main concepts that build the dominant framework of theory within biology. Recall, for instance, how important problems have been raised regarding the structure of evolutionary theory (Sober, 1984). In the first chapter of this book, Richard Lewontin offers a critique of the evolutionary process as one of engineering design. The idea that organs and biological structures have clearly identifiable functions—hands are made to hold, the Krebs cycle is made to degrade carbon sources—is already a questionable beginning for theory construction and has important consequences, as Lewontin shows. Much later in this book, Boris Magasanik illustrates how difficult it is for us to reconstruct, step-by-step, the origin of a complex interrelated system. "Which came first, the egg or the chicken?" seems to reflect our human limitations in considering the origin of organisms.

The study of the very rich internal structure of organisms and plausible avenues to unifying views for such structure and its associated dynamical properties at different levels of analysis is the subject of this book. The intuition that general atemporal rules must exist that partially govern the organization and functioning of biological organisms has supported a school rooted in the history of biology, for which the structure and its description has been the main concern. This school has been much less dominant since the emergence of evolution as the main framework of integration in biology (see Kauffman, 1993). A good number of perspectives addressed in this book can be considered part of such a tradition in biology.

One formal (at least more formal than biology) discipline that currently is more seriously dominant in the multifaceted marriages with other disciplines of which biology is capable is computer science (perhaps in part because of the underlying metaphor discussed in chapter 1). It should be clear, however, that this is not a book devoted to computational approaches to molecular biology. In fact, a sizable number of computational approaches currently applied to molecular biology are not represented in this book. (For an account of artificial intelligence in molecular biology, see Hunter, 1993.) How are the science of the artificial and the science of complexity (as computer science and artificial intelligence are self-identified) going to enrich molecular biology? A formal discipline that studies complex systems appears an attractive one to apply to biology, although promises in artificial intelligence—not only in molecular biology but also in the neurosciences and cognition—sometimes are too expansive and, historically, their effective goals have been modified over time (Dreyfus, 1992). Questions related to such issues are discussed in chapter 15 of this book by Robert Berwick, who draws on lessons from computational studies of natural language.

We do not attempt, with this book, to provide a complete account of integrative approaches to molecular biology. This text is the outgrowth of a meeting held at the Center for Nitrogen Fixation at the National Autonomous University of Mexico, in Cuernavaca, in February 1994. Sponsors for this workshop were the US National Science Foundation, the National Council for Science and Technology (México), and the Center for Nitrogen Fixation. Unfortunately, not all contributors to this book were at the meeting and not all participants in the workshop are represented in the book. Theoreticians, computer scientists, molecular biologists, and science historians gathered to discuss whether it is time to move into a more integrated molecular biology (and if so, how). As it was at the workshop, the challenge of this book is to show that different approaches to molecular biology, which employ differing methodologies, do indeed address common issues.

This book represents the effort of many people. We want to acknowledge the work of contributors as well as other colleagues who participated in correcting others' work and advising authors about their contributions. We also acknowledge Concepción Hernández and especially Heladia Salgado for their help in editing the book. Julio Collado-Vides is grateful to his wife, María, and sons, Alejandro and Leonardo, for their support and enthusiasm during the workshop and compilation of this book.

Integrative Approaches to
Molecular Biology

1 Evolution as Engineering

Richard C. Lewontin

All sciences, but especially biology, have depended on dominant metaphors to inform their theoretical structures and to suggest directions in which the science can expand and connect with other domains of inquiry. Science cannot be conducted without metaphors. Yet, at the same time, these metaphors hold science in an iron grip and prevent us from taking directions and solving problems that lie outside their scope. As Rosenbleuth and Weiner (1945) observed, "The price of metaphor is eternal vigilance." Hence, the ur-metaphor of all of modern science, the machine model that we owe to Descartes, has ceased to be a metaphor and has become the unquestioned reality: Organisms are no longer *like* machines, they *are* machines. Yet a complete understanding of organisms requires three elements that are lacking in machines in a significant way. First, the ensemble of organisms has an evolutionary history. Machines, too, have a history of their invention and alteration, but that story is of interest to only the historian of technology and is not a necessary part of understanding the machine's operation or its uses. Second, individual organisms have gone through an individual historical process called *development*, the details of which are an essential part of the complete understanding of living systems. Again, machines are built in factories from simpler parts, but a description of the process of their manufacture is irrelevant to their use and maintenance. My car mechanic does not need to know the history of the internal combustion engine or to possess the plans of the automobile assembly line to know how to fix my car. Third, both the development and functioning of organisms are constant processes of interaction between the internal structure of the organism and the external milieu in which it operates. For machines, the external world plays only the role of providing the necessary conditions to allow the machine to work in its "normal" way. A pendulum clock must be on a stable base and not subject to great extremes of temperature or immersed in water but, given those basic environmental conditions, the clock performs in a programmed and inflexible way, irrespective of the state of the outside world. Organisms, on the other hand, although they possess some autoregulatory devices like the temperature compensators of pendulum clocks, generally develop differently and behave differently in different external circumstances.

Despite the inadequacy of the machine as a metaphor for living organisms, the machine metaphor has a powerful influence on modern biological research and explanation. Internal forces, the internal genetic "programs," stand at the center of biological explanation. Although organisms are said to develop, that development is the unconditional unfolding of a preexistent program without influence of the environment except that it provides enabling conditions. Individual differences are regarded as unimportant, as are evolutionarily derived differences among species. The homeobox genes are at the center of modern developmental biology precisely because they are supposed to reveal the universal developmental processes in all higher organisms. Individual and evolutionary histories and the interaction of history and environment with function and development are regarded as annoying distractions from the real business of biology, which is to complete the program of mechanization that we have inherited from the seventeenth century.

The dominance of metaphor in biology is not only at the grand level of the organism as machine. There are what we may call *submetaphors* that govern the shape of explanation and inquiry in various branches of biology. Evolutionary theory, in particular, is a captive of its own tropes. The evolution of life is seen as a process of "adaptation" in which "problems," set by the external world for organisms, are "solved" by the organisms through the process of natural selection. One of the most interesting developments in the history of scientific ideas has been the back-transfer of these concepts into engineering, where they originated. The idea that organisms solve problems by adaptation derives originally, metaphorically, from the process by which human beings cope with the world to transform it to meet their own demands. This metaphorical origin of the theory of adaptation has been forgotten, and now engineers believe that the model for solving design problems is to be found in mimicking evolutionary processes since, after all, organisms have solved their problems by genetic evolution. Birds solved the problem of flying by evolving wings through the natural selection of random variations in genes that behave according the rules of Mendel, so why can we not solve similar problems by following nature? Most important, natural selection has solved the problem of solving problems, by evolving a thinking machine from rudiments of neural connections; yet we have not solved the problem of making a machine that thinks in any nontrivial sense, so perhaps we should try the method that already has worked in blind nature. The invention of genetic algorithms as a tool of engineering completes the self-reinforcing circle in the same way that sociobiological theory derives features of human society from ant society, forgetting entirely the origin of the concept of society. Nonetheless, genetic algorithms have been singularly unsuccessful as a technique for solving problems, despite an intense interest in their development. If nature can do it, why can't we? The problem lies in the inadequacy of the metaphor of adaptation: Organisms do not adapt, and they do not solve problems.

THE INFORMAL MODEL

The model of adaptation by natural selection goes back to Darwin's original account (Charles Darwin, *Origin of Species*, 1859), which has been altered only by the introduction of a correct and highly articulated description of the mechanism of inheritance. It begins with the posing of the problem for organisms: The external world limits the ability of organisms to maintain and reproduce themselves, so that they engage in what Darwin called a "struggle for existence." This struggle arises from several sources. First, the resources that are the source of metabolic energy and the construction materials for the growth of protoplasm are limited. This limitation may lead to direct competition between individuals to acquire the necessities of life but exists even in the absence of direct competition because of the physical finiteness of the world. Darwin writes of the struggle of a plant for water at the edge of a desert even in the absence of other competing plants. Second, the external milieu has physical properties such as temperature, partial pressures of gases, pH, physical texture, viscosity, and so forth, that control and limit living processes. To move through water, an organism must deal with the viscosity and specific gravity of the liquid medium. Third, an individual organism confronts other organisms as part of its external world even when it is not competing with them for resources. Sexual organisms must somehow acquire mates, and species that are not top predators need to avoid being eaten. The *global* problem for organisms, then, is to acquire properties that make them as successful as possible in reproducing and maintaining themselves, given the nature of the external milieu. The *local* problem is to build a particular structure, physiological process, or behavior that confronts some limiting aspect of the external world without sacrificing too much of the organism's ability to cope with other local problems that have already been solved. In this view, wings are a solution to the problem of flight, a technique of locomotion that makes a new set of food resources available, guarantees (through long-distance migration) these resources' accessibility despite seasonal fluctuations, and helps the organism escape predation. In vertebrates this solution was not without cost, because they had to give up their front limbs to make wings and so sacrificed manipulative ability and speed of movement along the ground. The net gain in reproduction and maintenance, however, was presumably positive.

Having stated the problem posed by the struggle for existence, Darwinism then describes the method by which organisms solve it. The degree to which the external constraints limit the maintenance and reproduction of an organism depends on the properties of the organism—its shape, size, internal structure, metabolic pathways, and behavior. There are processes, internal to development and heredity and uncorrelated with the demands of the external milieu, that produce variations among organisms, making each one slightly different from its parents. Hence, in the process of reproduction, a cloud of variant types is produced, each having a different ability to maintain and reproduce itself in the struggle for existence. As a consequence, the ensemble of organisms in any generation then is replaced in the next generation by a new ensemble that is enriched for those variants that are closer to the solution. Over time, two processes occur. First, there is a continual enrichment of the population in the proportion of any new type that is closer to the solution, but this process alone is not sufficient. The end product would be merely a population that was made up entirely of a single type that was a partial solution to the problem. Therefore, second, the entire process must be iterative. The new type that is a partial solution must again generate a cloud of variants that includes a form even closer to the solution so that the enrichment process can proceed to the next step. Whether the enrichment and novelty-generating process go on simultaneously and with the same characteristic time (gradualism), or the waiting time to novelty is long as compared with the enrichment process (punctuation) is an open question, but not one that makes an essential difference. What is critical to this picture is that the cloud of variants around the partial solution must include types that both are closer in form to the ultimate one and provide an incrementally better solution to the problem of reproduction and maintenance. The theory of problem solving by natural selection of small variations depends critically on this assumption that being closer to the type that is said to be the solution also implies being functionally closer to the solution of the struggle for existence. These two different properties of closeness become clearer when we consider a formalized model of the natural selective process.

THE FORMAL MODEL

The model of adaptation and problem solving by natural selection can be abstracted in such a way as to clarify its operating properties and also to make more exact the analogy between the supposed processes of organic evolution and the proposed technique of solving engineering problems. The formal elements of the system are (1) an ensemble of organisms (or other objects, in the case of a design problem), (2) a state space, (3) variational laws, and (4) an objective evaluation function.

A State Space

State space is a space of description of the organisms or objects, each object being represented as a point in the multidimensional space. For organisms, this state space may be a space of either the genotypical or phenotypical specification. In some cases, it is necessary to describe both phenotypical and genotypical spaces with some rules of mapping between them. The laws of heredity and the production of variations between generations operate at the level of the genotype. Unless the relation between the genotypical state and the phenotypical state are quite simple so that there is no dynamical error made in applying the notions of mutation and heredity to phenotypes, the fuller model must be invoked. In the case of the engineering analogy, the distinction is unnecessary.

Variational Laws

It is assumed that the ensemble of objects in the state space at any time, t, will generate a new set of objects at time $t + 1$ by some fixed rules of offspring production. In the simplest case, objects may simply be copied into the next generation. Alternatively, during the production process, the new objects may vary from the old by fixed rules. These may be simply mutational laws that described the probability that an imperfect copy will occupy a particular point in the state space different from the parental organism, or they may be laws of recombination in which more than one parental object participates jointly with others in the production of offspring that possess some mixture of parental properties. For sexually reproducing organisms, these are the laws of Mendel and Morgan. The set of objects *produced* under these laws is not the same as the ensemble of objects that will come to characterize the population in this generation as there is a second step, the selection process, that differentially enriches the ensemble for different types.

Objective Evaluation Function

Corresponding to each point in the state space—that is, to each different kind of organism—there is a score that is computable from the organism's position in the state space. Every point does not necessarily have a different score, and the score corresponding to a point in the space may not be a single value but may have a well-defined probability distribution. In evolutionary theory, these scores are so-called fitnesses. The score or fitness determines the probability that an object of a given description will, in fact, participate in the production of new objects in the next generation and to how many such offspring it will give rise. It is at this point that the connection is made between the position of the points in the state space of description and the notion of problem solving. The fitness score is, in principle, calculated from a description of the external milieu and an analysis of how the limitations on maintenance and reproduction that result from the constraints of the outside world are a function of the organism's phenotype. The story for wings would relate the size and shape of the wings to the lift and energy cost of moving them, coupled with calculations of how the flight pattern resulted in a pattern of food gathering in an environment with a certain distribution of food particles in space and time. Does the food itself fly, hop, or run? Is it on the ends of tree branches? How much energy is gained and lost in its pursuit and consumption?

The fitness scale can be regarded as an extra dimension in the space of description, producing a fitness surface as a function of the descriptive variables. The process of problem solving is then a movement along the fitness surface from a lower to a higher point. More precisely, the population ensemble is a cloud in the space of description that maps to an area on the fitness surface, and the evolutionary trajectory of the population is traced as a movement of this area from a region of lower to a region of higher fitness, ultimately being concentrated around the highest value. Alternatively, this trajectory can be pictured as a line on the surface corresponding to the historical trajectory of the average fitness of the ensemble or, of more interest to the engineering model, as the line giving the historical trajectory of the fit test type in the ensemble. Using this picture, we can now explore the analogies and disanalogies between the process of organic evolution and the process of problem solving.

PROSPECTIVE AND RETROSPECTIVE TRAJECTORIES

Problem solving in the usual sense is a goal-seeking process carried on by a conscious actor who knows both the final state to be achieved and the repertoire of possible starting conditions. The question then is, "Can I get there from here and, if so, how?" We assume that the problem cannot be solved in closed form so that some search strategy is needed. For example, if the problem is to make a machine that flies, starting from nuts, bolts, wires, and membranes, we suppose the solution cannot be arrived at simply by the application of some complete theory of aerodynamics and the strength of materials. One approach is a random or exhaustive search of all the possibilities in state space until the desired outcome is achieved. For all but the simplest problem, this clearly is out of the question. One cannot solve, by exhaustive enumeration, the traveling salesperson's problem for a reasonably sized case. The alternative, then, is to find an algorithm that, when iteratively applied, eventually will (in a reasonable number of steps) arrive at the final state. Such an iterative procedure requires an objective function that can be evaluated at the final state and at alternative intermediate states to test the closeness of each step to the desired end: That is, there must be some metric distance from the final state that is reduced progressively at each iteration and, if an iteration fails to reduce the distance, then a corrective routine must be applied. One may go back a step and try another path, or take a small random step and try the algorithm again but, whatever the rule, the criterion of stepwise success is always the distance from the final state on some measure.

However, the calculation of such a distance requires that the final state be known in advance: That is, problem solving is a prospective process. This is the first disanalogy with organic evolution. There is no natural analog to a knowledge of the final state. There is no evolutionary action at a distance. Given an organism without wings, flying is not a problem to be solved or, alternatively, we might claim that flying is a problem to be solved by all organisms including bacteria, earthworms, and trees. At what stage in the evolution of the reptilian ancestors of birds did flying become a problem to be solved, so that the fitness of an ancestral organism could be evaluated as a function of its distance from the winged state? The confusion between the prospective and retrospective nature of problem solving and evolutionary change has resulted in a biased estimate of the efficacy of natural selection as a method for problem solving. If we define as problems to be solved by evolution only those final states that are seen *retrospectively* to have actually been reached, then it will appear, tautologically, that natural selection is marvelously efficient at problem solving and we ought to adopt its outline for engineering problems. The mechanisms of evolution have, indeed, produced every result that has appeared in evolution, just as past methods of invention have indeed produced everything that has ever been invented. It is a vestige of teleological thinking, contained in the metaphor of adaptation, that problems for organisms precede their actual existence and that some mechanism exists for their solution.

THE SHAPE OF FITNESS SURFACES

The claim that organisms change by natural selection from some initial state to some final state implies that the fitness surface has a special property. It must be possible to draw a trajectory on the fitness surface that connects the initial and final state such that the fitness function is monotonically increasing along the trajectory: That is, every step along the way must be an improvement. This implies a strong regularity of the relationship between phenotype and fitness. If the fitness surface is very rugged, with many peaks and valleys between the initial state and some other state that has a very high fitness, the species is likely never to reach that ultimate condition of highest fitness. Evolution by mutation and natural selection is a form of local hill climbing, and the result is that fitness is maximized locally in the phenotypical state space but not necessarily globally. There is a further constraint and two possible escapes from local maxima. The added constraint is that the laws of Mendel and the patterns of mating imply a specific dynamic on the fitness surface, so that fitness must not only increase along the trajectory through state space but must also increase in conformity with fixed dynamical equations. This extra constraint almost excludes passage between two states that are not connected by a simple monotonic slope of the fitness surface. It is possible to escape from the local maximum if new mutations are sufficiently drastic or novel in their phenotypical effects or if chance variations in the frequency of types in finite populations push the ensemble down the fitness surface. The first of these phenomena is surely very rare. The second is extremely common, but its effectiveness depends on how rugged and steep is the fitness surface. It is equivalent in natural selection to simulated annealing in algorithmic problem solving, but without the possibility of tuning the parameters of the annealing, so it may be counterproductive. The existence of rugged fitness surfaces means that most evolutionary processes are best thought of as satisficing rather than optimizing, reaching only local optima.

A special difficulty arises in the selection of novelties, which is therefore particularly apposite in the analogy with problem solving. The fitness surface may be essentially without any slope for variations that, in retrospect, appear as rudimentary stages of an adaptation. This was a problem recognized by Darwin, who devoted special attention to what he called the "incipient stages" of a novel structure. The evolution of the camera eye, which has occurred independently in both vertebrates and invertebrates, with a light-receptive retina, a focusing lens, and a variable aperture, began as a small group of light-sensitive cells with associated neural processes that could enervate muscles directly or indirectly. Darwin argued that even such rudimentary eyespots are of selective advantage and the rest of the apparatus of the camera eye was an improvement on an already adaptive structure. This argument will not work, however, for incipient wings. Small flaps of tissue provide no lift at all because of the extreme nonlinearity of aerodynamic relations. (The reader may verify this by holding a ping-pong paddle in each hand and waving them up and down vigorously to see what effect is produced.) The present theory is that wings, in insects at least, were selected as heat collectors (butterflies regularly orient their wings parallel or at right angles to the sun as a form of heat regulation), and that only as the wings grew larger did they incidentally allow some flight. If wings are a solution to the problem of flight, they are an example of a problem being created by its own solution.

The recruitment of already existent structures for novel functions is a common feature of evolution: Front legs have been recruited for wings in birds and bats, the jaw suspensory bones of reptiles have been recruited to make the inner-ear elements of mammals, motor areas of the primate brain have become speech areas in the human cerebral cortex. Such recruitment has been possible only because the former function could be dispensed with or because it could be taken over by other structures. Redundancy of already existing complex structures can then be a precondition for the evolution of new problems and their solutions.

THE EVOLUTION OF PROBLEMS

The deepest error of the metaphor of adaptation, and the greatest disanalogy between evolution and problem solving, arises from the erroneous view of the relationship between organisms and their external milieu. *Adaptation* implies that there is a preexistent model or condition to which some object is adapted by altering it to fit. (That is why organisms are said to be highly "fit.") The idea of adaptation is that there is an autonomous external world that exists and changes independent of the organisms that inhabit it, and the relationship of organisms to the external world is that they must adapt to it or die—in other words, "Nature, love it or leave it." The equations of evolution then are two equations in which organisms change as a function of their own state and of the external environment, whereas environment changes only as a function of its own autonomous state.

The truth about the relation between organisms and environment is very different, however. One should not confuse the totality of the physical and biotic world outside an organism with the organism's *environment*. Just as there is no organism without an environment, there is no environment without an organism. It is impossible to describe the environment of an organism that one has never seen, because the environment is a juxtaposition of relevant aspects of the external world by the life activities of the organism. A hole in a tree is part of the environment of a woodpecker that makes a nest in it, but it is not part of the environment of a robin who perches on a branch right next to it. The metaphor of adaptation does not capture this action of organisms to sort through and structure their external world and would be better replaced by a metaphor such as construction. If problems are solved by organisms, it is because they create the problems in the first place and, in the act of solving their problems, organisms make new problems and transform the old ones.

First, organisms select and juxtapose particular elements of the external world to create their environments. Dead grass and small insects are part of the environment of a phoebe, which makes nests out of the grass and eats the insects. Neither the grass nor the insects are part of the environment (or of the problems to be solved) for a kingfisher, which makes a nest by excavating a hole in the earth and which eats aquatic animals. Nor do organisms experience climate passively. Desert animals live in burrows to keep cool, and many insects avoid direct sunlight, staying in the shade to avoid desiccation. By their metabolic activity, all terrestrial organisms, including both plants and animals, produce a boundary layer of moist warm air that surrounds them and separates them from the outer world only a few millimeters away. It is the genes of lions that make the savannah part of their environment, just as the genes of sea lions make the sea part of theirs, yet both had a common terrestrial carnivore ancestor. When did living in water become a problem posed by an external nature for sea lions?

Second, organisms alter the external world as they inhabit it. All organisms consume resources and excrete waste products that are harmful to themselves or their offspring. White pine is not a stable part of the flora of southern New England because pine seedlings cannot grow in the shade of their own parental trees. However, organisms also produce the conditions of their own existence. Plants excrete humic acids and change the physical structure of the soil in which they grow, making possible the growth of symbiotic microorganisms. Grazing animals can actually increase the rate of production of vegetation on which they feed. The most striking change wrought by organisms has been the creation of our present atmosphere of 18% oxygen and only trace amounts of carbon dioxide from a prebiotic environment that had virtually no free oxygen and high concentrations of carbon dioxide. Photosynthesis has produced the oxygen, whereas the carbon dioxide was deposited in limestone by algae and in fossil fuels. Yet the current evolution of life must occur within the conditions of the present atmosphere: That is, natural selection occurs at any instant to match organisms to the external world, but the conditions of that world are being recreated by the evolving organisms.

Third, organisms alter the statistical properties of environmental inputs as they are relevant to themselves. They integrate and average resource availability by storage devices. Oak trees store energy for their seedlings in acorns, and squirrels store the acorns for the nonproductive seasons. Mammals store energy in the form of fat, averaging resource availability over seasons. Beavers buffer changes in water level by building and altering the height of their dams. Organisms are also differentiators, responding to rates of external change. For example, Cladocera change from asexual to sexual reproduction in response to sudden changes in temperature or oxygen concentration in either direction, presumably as a way of mobilizing variation in an unpredicted environment.

Finally, organisms transduce into different physical forms the signals that come in from the external world. The increase of temperature that comes into a mammal as a form of thermal agitation of molecules is converted into changes in chemical concentration of hormones regulating metabolic rates that buffer out the thermal changes. The outcome of the constant interpenetration of organisms and their external milieu is that living beings are continually creating and recreating their problems along with the instantaneous solutions that are generated by genetic processes. In terms of the formal model, the fitness surface is not a constant but is constantly altered by the movement of the ensemble in the space. The appropriate metaphor is not hill climbing but walking on a trampoline. The reason that organisms seem to fit the external world so well is that they so often interact with that world in a way dictated by their already existing equipment. This in no way nullifies the importance of natural selection as a mechanism for further refining fitness relations. There were undoubtedly genetic changes more akin to local hill climbing in a small region of a nearly fixed fitness surface that further refined the musculature and behavior of skates and rays once they were committed to propelling themselves through water by flapping and flying motions, as opposed to the side-to-side undulations of their shark relatives.

If genetic algorithms are to be used as a way of solving engineering problems by analogy to the supposed success of natural selection in producing adaptation, then they must be constructed for the limited domain on which that analogy holds. The alternative is to evolve machines and later to find uses for them to which they are preadapted, a process not unknown in human invention. Digital computers were not invented so that we might see winged toasters flying by on the screen, yet they seem extraordinarily well-adapted to solving that problem. Organisms fit the world so well because they have constructed it.

I Computational Biology

The accumulation of large amounts of information in molecular biology in recent decades brings back into focus the question of how to deal with structure, its dynamical properties, and its description. Understanding molecular sequences and their three-dimensional (3-D) structure, understanding physiology and gene organization as well as cell biology and even higher levels of organism and ecological organization, can be accomplished to a certain extent without regard, for a moment, for their evolutionary history. Certainly, to achieve a synthetic description of these biological structures represents a paramount challenge. The computational infrastructure to support adequate organization, for easy retrieval and visualization, of the ever-increasing amount of data (genome projects included) in molecular biology is fundamental if such data are to be fully exploited in the field. The organization of this book emphasizes that this is the first necessary step toward a new integrative molecular biology.

Once the various genome projects, including that of humans, are understood and we have full knowledge of the completed sequence of the DNA contained in an organism, we will return to the science of biology with new tools and new questions. The types of questions and problems that might arise in this aftersequence period are illustrated by Elizabeth Kutter in chapter 2.

Chapter 3, by Temple Smith, Richard Lathrop, and Fred Cohen, a very useful review of the methodology around pattern recognition in proteins, providing a look at the mathematical, computer scientific, and other formal methods that currently are being used extensively to decipher 3-D structure and function from the primary sequence of the molecules. By its nature, this subject has required the early interdisciplinary work of computer scientists, molecular biologists, chemists, and physicists, who have formed work teams.

The more established interdisciplinary character of this type of research, as compared to the much more recent research on computational representations of gene regulation and metabolism (see part II), is reflected by the type of questions addressed in the chapters in this part. For instance, in chapter 4, Robert Robbins looks at how databases will be integrated into a federal infrastructure and the consequences, once genome projects are completed, of comparative studies of chromosome organization. This may well be recognized as a "higher-generation" database problem, whereas in gene regulation, physiology, and metabolism, the scientific community nowadays is working on what can be called "first-generation" types of problems.

Antoine Danchin, in chapter 5, provides an overall account of alternative formal methodologies, centered on different versions of information theory, that might help us to devise a better integration and theory construction in molecular biology—more specifically, the molecular biology of large amounts of DNA and protein linear sequences.

Historically, these approaches and ideas can be traced back to the prediction by Erwin Schrödinger that DNA is an aperiodic crystal. DNA is a message to be studied by information theory; it codes for hereditary information and thus usually is conceived as the physical container of a developmental program, connecting molecular biology to computer science. However, DNA is also a language, amenable to study within linguistic theories. One single object, DNA, (or two, if we include protein sequences) gives rise to no fewer than three differing attempts to apply formal disciplines to illuminate biology.

According to Claude Bernard (1865), biology is a science in which it is common to search for ideas—and methods—from more formalized sciences. This is a risky enterprise if we want to go beyond building analogies: Methods are to be used and tested within biology. They have to fit within the biological framework of understanding. The main risk and source of misconceptions is the assumption that ideas and principles that adequately explain domains within other disciplines will conserve their applicability and meaning within biology.

2 Analysis of Bacteriophage T4 Based on the Completed DNA Sequence

Elizabeth Kutter

The T-even bacteriophages are intricate molecular machines that show many of the complexities of higher organisms—developmental regulation, morphogenesis, macromolecular metabolic complexes, and introns—with a genome of only 168,895 base pairs (Kutter et al., 1994b). They have been major model systems in the development of modern genetics and molecular biology since the 1940s, with investigators taking advantage of the phages' useful degree of complexity and the ability to derive detailed genetic and physiological information with relatively simple experiments. This work has been fostered by the viruses' total inhibition of host gene expression (made possible in part through the use of 5-hydroxymethylcytosine rather than cytosine in the viruses' DNA) and by the resultant ability to differentiate between host and phage macromolecular synthesis. For example, T4 and T2 played key roles in demonstrating that DNA is the genetic material; that genes are expressed in the form of mRNA; that a degenerate, triplet genetic code is used, which is read from a fixed starting point and includes "nonsense" (chain termination) codons; that such stop codons can be suppressed by specific suppressor tRNAs; and that the sequences of structural genes and their protein products are colinear. Analysis of the assembly of T4's intricate capsid and of the functioning of its nucleotide-synthesizing complex and replisome have led to important insights into macromolecular interactions, substrate channeling, and cooperation between phage and host proteins within such complexes. The T-even phages' peculiarities of metabolism give them a broad potential range of host and environment and substantial insulation against most host antiviral mechanisms; these properties make them very useful for molecular biologists, and the phages even produce several enzymes that have important applications in genetic research.

The vast amount we have learned from studying the large lytic phages is, in part, a tribute to the vision of Max Delbrück in the early 1940s (Cairns, Stent, and Watson, 1966). He first convinced the growing group of phage workers to concentrate their efforts on one bacterial host—*Escherichia coli* B—and seven of its phages. He also organized the Cold Spring Harbor phage courses and meetings to bring together strong scientists from a variety of disciplines—in particular, to draw outstanding physicists, physical chemists, and biochemists into the study of fundamental life processes. From that time, phage work has emphasized the union of techniques from genetics, physics, microbiology, biochemistry, mathematics, and structural analysis, and has been characterized by very open communication and widespread collaboration. Delbrück succeeded in galvanizing a generation of old and young scientists from these many disciplines to work together to think about biological processes in new ways, and this legacy remains strong in the phage community.

Despite the intensive study of T-even phages over the last 50 years, major puzzles still remain in analyzing their efficient takeover of *E. coli*; many such puzzles are discussed in the new book *Molecular Biology of Bacteriophage T4* (Karam, 1994). For example, although the major phage proteins involved in shutting off host replication and transcription have been identified, the mechanisms of the rapid and total termination of host-protein synthesis (figure 2.1) are not at all clear. Also we do not understand the mechanism of lysis inhibition induced by attachment of additional phage to already-infected cells, or of the specific stimulation of phosphatidyl glycerol synthesis after infection. Few studies have looked at the physiology of infected cells under conditions experienced "in the wild," such as anaerobic growth or the energy sources available in the lower mammalian gut. Even less is known about the surprising, apparent establishment of a "suspended-animation" state in stationary-phase cells infected with T4 that still allows them to form active centers on dilution into rich medium.

Resolution of many of these puzzles should be facilitated by the recent completion of the sequence of T4. Since 1979, our group at Evergreen has been largely responsible for organizing genomic information from the T4 community, in collaboration with Gisela Mosig at Vanderbilt University, Nashville, TN, and Wolfgang Rüger at Ruhr Universitaet Bochum, Bochum, Germany. While on sabbatical with Bruce Alberts at the University of California at San Francisco, I worked with Pat O'Farrell to produce a detailed T4 restriction map correlated with the genetic map. At the same time, Burton Guttman completed the initial draft of the integrated map in figure 2.2, a large version of which can be found hanging in laboratories from Moscow and Uppsala to Beijing and Tokyo. (As we think about elaborate computer databases, it is important not to lose sight of the usefulness of such simple, accessible, detailed visual representations.)

Most of the T4 sequence was determined a number of years ago, taking advantage of the detailed genetic and restriction-map information available, but the last 15 percent turned out to be very challenging. The job has recently been completed, with Evergreen students and visitors from Tbilisi (Georgia) and Moscow responsible for the final difficult segments and for the integration of data from the worldwide T4 community. Bacteriophage T4 is now the most complex life form for which the entire sequence is available in an integrated and well-annotated form. There have been many surprising results, the analysis of which will be possible only by the combined

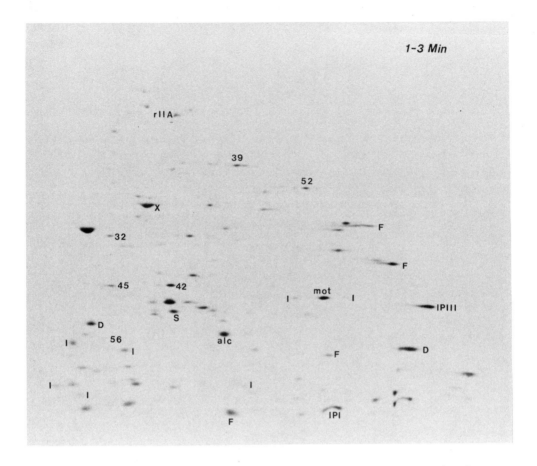

Figure 2.1 Proteins labeled 1–3 min after T4 infection of *E. coli* B at 37°C. The infection was carried out as described by Kutter et al. (1994a). Only minor traces of host proteins are still being made. *I* indicates otherwise unidentified immediate to early proteins. The genes whose products are explicitly identified can be characterized using the map in Figure 2.2. Those labeled F, P, D, and S are missing in phage carrying certain large-deletion mutations and thus are nonessential under standard laboratory conditions; many are lethal to the host when efforts are made to clone them, but their functions are otherwise unknown.

approaches of biochemistry, biophysics, molecular biology, microbiology, microbial ecology, and informatics—as will be true for all genomes. We can now approach a number of questions that can best be addressed for entities for which we have complete sequence information, questions related to genomic organization, redundancies, control sequences, roles of duplication, and prevalence of "exotic passengers."

In recent years, we have become very interested in the broader challenge of integrating and presenting many kinds of information about T4, including its physiology, genetics and morphogenesis, DNA and protein sequences, protein structures, enzyme complexes, and data from two-dimensional protein gels. Our detailed knowledge of T4 genetics and physiology, combined

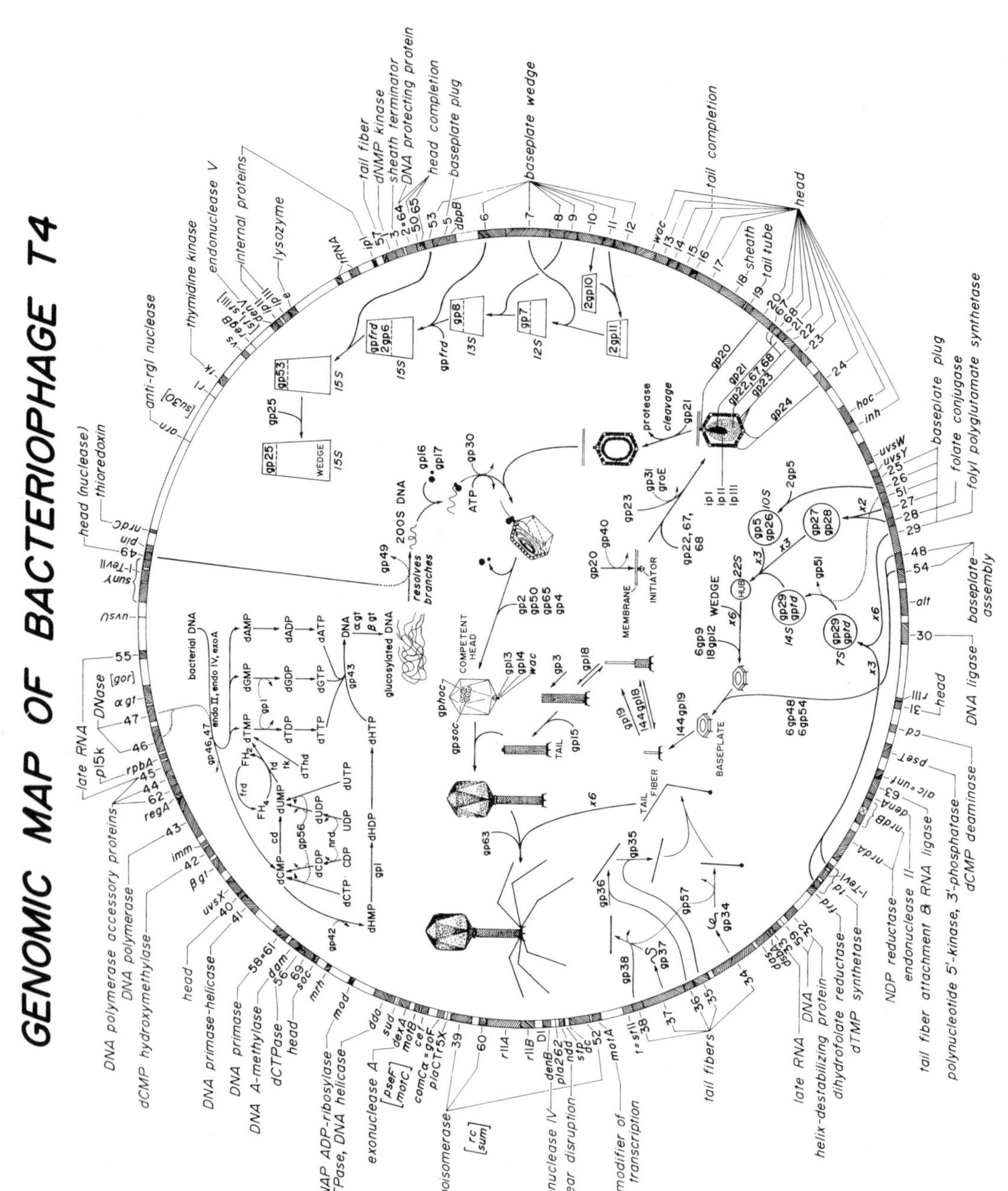

Figure 2.2 Map of the characterized genes of bacteriophage T4, with the spacing based on the complete sequence of the genome. The various pathways and products are indicated in juxtaposition with the genes (Reprinted from Kutter et al., 1994b.)

with the great additional advantage of having the complete DNA sequence, makes this phage a useful, manageable system for testing some of those analytical methods by which "whole-genome data sets can be manipulated and analyzed," as discussed by Robbins in chapter 4.

THE GENES OF T4

T4 has nearly 300 probable genes packed into its 168,895 nucleotide pairs—twice the number of genes expected not many years ago, and nearly four times the reported gene density of herpesvirus and yeast chromosomes III and VII (Koonin, Bork, and Sander, 1994). Most appear likely to be expressed, on the basis of such criteria as relationship to promoters and other genes, the presence of apparent translation initiation regions, a "correlation coefficient" that compares base frequencies at each codon position to those of a set of known T4 genes, and linguistics-based analyses such as GenMark (all further discussed later). This number reflects both the small size of many T4 genes and the fact that most of the available space is used efficiently. There are very few regions of apparent "junk" of any significant length, and even regulatory regions are compact or overlap coding regions; it appears that a total of only approximately 9 Kb does not actually encode either proteins or functional RNAs, and much of that includes regulatory sequences. In 42 cases, the end of one gene just overlaps the start of the next using the DNA sequence ATGA, where TGA is the termination codon of one gene and ATG is the initiation codon of the other gene; 34 additional genes actually overlap (usually by 4 to 16 bases). Perhaps more surprising, it has been clearly shown that one 8.9-kDa protein (30.3') is read out of frame within another coding region (Nivinskas, Vaiskunaite, and Raudonikiene, 1992; Zajanckauskaite, Raudonikiene, and Nivinskas, 1994) and the 6.1-kDa protein 5R is read in reverse orientation within gene 5 (Mosig, personal communication).

Only 70 of T4's genes are "essential," as determined by mutants that are lethal under standard growth conditions. These key genes use almost half the genetic material and mainly include elements of the replisome and nucleotide-precursor complex, some transcriptional regulatory factors, and the proteins that form the elaborate phage particle. Approximately 70 more genes have been functionally defined and encode such products as enzymes for nucleotide synthesis, recombination, and DNA repair; nucleases to degrade cytosine-containing DNA; eight new tRNAs; proteins responsible for excluding superinfecting phage, for lysis inhibition under conditions of high phage density, and for some other membrane changes; and inhibitors of host replication and transcription, and of the host Lon protease.

In our own analysis of the genetic information, we have been particularly focusing on the surprisingly large fraction of T4 genes apparently devoted to restructuring the host "factory": A large fraction of the nearly 150 otherwise uncharacterized T4 open reading frames (ORFs) seem, by a variety of criteria, to be involved in the transition from host to phage metabolism. They are located just downstream of strong promoters active immediately after infection, and they are lethal or very deleterious when cloned in most vectors. (This is one factor that made completing the sequencing so difficult.) Some of them are very large, but most encode proteins of less than 15 kDa, emphasizing the importance of not ignoring small potential ORFs; the smallest well-characterized T4 protein, Stp, has only 29 amino acids. Many of these genes are in regions that can be deleted without seriously affecting phage infection under usual laboratory conditions, suggesting that they are necessary only for certain environments or for infecting alternative hosts or that there is some redundancy in their functions. At the same time, their functions are important enough to the phage that, despite this apparent deletability, they have been retained in T4 and also in most related phages (cf. Kim and Davidson, 1974).

Most of the 37 promoters that function immediately after infection are in these deletable regions, which are very densely packed with ORFs, few of which have yet been defined genetically. However, their protein products—or at least those exceeding about 9 kDa—can be identified on two-dimensional gels of proteins labeled after infection, by comparing wild-type T4 with mutants from which known regions are deleted (Kutter et al., 1994a). It can thus be seen that these proteins are characteristically produced in large quantities just after infection. Most of these new, immediate early genes show very little homology with other genes. The fact that they are so deleterious to the host when cloned reinforces our belief that their proteins specifically inhibit or redirect important host protein systems, and a number may be useful in studying these host proteins in their active, functional state.

A prime example is the Alc protein, which specifically terminates elongation of transcription on cytosine-containing DNA. Alc seems to recognize selectively the rapidly elongating form of the RNA polymerase complex present at physiological nucleotide concentrations. A variety of evidence now strongly supports an "inchworm" model of polymerase progression, in which the polymerase inserts up to 10 nucleotides before moving ahead to a new site on the DNA (Chamberlin, 1995), with at least two binding sites each for RNA and DNA to prevent premature termination of transcription. Alc is potentially a very valuable tool for studying the dynamic structural changes that apparently occur in the polymerase; all other current approaches can look only at the polymerase paused at particular sites and infer its behavior from the resultant static picture.

One may well expect to find the same kind of specificity for particular active states of other enzymes. This would be especially useful for considering mechanisms with which other T4 proteins rapidly subvert host functions. Furthermore, some of these proteins eventually may suggest new approaches to making antibiotics, and may also prove useful for viewing the details of evolutionary relationships and protein-protein interactions.

COMPLEX PROTEIN MACHINES

Most known T4 proteins do not function alone but rather as part of some tight macromolecular complex. This is true not only for the elegant and complex capsid, with its six tail fibers and contractile tail, but also for most of its enzymes and the other early proteins that redirect host metabolism. Chemical equations, concentrations, and kinetic constants are only part of the story. Understanding such metabolic pathways requires not only work with purified enzymes and the kinds of analyses discussed by Mavrovouniotis in chapter 11, but also consideration of the convoluted interactions in such tightly coupled protein machines, which may turn out to be the rule rather than the exception in nature.

The best-understood of these enzymatic machines is T4's nucleotide precursor complex (reviewed by Matthews, 1993; Greenberg, He, Jilfinger, and Tseng, 1994). It takes both cellular nucleotide diphosphates (NDPs) and the nucleotide monophosphates (dNMPs) from host DNA breakdown and converts them into nucleotide triphosphates (dNTPs), in exactly the proper ratios for T4's DNA (that being that A and T together constitute two-thirds of the sequence). The synthesis occurs at the appropriate rate for normal T4 DNA production, even when DNA synthesis is otherwise blocked, implying that the regulation is somehow intrinsic, not a consequence of feedback mechanisms. Proteins of the nucleotide-precursor complex undergo further extensive protein-protein interactions as they funnel nucleotides directly into the DNA replication complex, which consists of multiple copies of nine different proteins (cf. Nossal, 1994). The interactions involved at all these levels have been documented by such methods as in vivo substrate channeling, intergenic complementation, cross-linking, and affinity chromatography, as well as by kinetic studies of substrates moving through the purified precursor complex. One consequence of the tight coupling is that dNTPs entering permeable cells must be partly broken down to enter the complex and must then be rephosphorylated to enter the DNA, so exogenous dNTPs are used severalfold less efficiently than are dNMPs or dNDPs. The complex, which includes two host proteins, has also been documented during anaerobic growth (Reddy and Mathews, 1978; Mathews, 1993, and personal communication), but the exact relationship of T4's two-component anaerobic NTP reductase to the other enzymes of the complex is not yet clear.

The replication complex, in turn, is strongly coupled to the complex of host RNA polymerase and phage proteins that transcribes the T4 late-protein genes, thus functioning, in effect, as a "mobile enhancer" to link the amount of phage capsid proteins to the amount of DNA being made to be packaged inside them (Herendeen, Kassavetis, and Geiduschek, 1992). Throughout infection, rapid transcription and replication are occurring simultaneously, with the replication complexes moving along the DNA at 10 times the rate of the transcription complexes, and both moving in both directions. One might expect frequent collisions between the two kinds of complexes, but recent evidence shows that T4's transcription and replication complexes can pass each other, with the polymerase changing templates when they meet head-on without interfering with the crucial total processivity of transcription (Liu, Wong, Tinker, Geiduschek, and Alberts, 1993; Liu and Alberts, 1995).

T-EVEN PHAGE EVOLUTION

There has long been interest in the origin of viruses, how they acquire their special properties and genes, and how they relate to one another. Botstein (1980) suggested that lambdoid phages are put together in a sort of mix-and-match fashion from an ordered set of modules, each of which may have come from a particular host, plasmid, or other phage. This concept has since been extended to other phages, including T4 (cf. Campbell and Botstein, 1983; Casjens, Hatfull, and Hendrix, 1992; Repoila, Tetart, Bouet, and Krisch, 1994).

T4-like phages, having complex tail structures and hydroxymethylcytosine rather than cytosine in their DNA, have been isolated all over the world, from places such as sewage treatment plants on Long Island (Russell and Huskey, 1974), the Denver zoo (Eddy and Gold, 1991), and patients with and phage preparations used to treat dysentery (Gachechiladze and Chanishvili, Bacteriophage Institute, Tbilisi, Georgia, unpublished data; Kutter et al., 1996). Studies in various laboratories have used genetic analysis, polymerase chain reactions, sequencing, and heteroduplex mapping to show that a large fraction of the phage genes are conserved; although some occasionally are lost or replaced, the general gene order also is conserved (cf. Kim and Davidson, 1974; Russell, 1974; Repoila et al., 1994).

Few T4 proteins, except those involved in nucleotide and nucleic acid metabolism, show substantial similarities to anything else under standard search protocols such as BLAST. Several of the similarities that have been found are to uncharacterized ORFs of eukaryotic and other prokaryotic viruses. T4 has, in fact, been accused of having had illicit sex with eukaryotes. This suggestion is based on sequence similarities between T4 and eukaryotic cells (Bernstein and Bernstein, 1989) and on the fact that mechanistic features of some of T4 enzymes of nucleic acid metabolism are much more similar to those of eukaryotes than to those of *E. coli*. This interesting similarity emphasizes how little is known about the origins of T-even phages or their relationships to other life forms. As discussed by Drake and Kreuzer (1994), T4's large genetic investment in "private" DNA metabolism may eventually provide insights into its ancestry as questions of horizontal or vertical transmission of genes are sorted out.

T4 DNA is approximately $\frac{2}{3}$ AT overall. If, indeed, it is put together from "modules" from various sources, one might expect different genes to have wide ranges of GC contents within the AT/GC composition of DNA. However, only 18 of the known and apparent genes have less than 60 percent AT, and only 4 have less than 58 percent. Interestingly, it is mainly the capsid proteins—presumably among the earliest to have developed—that have lower AT/GC ratios, closer to the AT/GC ratio of *E. coli*. Gene 23, the major head protein, is the lowest, at 55 percent. Also, there seems to be a substantial bias toward G and against C: Only 4 genes have more than 20 percent C, whereas approximately 130 have more than 20 percent G, and 37 have more than 22 percent.

Only one group of 13 T4 genes seems to show clear evidence of horizontal transfer (Sharma, Ellis, and Hinton, 1992; Gorbalenya, 1994; Koonin, personal commununication); a substantial fraction of the genes unique to T4 seem to be in this class. The group consists of apparent members of all three mobile nuclease families first identified in eukaryotic mitochondrial intron genes. It includes the genes for two enzymes that can impart specific mobility to the introns in which they reside (Shub, Coetzee, Hall, and Belfort, 1994). Not yet clear is how many of the others still are expressed or whether any of them have been coopted to perform functions useful to the phage. One is situated in reverse orientation to the genes in its region, with no apparent promoter from which it could be expressed; at least two others—including the gene in the third T4 intron, *nrdB*—seem to be pseudogenes, the nonfunctional residues of a genetic invasion. In general, T4's tight spacing of genes may help discourage invasion by such external DNA.

Some of the small proteins that have been studied in detail are especially highly conserved at the protein level, presumably reflecting their tight and complex interactions with multiple cell components; for example, the *alc* gene in one of the T-even phages differs from that in T4 by 17 nucleotides, but the proteins differ by only one amino acid (Trapaidze, Porter, Mzhavia, and Kutter, unpublished data). Other regions show very complex patterns of high regional conservation, variability, and large-block substitution that may help us better understand T-even gene origins and commerce among the phages (Poglazov, Porter, Mesyanzhinov, and Kutter, manuscript in preparation). Such studies also are potentially very helpful in sorting out the functions of at least some of these genes and in enhancing our understanding of the takeover of host metabolism by these large lytic phages.

ANALYSIS "IN SILICO"

In collaboration with Judy Cushing, an Evergreen colleague expert in object-oriented databases, and with help from the National Science Foundation-IRI database and expert systems program, we are now using our T4 genomic analysis to explore ways to integrate a large variety of structural, genetic, and physiological information. We are working closely here with developers of several promising systems: Tom Marr's object-oriented Genome Topographer database; Jinghui Zhang, Jim Ostell, and Ken Rudd's sophisticated Chromoscope viewer; Eugene Golovanov and Anatoly Fonaryev's hypertext Flexiis genomic encyclopedia; Monica Riley's FoxPro database of *E. coli* physiological data; and Peter Karp's expert system application EcoCyc for biochemical pathways.

Our aim is to assemble a complete T4 database with its associated computing tools as a model for genetic databases in general, which will be useful for teaching in addition to T4 applications. Another major goal is the development of students who are highly skilled in both molecular biology and relevant computer sciences, this is crucial for implementing the integrative ideas discussed extensively in this symposium.

Several major challenges arise in populating this sort of complete genomic database, challenges that can be facilitated by using the database itself in an interactive process. These include accurately identifying the coding regions; determining the functions and structure-function relationships of the encoded proteins, including their interactions with other phage and host proteins in macromolecular complexes; and determining the sources and evolutionary relationships of the various genes.

IDENTIFYING GENES

Work with T4 makes it clear that identifying regions encoding proteins is a good deal more complex than just locating the codons for translation initiation, ATGs, with a reasonable Shine-Dalgarno sequence followed by an extended open reading frame. Determining the start(s) and even, occasionally, stops of genes turns out to be highly complex and emphasizes the concern expressed by Robbins (chapter 4) about the meaningfulness of the term *gene*.

In general, recognizing gene ends—the stop codons—is relatively straightforward. However, several factors can affect whether a particular stop codon really is the end of the expressed protein, in addition to the frequent presence of suppressor tRNAs, which can mediate read-through of stop codons with varying efficiency and which may have normal cellular functions beyond their usefulness to molecular biologists.

1. There can be intron splicing: This is rare in prokaryotes but occurs in at least three T4 genes. For example, what were earlier called *ORFs 55.11 and 55.13* are now known to encode the protein NrdD.

2. There can be ribosomal frameshifting, which shifts translation by one base into a different reading frame at specific sites. It has not yet been confirmed in T4 but is suggested as a possibility in some places and is clearly demonstrated for two genes in T7 by Dunn and Studier (1993). Frameshifting can also occur by folding out a piece of mRNA in a very stable structure, such as the 50-bp segment in T4 gene 60 (Huang, Ao, Casjens, Orlandi, and Zeikus, 1988).

3. As recently discovered, the stop codon UGA can, in the proper very extended context, encode a twenty-first amino acid, selenocysteine (Bock et al., 1991). It is not yet clear whether T4 has any such sites, but they seem important in some viruses, such as the human immunodeficiency virus (Taylor, Ramanathan, Jalluri, and Nadimpalli, 1994), as well as in *E. coli*.

Initiation of protein synthesis depends on the ability of a stretch of the mRNA to bind to the 30S ribosomal subunit so as to position an initiator codon appropriately and interact with the initiator fMet-tRNA and initiation factors, followed by binding the 50S subunit. In prokaryotes, this was classically thought to involve an AUG (or occasionally GUG) codon as well as a Shine-Dalgarno sequence a few nucleotides upstream, which provides complementarity to a certain stretch of the ribosomal RNA. Schneider and colleagues have taken a different approach, using their Perceptron algorithm, as discussed in chapter 5. By lining up the starts of well-characterized genes from a given organism, they calculate the distribution of bases at each position from -20 to $+18$ relative to the AUG/GUG, covering the entire region protected by the ribosome during initiation. This defines a matrix of values that can then be compared to other putative start sites. A translation initiation region (TIR) or ribosome-binding site (RBS) value, expressed as bits of information, can thus be determined. One can then ask questions about the correlation between this value and experimental evidence suggesting whether this is a translational start site; are also can look for factors that may affect the efficiency of initiation. Neural net approaches to the same problem are being developed.

Miller, Karam, and Spicer (1994) discuss many of the issues in identifying T4's translational start sites and coding regions. More are continually revealed as we examine the full T4 sequence. Ten T4 genes use GTG as their start codon. At least one, 26', uses ATT (Nivinskas et al., 1992), whereas *asiA.6* uses a pair of ATAs (Uzan, Brody, and Favre, 1990). We have been working with Schneider, Alavanja, and Miller to characterize T4 TIRs more thoroughly, in the hope of helping to explain the switch from host to phage translation. T4 TIRs do appear to have a higher number of bits of information on average than those of *E. coli*, and TIRs for each score much lower generally when tested with parameters derived from the other, but there are too many overlaps and discrepancies among the values for this to account for the rapid and total shutoff of host translation (see figure 2.1). Furthermore, scanning of the genome using the T4 "early" pattern turns up several times as many "high-probability" apparent TIR sites as do probable genes, even with a TIR value cutoff that does not pick up a third of the known gene starts. This number is far greater than would be expected on a random basis, and most of these "pseudo-TiRs" are in the transcribed strand, further emphasizing their nonrandomness. At least some such apparently "nonfunctional" TIRs are able to direct translation of reporter genes cloned adjacent to them, even though in T4 they would encode only peptides a few amino acids long (Mosig, personal communication; Krisch, personal communication).

Five T4 genes and several other ORFs have already been shown to have functional internal starts, with good evidence in several cases that the shorter protein has a distinct functional role. In addition, seven genes have two closely spaced start codons with equally strong TIR values. It will be very interesting to determine whether both are used, and why. (In bacteriophage lambda, there is an example of one such pair of proteins that have opposing functions and differ by only two amino acids in length: One makes the pore to permit access by lysozyme to the peptidoglycan layer, whereas the other delays pore formation. The regulation of the two is not understood at all.)

Defining Genes by Information Content

One can define probable coding regions surprisingly well by looking at various statistical properties of the stretch of DNA in question as compared to those of known sets of genes from the organism in question. Stormo has developed an algorithm using a correlation coefficient that compares the base preference at each codon position of a stretch of DNA with that of the known genes in an organism to determine its coding potential (Selick, Stormo, Dyson, and Alberts, 1993). The codon base preference is a matrix that counts the occurrences of each base at each codon base position. In coding sequences of T4, for example, one finds that more than half of the Gs are in the first position, and nearly half of its Ts are in the third position. The correlation coefficient can range in value from -1 to $+1$; most T4 genes have values in excess of 0.85, as do most of the unassigned ORFs.

Higher-order Markov chain models have also been used successfully for several organisms (cf. Borodovsky, Rudd, and Koonin, 1994) and work very well in predicting T4 coding regions. These are related to, though quite distinct from, the codon usage analysis and the linguistic, grammatical approaches discussed in chapter 5. Both these approaches have made some interesting, testable predictions and raised important questions about the assignments of probable genes in T4.

For seven known genes and six ORFs, a combination of TIR with other analyses suggested questions about the hypothesized coding region. For several, this has led to identifying sequencing mistakes; in others, it still raises interesting possibilities of unusual start codons or RNA secondary structure. For example, the gene-38 Shine-Dalgarno sequence is located 23 bases away, with a stable stem-loop structure then bringing it to five bases from the start codon (see Miller et al., 1994). The existence of a 50-bp foldout intron in the middle of gene 60—a phenomenon not yet reported for any other organism—emphasizes the need to look very carefully at anomalies rather than just dismissing them as mistakes.

Expediting Genomic Analysis

Genomic analysis is still a very laborious process, even for a genome the size of T4, requiring a great deal of hand work in exploring the databases, keeping track of the results, integrating them, and then going back to explore new predictions. As discussed in general in chapter 4, techniques for effectively and efficiently mining the data are badly needed and are being developed. The following is a summary of questions to be asked about any genome, with applications to T4.

Where the "Genes" Start and Stop?

1. What regions of a piece of DNA are predicted to be coding regions using Markov chain models, such as GeneMark, trained to known genes of this organism (or to a structurally defined subset of genes from the organism)? Note: These analyses are available on-line for *E. coli* and several other organisms, at genmark@ford.gatech.edu. The graphical version is needed along with the table of values for their predicted start sites, as the latter assumes an AUG or GUG start codon; no other assumptions are made about ribosome-binding sites.

2. What are the TIR values (or other measures of ribosome-binding sites, such as Shine-Dalgarno sequences and their spacing) for the potential start sites predicted by GenMark or other more traditional forms of analysis? Are there apparent ORFs where some other sort of start must be involved if they are to be functional?

3. What promoters, terminators, and thus probable transcripts are predicted from the sequence for this region or are known from experimental data?

4. Are there regions of abrupt reading frameshift that might reflect sequencing errors, translational frameshifting, or pseudogenes differing from some original version by a frameshift mutation or longer deletion? (A pseudogene is seen for one of the three intron genes, *I-TevIII*, which is intact in the related phage RB3, as shown by Eddy and Gold, 1991).

A combination of these forms of analysis enabled us to predict a number of sequencing errors around the genome that were indeed found on experimental testing of genomic DNA; this has been very helpful to the scientists working in the regions in question. In general, sequencing from related phages, cloning and overexpressing the putative gene, and gel identification of the product are needed to resolve questions. Here, complex computer work helps suggest and inform interesting laboratory experiments.

Computer Efforts at Predicting the Functions of New ORFs

5. For those ORFs that probably are transcribed (see question 3) and appear likely to be expressed (see questions 1 and 2), what significant sequence similarities can be identified with proteins in the current databases, using various parameters with the variety of analytical programs available?

6. Which of these ORFs appear to encode known functional motifs, such as those from the ProSite library: nucleotide-binding sites, zinc fingers, phosphorylation sites, or probable membrane-spanning domains? Similarly, what sorts of secondary structure are predicted, and how consistently, by different algorithms?

7. Which of these similarities are to proteins of known three-dimensional structure, and are there any predictions one can make using new programs or databases suggesting structure-sequence alignments?

8. Where have parallel regions been sequenced in related phages, and what can be inferred from the degree of conservation? Is the gene or region found in all of them? Are most sequence differences in the third codon position, indicating that the protein sequence is very highly conserved though it has evolved at the DNA level?

Occasionally, simple sequence comparisons excitingly identify new proteins, as with both parts of the T4 anaerobic ribonucleotide reductase, which are 60 percent identical with host homologs. This gave the first proof of anaerobic-state T4 enzymes and also identified the function of the last T4 intron-containing gene, *sunY*, the other two also being in genes of nucleotide metabolism.

More often, the relevant information needs to be teased from large amounts of statistical noise by using sophisticated analyses and cross-comparisons. Phylogenetic comparisons can define key conserved elements of proteins; very strong protein conservation despite many DNA changes suggests that the protein interacts with a number of other components and cannot even tolerate many conservative changes (as is true in T4 for Alc and the gp32 single-stranded DNA binding protein, for example.)

Mining Sequences for New Kinds of Data

9. What other strong apparent TIR/RBS sites are predicted in this region by computer analysis? May these be reflecting alternate start sites or genes within genes, or could they have some interesting regulatory role in translation of those regions? For instance, there are indications that certain sequences with a Shine-Dalgarno sequence shortly before a stop codon are involved in ribosomal frameshifting.

10. Can one find any other possible frameshifted genes, pseudogenes, regulatory sites, or other sign of origin by applying programs such as BLASTX to the few extended stretches between probable ORFs?

11. What can one suggest about possible evolutionary relationships or different coding regions by using a GenMark analysis trained for a different set of genes? For example, GenMark trained for the main chromosomal genes (class 1) of *E. coli* identifies very few (parts of) T4 genes; GenMark trained for several other bacteria reveals a somewhat larger, partially overlapping set. Many, interestingly, show up if GenMark has been trained using known yeast or GC-rich human genes.

12. What possible secondary or tertiary structures are predicted for the RNA that might affect the probability or extent of translation initiation at the predicted TIRs or that might suggest other phenomena such as frameshifting, the use of selenocysteine, or additional stable RNAs? Examples would be the

obligatory gene-38 TIR region discussed earlier that brings the Shine-Dalgarno into appropriate position; stem-loops that occlude the TIRs on early transcripts of such T4 genes as *soc* and *e*, while they are translated effectively from transcripts initiated at late promoters within the stem-loop region; and T4's two still-missing short stable RNAs.

Analyses Integrating Various Other Kinds of Laboratory Data with Sequence Information

13. How well can one correlate both genes and potential ORFs with two-dimensional protein gel data, and what can one then say about timing and levels of expression under various conditions (cf. Cowan, d'Acci, Guttman, and Kutter, 1994; see also chapter 7)?

14. How can we best integrate and use metabolic pathway data—for example, to understand the effects of phage infection on host metabolism and the complexities of the phage-infected cell, under various conditions? There appears to be enormous promise here in overlaying host- and phage-induced pathways and known phage alterations in a program such as Karp's EcoCyc, and in setting up straightforward links between that and other databases.

15. What is known about probable protein-protein interactions? How can knowledge from affinity chromatography, cross-linking, genetic studies, and the like be integrated and applied to demonstrate interactions in the cell and to suggest new experimental approaches?

It is clear that any real understanding of what is happening in living cells must take the complex interactions between macromolecules into consideration, along with individual enzyme properties and the effects of intracellular concentrations of ions and other small ligands. As discussed previously, most known T4 proteins function as part of some tight macromolecular complex: the nucleotide precursor complex, the DNA replication complex, the phage capsid itself. Also, many of the new small proteins that seem to be involved in subverting host function are likely to interact in interesting and dynamic ways with host proteins, as described for Alc, which interacts with a specific active form of polymerase to terminate host transcription. It is thus crucial to develop better ways to measure all these parameters, to keep track of the many kinds of evidence in readily accessible formats, and to display potential structural interrelationships in ways that facilitate new modes of thinking about the living cell.

More three-dimensional protein structures are becoming known, and comparison and modeling programs are becoming more sophisticated, including the addition of tools such as Sculpt, with which one can change parts of molecules from those indicated by crystal structures and examine the implications. How soon will we be able to obtain additional biological insights by playing with possible interactions and shapes on the computer? Such capability could help us design future experiments more knowledgeably and productively, as well as help students develop flexibility, insight, and creativity. (Let us remember, after all, the enormous impacts of the double-helix DNA structure and the fluid mosaic cell membrane model on advances in biology.) The phage-infected bacterial cell seems to be one area in which such approaches are particularly promising.

CONCLUSIONS

A kind of ideal of experimental biology is to know everything about one organism—in mathematical terms, to solve one complete system. As expressed in many ways at this conference, this ideal seems closest to being attainable for *E. coli*, but there is still an enormous amount to learn. One major step along the way would be to gain a thorough understanding of *E. coli* infected with a phage such as T4, which completely shuts off so many host functions. We have here outlined the experimental and analytical advantages of using T4, especially the extensive knowledge of its genetics and physiology; its complete, well-annotated DNA sequence; growing information about related phages; extensive information about the expression patterns, functions, and interactions of a large fraction of its proteins; and the rapidly growing knowledge about its bacterial host and intense work on tools to manipulate that information. At the same time, it is complex enough while retaining enough mystery to be a very good system for many of the new analytical tools. Computers have become powerful instruments in the genetic toolbox and can now be used to greatly enhance further analysis and identification of the remaining phage gene functions and to help inform the experiments that can lead ultimately to that goal of a reasonably complete understanding for this one "simple" system. With such an understanding and the analytical tools refined through this exercise, we can then apply the information and methods to the more complex challenges that remain.

SUGGESTED READING

Borodovsky M., Koonin E., and Rudd K. (1994). New genes in old sequence: A strategy for finding genes in the bacterial genome. *Trends in Biochemical Sciences, 19,* 309–313.

Casjens, S., Hatfull, G., and Hendrix, R. (1992). Evolution of dsDNA tailed-bacteriophage genomes. *Seminars in Virology, 3,* 383–397.

Karam, J. D. (Ed.). (1994). *Molecular biology of bacteriophage T4.* Washington, DC: American Society for Microbiology.

Many chapters of this excellent reference, in addition to those specifically cited, will help in exploring ideas discussed in this chapter.

Mathews, C. (1993). The cell—bag of enzymes or network of channels? *Journal of Bacteriology, 175,* 6377–6381.

Matthews, C. (1993). Enzyme organization in DNA precursor biosynthesis. *Progress in Nucleic Acid Research and Molecular Biology, 44,* 167–203.

3 The Identification of Protein Functional Patterns

Temple F. Smith, Richard Lathrop,
and Fred E. Cohen

There is a vast wealth of amino acid sequence data currently available from a great many genes from many organisms. An even greater number is anticipated in the near future. However, our ability to exploit the data fully is limited by our inability to predict either a protein's function or its structure directly from the knowledge of its amino acid sequence. Interestingly enough, it currently is easier to predict function rather than structure for the average newly determined sequence. This is true even though knowledge of the folded structure of the amino acid chain is, in principle, essential to understanding the steric constraints and side-chain chemistry that determine its biochemical function(s).

Anfinsen's experiments with bovine ribonuclease in the 1950s (Anfinsen, 1973) indicated that the tertiary structure of a protein in its normal physiological milieu is one in which the free energy of the system is lowest. Hence, theoretically, the sequence of amino acids completely specifies a protein's structure under physiological conditions. Although this view may not be completely universal given the involvement of chaperones, in the folding of large proteins (Anfinsen, 1973), it still is the general rule. The direct static or molecular dynamic minimum-energy approaches to the protein folding problem have all proven to be computationally intractable for sequences over a few tens of amino acids in length. The difficulties are well recognized and are due to the combination of the exceedingly large conformational space available and to the "rough" nature—many complex local minima—of the energy landscapes defined over the available conformational space. How, then, is it that we can so often accurately predict a protein's function and its basic structure?

It is the very wealth of available data that provides the answer. As has been pointed out many times, the evolutionary history of life on earth allows us to recognize proteins of similar function because of the conserved similarity of their sequences. This is true even when the evolutionary distances are large, resulting in only a few amino acids being shared in common. That is because evolution, once having solved a problem, needs to conserve only those aspects of the sequence most functionally "descriptive" of that solution! Thus, while only a limited number of proteins have had their functions determined by either detailed biochemistry or genetic analyses, any protein that has determined function provides information on all sequence-similar proteins. Comparative sequence analysis therefore provides modern biology with its own Rosetta stone. It is believed that as the number of determined structures increases, analogous approaches such as homologous extension modeling will be refined to allow the accurate prediction of structure from sequence similarities as well (Ring and Cohen, 1993). Appropriate to the title of this book, the review that we present will show that current and envisioned methods of identifying protein structure and function are highly integrative: They depend on methodologies drawn from mathematics, computer science, and linguistics, as well as physics, chemistry, and molecular biology.

Related enzymatic functions generally are encoded by similar structures, again as a result of common origins. It is not surprising, therefore, that patterns of conserved primary sequence, secondary structure, or other elements can be associated with a particular protein function. These patterns represent a complex mix of history and conserved physical chemistry. As with standard sequence comparison (Kruskal and Sankoff, 1983), pattern matching has proven useful in identifying the probable function of newly sequenced genes. More importantly, it can be used to focus site-directed mutagenesis experiments. Their utility may prove to be even more general and allow the accurate prediction of some protein substructures and subfunctions with common biochemical and structural constraints. Nearly all the current methods of discovering patterns begin from a similar point, a defining set of related sequences of known function. From such sets, patterns common to all set members are sought. These patterns have been represented as consensus sequences, regular expressions, profile weight matrices, structural motifs, neural net weights, or graphic structures.

In this chapter, we outline a number of protein pattern methodologies, indicate their similarities and differences, and discuss evaluation methods and likely future developments. The study is organized into four sections: pattern representation, pattern matching, pattern discovery, and pattern evaluation. It is recognized that the choice of pattern representation is a function of the type of information to be included and is therefore a function of the method of discovery. Furthermore, the representation also influences the choice of search and match method, which in turn has an impact on statistical measures of diagnostic ability. Thus, the four sections of this chapter are not fully independent but have been chosen to structure the presentation and disentangle the concepts as much as possible.

PATTERN REPRESENTATION

There are two representation languages involved in describing protein sequence function-structure correlations. First is an *instance language*, in which the universe of known proteins is described. Second is a *description language*, in which the patterns are described. To understand the importance of the

Smith, Lathrop, and Cohen

distinction, one needs to recall that a language contains both syntactical and semantic constructs. The grammatical rules (syntax or algebra) used to describe a particular instance, a given protein, can be very different from those used to describe a pattern identifiable within that same protein. For example, the syntax for describing a protein might allow only a linear sequence of symbols (primitive tokens) representing the amino acids. On the other hand, the syntax for describing a pattern might admit parentheses for grouping, or Boolean operators such as AND, OR, and NOT. The meaning or referent of terms (semantics) can also differ considerably between the languages. This is because the instance language refers to particular components of a specific individual, whereas the description language refers to classes (the distinction between "that brown easy chair in my living room" and "the notion of a chair"). Obviously, the instance and description languages are not independent. For example, if we are to identify the instances containing the pattern, the symbolic alphabet used to describe a pattern must have a well-defined mapping to the alphabet used to express each protein or instance sequence.

The instance language used to represent protein sequence information is normally an annotated linear string of characters chosen from the 20 amino acid symbols. The degree of annotation is a function of the level of our knowledge. For example, if a protein's complete three-dimensional structure is known, the instance representation includes a set of atomic three-dimensional coordinates as the annotation of each amino acid. If substrate cocrystallization information is available, the annotation might include information as to which amino acid atoms interact with the substrate. Any individual amino acid may be annotated with individual, local, or global properties, and these may be either measured or predicted. Properties specific to an individual amino acid, such as its size, phosphorylation potential, tolerance to mutation, or hydrophobicity, are individual properties. Secondary structure designation, charge clusters, or the specification of hydrogen bond pairs are local properties (because they apply to a neighborhood larger than an individual amino acid but smaller than the whole protein). A protein's tertiary structure classification, tissue or compartment location, and catalytic functions are global properties of the whole protein. Local and global annotation are of particular interest, as they often can be calculated or estimated through comparison with homologous proteins or models. In summary, while the instance language is generally a simple linear string of characters denoting amino acids, it can contain considerable context-dependent information attached to any or all of those elements.

The simplest pattern representations are linear strings expressed in the symbol alphabet of the 20 amino acids, plus *place holders* as nulls, *wild cards*, and *indels* (insertions and deletions). Linear string patterns have a very restricted syntax or algebra that specifies only the order of the symbols. The simplest such patterns contain consensus information only. They are generated directly from observed common substrings or conserved regions among an aligned defining set of sequences. Each position in the pattern is assigned the amino acid that occurs most frequently in that relative position among the set of defining sequences. Because consensus patterns are the least sophisticated, and their representation is a restricted form of the more regular expression representation, they will not be discussed further.

Most pattern representations attempt to include at least the observed type and extent of variation at each equivalent position in the defining set and any variation in spacing between the pattern elements. The need to incorporate this variability arises from at least three facts. First, nearly identical structures can be formed, or functions carried out, by very different amino acid sequences. Thus the pattern language should allow various amino acid groupings. Second, in the folded protein, conserved residues that are proximal in space may be dispersed along the primary sequence, separated by regions variable in both length and composition. Thus the pattern language must allow variable spacing between pattern elements. Finally, our knowledge is always restricted to a small sample or representative set of all proteins of any given function or structural type. This suggests that useful patterns should allow for both observed and anticipated variability.

Attempts to model more of the natural variability found in proteins led the development of pattern languages in two directions. In one, the pattern syntax or algebra was extended by adding more flexible and expressive language constructs, permitting more complicated patterns (Chomsky, 1956). This reflected the intuition that proteins form complicated interdependent structures. In the other direction, the pattern semantics were extended. Here, for example, the pattern symbol alphabet is extended by adding numerical weights to the pattern elements, permitting some parts of a pattern to be more important than others. This reflected the intuition that the components of a biological structure are not all of equivalent importance in determining biological response. The first direction gave rise to regular expressions, the second to weight matrices or profiles. The combination has led to various hybrid approaches.

Regular Expressions

Formally, regular expressions are linear strings of a given symbol alphabet —the 20 amino acids—recursively linked by the operators OR (+), CONCATENATE (&), and INDEFINITE-REPEAT (*) (Aho, Hopcroft, and Ullman, 1983). This minimalist formal syntax usually is extended by additional language constructs. A full biosequence pattern-matching program, QUEST, based on the UNIX regular expression language and extended with THEN, AND, and NOT operators on patterns, was first introduced by Abarbanel, Wieneke, Mansfield, Jaffe, and Brutlag (1984). Because regular expressions can be nested to an arbitrary depth, extremely complex patterns can be stated concisely. For clarity, we will distinguish between simple regular expressions, as generally used for genetic pattern representation, and the general fully formal definition previously given. In particular, a simple regular expression does not allow indefinite repeats for any pattern element of sub-pattern with the exception of the null or wild-card character. In addition, arbitrary deep nesting of subpatterns is not allowed.

Whether recognized or not, when we express a primary sequence pattern as an English sentence, it is usually equivalent to such simple regular expressions. The generalized guanine nucleotide-binding domain pattern provides an example (Walker, Savaste, Runswick, and Gray, 1982; Bork and Grunwald, 1990). This can be described as a sequence of a glycine, serine, or alanine; 4 amino acids of "nearly" any type; a glycine; either an arginine or lysine; a serine or threonine; and 6 to 13 amino acids of nearly any type, followed by an aspartic acid. This has a regular expression representation: [G, S, or A] X{4} G [K or R] [S or T] X {6 to 13} D. Here the concatenation between each sequential symbol is implicit: the X is a wild card equivalent to an OR among all 20 amino acid symbols, and the numbers in braces indicate the range of times the preceding element is to be repeated. The particulars of the actual syntax used in the representation are not important. Any such expression that can be reduced to a simple list of alternative linear strings can be written as a regular expression. This ranges from simple majority-rule consensus patterns to the explicit listing of all known example sequences of interest. However, neither of those extremes exploits the full power of the regular expression representation even in its simple restricted form.

The nucleotide-binding pattern regular expression represents a very large set of alternative amino acid strings of length 15 to 21, much longer than the number of distinct proteins composing a human. In fact, it represents more than 10^{20} subsequences. That number is clearly many times larger than the number of distinct nucleotide-binding proteins likely to have ever existed on earth, many of which would surely not form functional binding sites. The power of the regular expression representation rests in the ability to represent a vast number of potential sequences concisely and explicitly. It allows the specification of a few "conserved sites" with the known natural variability, while ignoring the intervening sites for which little restrictive information is available. We know, for example, that not all the combinations allowed by the four wild-card positions between [G, S, or A] and G will be compatible with the loop or beta turn structure anticipated for this region, but our limited knowledge of protein structure constraints is represented by this ambiguity.

Regular expressions per se do not have weights associated with the different alternatives. One of their major limitations is that no notion of differential similarity can be explicitly represented: That is, the pattern itself does not contain the information that mismatching some elements is less important than missing others. It is important not to confuse such *pattern representation* limitations with the *pattern matching* or occurrence search methods, such as dynamical programming (to be discussed later). There, different scores can be associated with the different pattern positions or elements, even though the pattern is represented as a simple regular expression.

Weight Matrices and Profiles

One logical direction in which to extend the expressive power of patterns is to specify weights reflecting the importance or preferences at each pattern position. This allows the pattern to reflect the observation that some positions are more strongly conserved than others and that even the different alternatives at a given position are not equivalent in their information content. The weight matrix provides a mechanism that implements this (Barton and Sternberg, 1987a,b; Gribskov, McLachlan, and Eisenberg, 1987; Boswell, 1988; Hodgman, 1989; Staden, 1989; Hertz, Hartzell, and Stormo, 1990; Bowie, Lüthy, and Eisenberg, 1991). A pattern is represented by a two-dimensional matrix. The pattern positions index the columns, and the rows are indexed by the alphabet of allowed elements (generally the 20 amino acids and the null or gap element). The value of each matrix element is a weight, or score, to be associated with the acceptability of a particular amino acid in that pattern position. Thus, a column (vector) of weights is attached to each pattern position, with large positive weights for acceptable or frequent amino acids and smaller or negative weights for less acceptable amino acids.

These matrix weights are often a transformation of the observed occurrence frequencies. From an analysis of the observed frequency of each amino acid at each position in an aligned set of defining protein sequences, an information or log-likelihood measure can be calculated (Hertz et al., 1990; Bowie et al., 1991). Sometimes a special quantity representing minus infinity is used as a score, permitting the identification of a "forbidden" amino acid at a particular position. Bork and Grunwald (1990) employed a reduced alphabet in which, instead of each position having 20 associated amino acid frequencies, there is a vector of 11 physicochemical properties. Bowie et al. (1991) extended this to include local structural environment information. However, the particular interpretation of the weights is not fundamental to this form of pattern representation and should be considered part of the match evaluation or pattern discovery methodologies discussed later.

Variable-length gaps can be represented in two ways within the weight matrix. A pattern can be represented as an ordered series of weight matrices, each representing a contiguous block of conserved amino acids separated by explicit gap indicators (Barton and Sternberg, 1990) (see discussion that follows on common word patterns). Alternatively, a gap or null-element weight can appear within each pattern position column. Such weights represent the relative cost of inserting a gap at that position in either the pattern or the sequence being searched for a match to the profile. The sum of the weights, one from each pattern position, defines a score for each alternative sequence represented by the matrix. This score is associated with a match having an instance sequence containing that particular alternative. The previous guanine nucleotide-binding domain pattern might have a profile defined as the negative log of the frequencies listed in table 3.1.

Table 3.1 Frequency table representing the potential information used to construct a profile or weight matrix of the nucleotide-binding pattern

Amino Acids	1	2	3	4	5	6	7	8
A	0.10	0.05	0.05	0.05	0.05	0.02	0	0
C	0	0.05	0.05	0.05	0.05	0	0	0
D	0	0.05	0.05	0.05	0.05	0	0	0
E	0	0.05	0.05	0.05	0.05	0	0	0
F	0	0.05	0.05	0.05	0.05	0	0	0
G	0.90	0.05	0.05	0.05	0.05	0.98	0	0
H	0	0.05	0.05	0.05	0.05	0	0.05	0
*	0	0.05	0.05	0.05	0.05	0	0	0
*	0	0.05	0.05	0.05	0.05	0	0	0
K	0	0.05	0.05	0.05	0.05	0	0.45	0
*	0	0.05	0.05	0.05	0.05	0	0	0
R	0	0.05	0.05	0.05	0.05	0	0.50	0
S	0	0.05	0.05	0.05	0.05	0	0	0.42
T	0	0.05	0.05	0.05	0.05	0	0	0.58
V	0	0.05	0.05	0.05	0.05	0	0	0
*	0	0.05	0.05	0.05	0.05	0	0	0

Note: Each column represents a pattern position. Asterisks indicate the rows for the unlisted amino acids.

The inclusion of weights is both a strength and a weakness of the profile or weight matrix representation over the regular expression. As discussed later, the meaningful assignment of these weights generally is constrained by the sample-size bias of the defining set. In the nucleotide-binding example, 441 weights must be assigned. Even with all null positions assigned a neutral weight of zero, there are still 300 weights for 15 potentially independent pattern positions. We rarely have sufficient data in the statistical sense to define that many parameters adequately. These problems are normally considered part of the pattern discovery procedures and will be discussed under that heading, but it must be remembered that profile matrix weights need not necessarily be associated with any particular profile-generating or pattern-matching algorithm. In the limiting case of equal weights associated with each allowed amino acid, the null elements assigned a weight of zero over fixed ranges, and negative infinity otherwise, the weight matrix can be equivalently represented as a simple regular expression representation.

One of the major limitations of weight matrix and some other representations is that correlations between alternatives at different pattern positions are not included. These can potentially be overcome in more complex representations. Our understanding of the close packing of internal residues in proteins and the shape specificity of most substrate binding suggests that

The Identification of Protein Functional Patterns

there must be many residue alternatives within such positions that are viable only if compensating alternatives exist in other positions (Ponder and Richards, 1987; Sibbald and Argos, 1990; Bowie et al., 1991). Thus, the "union" of two patterns, each representing correlated alternatives of A_i if B_j and C_i if D_j, is not properly represented by the simple Boolean combinations, A_i or C_i and B_j or D_j. However, it is by (A_i and B_j) OR (C_i and D_j). The inclusion of this "global" OR operator between alternate patterns creates serious problems for most of the pattern discovery methods, as discussed later. It is sufficient, at this point, to note that in the limit, a list of all known amino acid sequences associated with a particular function ORed together is a legitimate regular expression but one containing no distillation of the information about how the function is encoded. Consequently, one generally restricts the use of OR to alternates at a given position.

Neural Networks

One of the more intriguing methods of representing a protein structural or functional pattern is in the weights and nodal interconnections of a neural net. Neural networks are members of the class of machine learning algorithms developed by Rosenblatt (1962) and Werbos (1974). They are conceptually networks of connected processing units or "neurons" with adjustable connection, combining, or signal-passing weights. The processing units can be designed to generate discrete or continuous output. The output is a function of the sum of weighted inputs plus some set output parameters. The input-to-output function must be considered part of the pattern's representation. In most cases, the output is a sigmoid function of the input, with the midpoint defining the mean output parameter. In the limiting step function case, there would be a fixed-value output if, and only if, the combined input exceeds the triggering or firing threshold.

The distinction between the pattern's representation and the associated match, discovery, and evaluation algorithms may seem the most difficult for neural nets. This is because neural networks are nearly always presented in combinations with the "learning" procedures and positive and negative training sets used to assign (adjusted) the various weights. However, the weights and connectivity or architecture that represents any particular pattern can be defined independently of any matching or training procedures.

At this juncture, only the simplest neural net matching procedure will be outlined to facilitate a conceptual understanding of the net's representation of a pattern (Pao, 1989). The identification of an instance match to a neural net–represented pattern begins by applying the position states of the instance sequence (amino acids or properties thereof) to a set of "input neurons," or the lowest level network nodes. These neurons fire depending on whether or not the state value is above the firing threshold. The resulting output signals are combined (as a function of the network connectivity and passing weights) to act as inputs to the next level of nodes. There the neurons fire if the combined weighted signal exceeds their firing thresholds. This continues until the top, or output-level, node(s) is reached, where a match, mismatch, or indeterminate signal is obtained. In the simplest case, there would be only one output neuron, the firing of which would indicate a match. It is worth noting that this representation is related to the profile-weight matrix. For example, the "perceptron" or "single-layer" neural net (Stormo, Schneider, Gold, and Ehrenfeucht, 1982b) is formally equivalent to a weight matrix. The inclusion of simply connected single hidden layers can code for logical ORs among different weight matrices. The exclusive ORs introduced in the common word representations can be accommodated within neural networks but only if at least one hidden layer is included.

Hidden Markov Model

Considerable research has taken place in the area of speech pattern recognition, using hidden Markov or discrete-state models. Although there have been limited applications to protein sequence analysis, the problems are conceptually very similar. Speech is a linear sequence of elements. Its pattern interpretations are highly complex and very context-dependent, not greatly different from the protein sequence's encoding of structure and function. Such models can be viewed as signal generators, or amino acid sequence generators in our case. The models often are represented as graphs. Each of the N nodes represents a model state. The connecting arcs represent the allowed transitions between the states. The set of states must include a subset of output states. Each output state defines an output on the distribution over some alphabet of symbols. Hidden Markov patterns are thus composed of an N-by-N state-to-state transition matrix and a set of output state distribution functions. A pattern represented by such a model is defined as the set of all possible sequences generated by the model along with their relative probabilities of generation. While hidden Markov models, in the same way as Markov chains, often are conceived as symbol sequence generators, they are employable in the reverse manner—that is, given a sequence of allowed symbols, the model can return the probability that it would have generated such a symbol string. When all the states of the models are output states, each representing a pattern position with the distribution of amino acids expected at that pattern position, similar to the columns in the weight matrix, an equivalence with the profiles or weight matrices can be obtained. White, Stultz, and Smith (1994) have developed a discrete-state model structure representation of the structural profiles similar to those of Bowie et al. (1991). Nonzero transition probabilities between nonadjacent modeled states allow for the "skipping over" of a modeled pattern position (a gap in the alignment), whereas nonzero transition probabilities between a modeled state and itself allow for the "skipping over" of an element (an alignment gap) in the sequence being searched for a match to the pattern.

The output states may be more complex, however, than just a sequence pattern position. They can represent a secondary structure, a supersecondary structure, a hydrophobic run, or even an entire structural domain. Each state may, in fact, be a discrete state model that contains many other states. One could construct a discrete state model of the pattern: an alpha-helix followed by either two hydrophobic amino acids or two to three positively charged amino acids with a two-thirds and one-third probability, respectively. We could connect an alpha-helical state with a hydrophobic run state and a positive-charge run state. The alpha-helical state would contain a set of output position states, each with the expected distribution of amino acids common to such helices.

The model's transition matrix and output-state probabilities, with which each pattern sequence is potentially generated, can be learned from some defining data set, or they can be constructed directly to represent a current level of knowledge. It is the latter that gives the discrete state model representation its potential power. As with profiles, if all the weights are assigned only from the limited set of known proteins of the function of interest, we might anticipate that the pattern representation may not be able to match new sequences of the same function, even though the novelty of the sequence could have been expected from our understanding of biochemistry.

Common Word Patterns

There are general pattern representations that include both the regular expression and profile representations within their vocabularies. We refer to them as *common word sets*. They are sets of "words" common to all members of the defining set of sequences and some set of relationships between the words, such as their linear order. The words themselves can range from a set of amino acid consensus sequences, simple regular expressions, profiles, and even neural nets. Any or all can be combined via the regular expression operators into a complex new pattern. Finally, the common word pattern can contain a set of weights, one associated with each word or subpattern. These could show some measure of the importance of each word in defining the degree of match. This, in turn, could be considered a higher-level weight matrix. Thus, the common word patterns could be represented by complex hierarchies of pattern representations.

The common word sets are a natural means of representing patterns that are composed of a number of short conserved amino acid subsequences embedded in very long and diverse proteins. In the work of Wallace and Henikoff (1992), Roberts (1989), and Smith and Smith (1990), the word sets are strictly ordered sets of short patterns, all of the same representational type. The order refers to the N-terminal to C-terminal order expected for any protein identified as containing the common word pattern.

Other Hybrid Strategies

A number of protein patterns can be best recognized at the secondary or supersecondary structure level. These patterns often are referred to as *structural motifs*. One example is the helical coiled coils (Cohen and Parry, 1986); a second is the helix-turn-helix motif of DNA-binding proteins (Dodd and Egan, 1987). Another is our previous example, the nucleotide-binding domain. This is the central beta-strand–alpha-helix motif of the Rossman fold (Rossman, Moras, and Olsen, 1974) in which a primary sequence pattern is embedded (see regular expression and weight matrix examples cited previously). Complex structures such as alpha-beta barrels have been described as an eightfold repeat of a simple beta-strand–turn–alpha-helix motif (Blundell, Sibanda, Sternberg, and Thornton, 1986). These motifs often are expressed as strings, but their three-dimensional information is not easily represented by simple regular expressions or profiles. For example, the helix-turn-helix DNA-binding motif ideally includes the spatial information about the angle between the helix axes and the relative orientation of their hydrophilic surfaces. This could be included through correlations between allowed amino acids at widely separated pattern positions. However, the representation of such motifs has not been uniquely formulated. These motifs are most often represented by a prototypical example including a set of alpha-carbon coordinates from the analysis of crystal structures. They need not be drawn from consecutive residues along the primary sequence. The Rossman fold, for example, consists of a set of three beta-strands linked by alpha-helices and short loops embedded in a five- or six-stranded beta-sheet. The central strand is connected through its C-terminus to the returning helix by the GXXXXG-containing loop. The motif definition does not, however, specify the strand connectivity. This is necessary as there are different interstrand connection topologies (the order of strands in the sequence is not correlated with their three-dimensional spatial layout) among the wide range of nucleotide-binding proteins containing this motif.

Structural motifs can be at least partially represented as linear strings, regular expressions, or weight matrices by expanding the pattern element alphabet to include secondary structure elements, length variation of the elements, and spacers. Such linear representations are generally of an unfolded structure only, and the spatial arrangement among the elements is represented only through the correlation of information between particular linear elements. A pattern representation may have the character of an object-oriented database, in which much of the pattern information is hierarchical or inherited though the properties attached to the elements as members of an object class. Hierarchical patterns allow the natural hierarchy of structural organization present in proteins to be reflected in the pattern representation. Thus, they permit complex representations to be assembled from simpler representations that may be drawn from a large library of pattern components.

A hierarchical graphic description can be used to represent a structural motif in which different description or element alphabets are used at each level. Different pattern element alphabets can thus be linked together, forming a pattern wherein the elements at one level of description are related to one another through their association with elements on a higher level. This description procedure uses the so-called hierarchical pattern descriptors (Lathrop, Webster, and Smith, 1987).[1] The result is a linkage relationship between the two sets of primary sequence elements via the secondary structure elements. Rather than a fixed range of intervening primary elements, the pattern contains two secondary structures of arbitrary primary element lengths. At each level in the linked hierarchy, match and mismatch weights can be assigned. The sum of those weights defines the degree of match of the pattern to a particular instance. Such representation of a complex pattern by a set of hierarchically linked objects with their associated lists of properties and weights is independent of the methods employed to locate such patterns in real sequences.

Through a more expressive syntax, including hierarchical pattern definitions, it has been possible to extend significantly the convenience and power of regular expression languages, as in the PLANS software of Abarbanel (Cohen, Abarbanel, Kuntz, and Fletterick, 1983). Here a pattern can include a complex set of rules that concern not only allowed combinations of subpatterns but also the instance language context properties governing which rules apply within that instance sequence context. The PLANS software pioneered an explicit knowledge-engineering approach to the organization of patterns as knowledge, interpreting patterns as modular rules encoding expertise about elements of protein structure. The ALPPS software (Cohen, Presnell, and Cohen, 1991; Presnell, Cohen, and Cohen, 1992) significantly extended this approach by adding a metaknowledge component that controls the underlying rule-based system. For example, matches to a subset of pattern elements could selectively control which other elements were "visible" and thus define the quality of the evidence on which an overall match was determined.

Searls and Dong (1993) persuasively argue that some biological sequence features (e.g., palindromes or tandem repeats) require more expressive power than regular expressions, possibly requiring grammars with context-free or greater expressiveness (Chomsky, 1959). Representing the pattern as production rules in a formal grammar naturally admits hierarchical pattern development in terms of nonterminal productions in the grammar. Searls's implementation admits as features parameter passing, procedural attachment, terminal replacement, a controllable degree of mismatching, and a method for attaching a scoring matrix to consensus patterns. All these features extend the expressive power of Searls's method beyond the context-free limit. (Where mechanisms such as explicit procedural attachment are admitted into the pattern representation, the distinction between pattern representation and pattern match algorithm becomes blurred.) The general implementation of a higher-order grammar as the syntactical structure for protein functional patterns must await future development.

In an annotated linear sequence, it generally is possible to represent the unfolded structure only, and the spatial arrangement among the elements can be represented only through the correlation of information between particular linear elements. It is as if the protein were picked up at each end of the sequence and gently stretched until the supersecondary motif had unfolded but before the secondary structures unraveled. Pieces of the motif that come from widely disparate portions of the chain generally cannot be associated unless each piece is distinctive enough to be recognized on its own. However in many cases, enough of the motif arises from a local region of the chain to be locally recognizable. For example, the primary sequence elements—G followed by K or R, then S or T—in the nucleotide-binding fold can be linked with the N-terminus of a nearby predicted alpha-helix. Next, the helix is linked to a predicted (following) beta-strand, which is linked to a following D. Constraints on allowable spacing variation may be imposed between every pair of adjacent elements. Weights may be associated with whether or not any particular element is observed. The result is a linkage between two sets of primary sequence elements via the spacing of secondary structure elements. Such a representation of a complex pattern by a set of linked objects is independent of the methods employed to locate such patterns in real sequences (although the match method employed will affect where and which matches are found).

Hierarchical Patterns

A hierarchical pattern is built recursively from simpler patterns. In turn, the patterns are built from still simpler patterns. Hierarchical patterns allow the natural hierarchy of structural organization present in proteins to be reflected in the pattern representation. This is useful because many of the biological structures of interest (e.g., supersecondary structure motifs) are most effectively expressed in terms of intermediate structures such as secondary structure and not directly at the lowest level of description (e.g., individual amino acids). Additionally, handling patterns in small modular pieces encourages selective pattern refinement and facilitates building a reusable library of commonly occurring lower-level patterns.

A hierarchical pattern language typically will divide into a number of primitive pattern elements and a method for combining pattern elements into a higher-level composite pattern. Once defined, the composite pattern becomes available for use in still higher-level patterns. The set of accessible patterns in a given hierarchical language is thus the transitive closure of the primitive pattern elements under the combining method. There is a correspondence between the patterns in a hierarchical pattern language, the pattern-action rules in a rule-based expert system, and the grammatical productions in a formal grammar. Hierarchical pattern languages often espouse an explicitly knowledge-engineering approach (Abarbanel, 1984) similar to that employed in a rule-based expert system, assuming that the researcher employing the system will craft the rules (patterns) to expose the specifics of the problem under study.

Hierarchical patterns often use an object-oriented methodology so successful in database applications (Won, 1990). This permits a clean separation between the data structures and the programmed match behavior. Objects are defined as belonging to various classes and are said to inherit those properties assigned to the class. New pattern elements with new match behavior can be easily added to the pattern language. This allows a potentially more complex (hence richer) representation than is possible in a simple linear string of characters. While the richness of the language (Chomsky, 1956) allows more of nature's complexities to be accurately modeled, a richer language may also permit more spurious coincidences to be detected (see under the heading, "Pattern Evaluation").

The same representational mechanism also makes it easy for hierarchical pattern languages to accommodate individual and local annotations of the protein, as described earlier. These might come from literature accounts of experimental data or from another program in a preprocessing step. For example, a program for detecting charge clusters might annotate the protein with specific clusters at specific locations. By an appropriate choice of interface, these can be recorded on the protein in such a way that they "look like" the results from a previous pattern match. They can then be used as a pattern-element in higher-level patterns, and the match process will function correctly because the underlying representation is the same.

PATTERN MATCHING

Most pattern discovery methods involve the identification of the best match between sequences and the developing pattern. Before discussing the various discovery methods, we will outline briefly the basic pattern-matching methods used. However, first we note that the verb *to match* has been used with two different meanings in the literature, generating some confusion: It is used to mean both "to attempt" to find an instance containing the pattern, and "to succeed" in that attempt! We will try to avoid this confusion. We would also note that there are a number of pattern-matching approaches that are purely statistical and are not considered in what follows. These include, at one extreme, the attempts to characterize protein-folding class patterns by gross statistical properties such as their amino acid composition. At the other extreme are the sophisticated statistical measures used to define various sequence amino acid "property runs" (Guibas and Odlyzko, 1980; Karlin, 1990; Karlin and Macken, 1991).

The three most prominent methods are finite-state machines, weight matrices or profiles, and dynamical programming, all of which have been exhaustively described in the literature (Wulf, Shaw, Hilfinger, and Flon,

Smith, Lathrop, and Cohen

1981; Stormo, Schneider, and Gold, 1982a; Sankoff and Kruskal, 1983; Waterman, Arratia, and Galas, 1984; Gribskov, McLachlan, and Eisenberg, 1987). We outline each briefly here.

Finite-state machines, or finite automata (Hopcroft and Ullman, 1979), are "devices" whose current state depends on some maximum (hence finite) number of past state values and the current input. These are used to identify a match to a regular expression. In the pattern-matching application (Dardel and Bensoussan, 1988), each sequence element sequentially causes the device to change state deterministically depending on the element value and the automata's current state, the current state being a function of the previous inputs and the initial state only. If, after processing an entire sequence, the device is left in any state but its final state, no match is considered found. Such devices often are represented as a matrix in which the rows are indexed by the input value and the column indexing is by state. The value indexed by any row and column pair then is the next state (column) to which the device is to move, given the current input and state. Such matrix indexing is straightforward to program on a computer and can provide a very fast search tool. It can be proven that the class of patterns (languages) representable as regular expressions is identical to the class recognizable by finite-state machines (Minsky 1968). With all its potential simplicity and speed, the finite-state machine is limited. For example, the set of all palindromes, including subsets composed of only a few symbols, cannot be found by any such device (Minsky, 1968). In their normal form, they cannot recognize a near or partial match to a regular expression pattern, although there have been recent extensions in this direction (Myers and Miller, 1989). For a particular pattern, a partial-match finite-state machine can be designed. However, there is no general algorithm that will construct an efficient machine for an arbitrary regular expression with an arbitrary degree of tolerated mismatch.

There are two popular weight matrix pattern-matching methods. These are the simple position weight sums and the use of dynamic programming. The former is extremely simple. Each column of the matrix is scalar "multiplied" (the dot product) by an instance position vector, sequentially for succeeding in the instance sequence. The instance vectors are the length of the pattern element alphabet and contain all zeros except at the element value corresponding to the amino acid occurring at that position, where they contain a one. The sum of all matrix column dot products represents the value of the match across those positions. A simple search over all starting positions in the instance sequence allows one to identify the maximum valued match. If the pattern definition includes a minimum match value, a match to any instance sequence is one in which the sum of matrix dot products equals or exceeds that value.

Dynamic programming algorithms provide a means of identifying the nearest or best match of various pattern representations to an entire sequence, or even to another pattern. Dynamic programming was introduced to molecular biologists as a minimal string edit measure of the number of replacements (substitutions) and indels required to convert one sequence into another (Needleman and Wunsch, 1970; Sankoff, 1972; Sellers, 1974). The basic idea is to find the minimum cost required to convert one sequence into the other in a symmetrical manner (a cost being associated with each element conversion). Given the minimal total cost, the dynamic programming algorithms allow construction of the associated alignment between the two strings. The alignment is a linear one-to-one mapping of each element of one string onto either an indel or an element of the other string. In the case where one of the strings is a linear pattern and the other is an instance protein sequence, the alignment identifies those elements in the sequence to be associated with each pattern element in the optimal match. The method has been generalized in a number of ways (Waterman et al., 1984). For example, dynamic programming can be used to find the most similar subsequence between any two strings (Smith and Waterman, 1981).

The power of the dynamic programming method is twofold: First, its mathematical properties are well understood, including a proof that the method always finds the minimum-cost optimal alignment. Second, very complex cost functions can be employed. They allow the pattern matching to include all possible amino acid substitutions or mismatches and indel weights. In addition, these cost functions can vary for different positions along the pattern. Such properties have allowed the use of dynamic programming for identifying the optimal match between many different pattern representations and any sequence. In particular, the optimal match between any regular expression (Smith and Smith, 1990), or profile-weight matrix (Bowie et al., 1991) and a sequence can be obtained. In the former, it allows one effectively to use regular expressions with mismatches. In the latter case, the optimal profile match is that which maximizes a weighted sum of profile column values.

A more complex pattern-matching method is the "time series" filtering method used for patterns represented as Markov, hidden Markov, or discrete-state models. As noted earlier, these models have been used in the domain of speech pattern recognition for many years (Moore, 1986). The filtering algorithms employed (White, 1988) provide a means of calculating the probability that any particular sequence (time series) was (could have been) generated by the discrete-state modeled pattern. This is done by properly summing the probabilities of generating the sequence along all possible paths through the model. The limitation is that such probabilities, being generally very small, and exponential function of pattern length are simply interpretable only as a relative likelihood. Thus, one must compare the match or likelihood of one model against another.

Neural nets identify a match between the pattern and a sequence by identifying those subsequences that, when taken as signals to the input nodes, cause the output layer "neuron" or node to fire. Normally, there is a range of inputs, given a network's connectivity, transmission, and threshold weights, that will allow a signal to propagate through to the output node. As noted

Smith, Lathrop, and Cohen

previously, hidden layers providing additional interconnectivity between the input and output nodes allow the net to incorporate much of the logic of finite-state automata and the complex weighting of the profile matching with dynamical programming. However, the net also may encode a global OR among the known or defining examples. The latter is referred to as *learning the training set*.

Finally, there are a number of hierarchical (Abarbanel et al., 1984; Lathrop et al., 1987; Taylor, 1988; Thornton and Gardner, 1989; Presnell et al., 1992) and grammatical pattern-matching methods. They may be viewed as hybrid strategies wherein the pattern-matching methods previously described are used to identify the lowest-level pattern elements, or primitives. The matched primitives then form a new string over which one searches for matches to successively higher-level pattern elements in a recursive fashion. Thus, a high-level element is matched to a protein sequence through its match to the required combination of lower-level pattern elements. The procedure of Abarbanel and Cohen (PLANS) (Cohen et al., 1983) employed a regular expression match at the lowest level, whereas higher-level pattern elements were matched to those constructed from the lower level using a rule-based expert system approach (Lindsay, Buchanan, Feigenbaum, and Lederberg, 1980; Davis and Lenat, 1982). Presnell and Cohen (ALPPS) (Presnell et al., 1992) extended this by adding a metalevel that provided explicit control of the match process. This permitted groups of patterns (rules) to be switched on and off, the results of previous pattern matches to be hidden or exposed, and so on. Lathrop et al. (1987) employed user-definable primitive elements at the lowest level. Their match behavior was determined by fragments of program code attached to the primitive elements' definition. Higher levels were matched using an A* search (Winston, 1992) that accommodates inexact matches and complex scoring schemes. Taylor (1986) employed weight matrices and dynamical programming.

The computational and conceptual power of these approaches resides in their ability to match very complex patterns in a descending hierarchy of decreasing complexity. The process is iterative, identifying matches and then forming allowed combinations of previously assembled intermediate matched elements into still higher-order elements. This can be continued until the final pattern match either is or is not found. The human language or grammar analogy is useful here. The English language can be decomposed into a hierarchical graph (Walsh and Walsh, 1939). At the top, one can define very complex classes of sentences as a pattern for which a match means the identification of the proper combination of particular categories of phrases. The phrases are identified (matched) with combinations of proper word categories—verb, noun, and so on. Finally, the actual instance words matching these categories are identified. The final pattern need never be described or defined in terms of the primitive English words (sequences of amino acids) if at each level only combinations of lower-level elements that can form acceptable next-level elements are kept. These hierarchical matching schemes generally are called *rule-based grammars*. Earlier characteristics of hierarchical pattern matching for protein include a set of primitive pattern elements with defined match semantics (behavior), a method for matching composite patterns by recursively matching the elements and combining the results, and the ability to treat an extended subsequence of the protein as a modular unit (e.g., a previous pattern match). In this last respect, pattern matches become identical to what we earlier called *local annotations*.

PATTERN DISCOVERY

Many of the currently recognized protein sequence patterns were discovered by direct (noncomputer) observation of common features among some sets of proteins. These involved the same logic used by the majority of computer-based methods. First, proteins with a common or suspected common function are identified as a "defining" set. Second, common features are sought that correlate with that function. This generally begins via sequence alignment methods. The important aspect of the logic is implicit in "correlate with function," which suggests that, in addition to the defining set, there is always a negative or control set, even if it is not explicitly defined such that the recognized features, while common to the defining set, are not so in other proteins.

Common Word Sets

Perhaps the simplest pattern discovery approach is that leading to a list of common words. One begins with some definition of a word. These definitions can be very complex, including classes of amino acids and regular expressions of amino acids or profiles, or as simple as any common four consecutive amino acids. A list then is constructed of words common to the protein sequences in a defining set or even some significant fraction thereof. Normally, the common word pattern then is defined as the maximum set of common words having the same relative order within each defining sequence. It is worth noting that the simplest form of such patterns is related to the current standard "hash-coding," sequence-similarity search methods used by FASTA (Pearson and Lipman, 1988) and BLAST (Altschul, Gish, Miller, Myers, and Lipman, 1990). In these, a dictionary of all words in two sequences is compared for a common set, and then subsets of the same linear order are identified.

The actual algorithms for efficiently identifying both the common words and their relative order can be difficult, even when they are simple symbol strings (Galas, Eggert, and Waterman, 1985; Mengeritsky and Smith, 1987; Grob and Stuber, 1988). Waterman and Jones (1990) find a succession of consensus word sets within a partial alignment, then construct from these an overall consensus sequence composed of a sequence of word sets, each work set being composed of similar words (e.g., one mutation apart). A consensus sequence then is generated by aligning the consensus words across all the protein sequences, shifting and concatenating them.

Smith, Annau, and Chandrasegaran (1990) devised a common-word method for cases wherein there is minimal primary sequence similarity (and hence wherein accurate multiple sequence alignments generally are not possible). Matches to all possible 3-amino acid patterns of the form aa_1-d_1-aa_2-d_2-aa_3, where the aa_i is amino acids and the d_i is spacings or null elements of 0 to 24, are accumulated into an array. Segments of the proteins containing those three patterns that occur most frequently are aligned. The block of neighboring amino acids then is scanned for similarities, permitting a consensus sequence or profile to be developed. The blocks then become the pattern-defining words. This method has been extended by Henikoff and Henikoff (1991) to produce an automated pattern discovery tool in which patterns are defined as an ordered set of "word blocks" in the form of weight matrices. This has been done for each of 437 defining protein sequence sets.

Karlin and coworkers have developed a number of methods for identifying statistically unexpected words or runs (Karlin, Blaisdell, Mocarski, and Brendel, 1989a; Karlin, Ost, and Blaisdell, 1989b; Karlin, 1990). Here the words are discovered as runs of some amino acid or class of amino acids that occur at some threshold above that expected for the composition of a reference set of sequences. This reference set need not be a negative control set, but merely a representative set of possible proteins. Sequences sharing sets of such words or runs are identified as having a common pattern. These methods are conceptually related to other attempts to identify runs and repeats (Rhodes and Klug, 1986).

Multisequence Alignment Methods

Many pattern discovery methods rely on a sequence-similarity alignment method. Normally, a local maximum-similarity dynamical programming algorithm (Smith and Waterman, 1981; Waterman and Eggert, 1987) is used to obtain all pairwise alignments or a multiple sequence alignment among a training or defining set of examples (Barton and Sternberg, 1987a,b; Henneke, 1989; Vingron and Argos, 1989; Smith and Smith, 1992). A pattern is constructed by assigning to each consecutive pattern position the consensus of a common property, or set, of amino acids observed in each aligned sequence position. This assignment normally is restricted to some maximally similar domain (e.g., a known amino acid physicochemical class). If the pattern is to be represented as a regular expression, the observed amino acid types in the aligned positions become alternatives (ORs) in the expanding pattern. If the pattern is to be represented as a weight matrix, the matrix column vectors can be derived from some transformation of the relative amino acid frequencies observed at the aligned sequence positions. If the pattern is represented as a minimal covering pattern (Smith and Smith, 1990), the smallest predefined class of amino acids containing all observed amino acids is assigned. Gaps in the defining alignments can be directly included as pattern-spacing variations, with the inclusion of indels in the regular expression or weights in the indel row of a profile.

There are several differences in profile discovery methods: the alphabet used, the function used to assign the matrix weights (given the frequency of observed element types at each position), and the determination of the threshold defining a match. The simplest alphabet is the 20 amino acids and a gap, or indel, symbol. In the physicochemical property vector representation of Bork and Grunwald (1990), whether each amino acid property is present in a particular position is indicated by *always*, *never*, or *sometimes* in the appropriate matrix element. This approach reduces the information detail from the defining set and thus reduces the sampling biases of the particular defining set. The simplest match threshold is the lowest sum of weights, one from each column associated with the observed element type, allowing a match to all members of the defining set. One of the most common weighting functions is the logarithm of the observed frequencies. In that case, the profile weights within a single column encode a measure of the information content of that pattern position. If the logarithm of the observed amino acid frequencies is divided by the "expected" frequencies, a log likelihood measure is obtained:

$$I_{ij} = -\log(f_{ij}/e_j)$$

Here f_{ij} is the observed frequency of the jth amino acid, class, or gap in the ith aligned position, and e_j is the expected frequency of jth amino acid, independent of position. This normally is viewed as a relative measure of the information in the defining positions as compared to a random sequence of similar composition. However, the e_j's could be calculated from a more conservative assumption—for example, the set of all sequences matching the pattern at a score between some lower threshold and that producing matches to the defining set.

These multisequence alignment methods may fail for defining sets having widely diverse primary sequences or multidomains, some of which are shared by all members of the set. Either no consistent alignment is found or the common pattern is not of statistical significance (e.g., too few positions). Additional information can often be used to constrain the alignments, such as secondary structure (restricting gap placement to surface loops [Smith and Smith, 1992]) or experimentally determined functional sites (Henneke, 1989; Fischel-Ghodsian, Mathiowitz, and Smith, 1990). However, in many cases, this still does not provide enough constraints to generate a statistically significant pattern directly from a multisequence alignment. Thus, many of the pattern discovery methods use an iterative approach or a statistical evaluation of potential pattern elements as an integral part of the discovery process. Some of these are outlined next. (As noted earlier, we will discuss the problems associated with pattern evaluation in the final section of this chapter).

Iterative Methods

One approach to overcoming the limits on multisequence alignment procedures is to use an iterative pattern discovery method. Taylor (1986), Patthy (1987), and Smith and Smith (1990) have developed such methods. They exploit the fact that a pattern induces a local multisequence alignment when it successfully matches more than two sequences. In the simplest case, the alignment associated with the two closest or most similar sequences is taken as a starting point. This alignment is used to generate a consensus regular expression, profile, or other pattern representation over the domain of highest similarity. In the approach of Patthy (1987) the pattern then is matched against the whole defining set, allowing a certain degree of inexact pattern element matches. The sequences matching this previous pattern best are used to construct the next pattern from the alignments with the previous pattern. This updated and more general pattern then is used for the next iteration until there is no change. In other approaches, such as that of Smith and Smith (1990), the new pattern is optimally aligned and updated with the next most similar sequence defined by an independent dendrographic (tree) relationship among the defining set. This process, which can be generalized in various ways, then is iterated until all the positive sequences have been included. Jones, Taylor, and Thornton (1992) have implemented an iterative scheme on the massively parallel Connection Machine computers using the Smith and Waterman local dynamical programming method (Smith and Waterman, 1981) to define the optimal similarity alignments at each iteration. Staden (1989) discusses several other related discovery iterative methods based on probability-generating functions.

The amino acid class-covering pattern procedure of Smith and Smith (1990) uses a predefined set of amino acid classes based on their physicochemical properties. The updated pattern is constructed from the alignment at each iteration by identifying the smallest amino acid class that covers (includes) the amino acids or previously defined covering class at each aligned position. The closely related iterative procedure of Taylor (1986) starts with amino acid templates (patterns of amino acids and classes of amino acids in association with structural features) initially derived from a small number of related sequences whose tertiary structures are known. Modified or refined templates are generated that represent ever more general features conserved in an increasingly larger set. These related sequences are of unknown structure but are believed to be homologous. The use of sequences with known tertiary structure as a starting point allows better control of the generalization process. The insertion of gaps in the alignment is restricted to regions that are not part of well-defined secondary structures. The work of Bowie et al. (1991) is closely related. The profile alphabet is a generalized set of known structural environments, and gap weights are a function of well-defined secondary structure positions.

One of the automated pattern-construction techniques without an explicit intermediate alignment step was developed by Stormo and colleagues (Stormo and Hartzell, 1989; Hertz et al., 1990). An initial defining sequence is subdivided into the set of all possible k-mers (subsequences of length k), where k is the nominal or expected size of a potential common pattern. These k-mers define an initial set of weight matrices. The optimal match for each of the k-mer matrices then is located among the remaining members of the defining set. Based on this match, each individual k-mer matrix is revised by updating amino acid occurrence frequencies or their transform with the alignment to the best, highest-scoring match. The defining set is reduced by removing the member optimally matched. The process then is iterated until no sequences remain. The matrix having the highest final sum of scores (in this particular implementation, the matrix having the highest information content) is considered the profile representing the common pattern. Their method was generalized to a log likelihood measure in an iterative framework by Lawrence and Reilly (1990) and was given a thermodynamical interpretation by Stormo (1990). The method appears very robust. If the defining set contains a small fraction of random sequences or nonmembers of the defining functional set, the final best profile is affected minimally. The major drawback in this procedure is that no negative control set is explicitly employed. Instead, the information measure used is a relative measure based on expected amino acid occurrence frequencies, equivalent to a random negative control.

None of these iterative methods test the updated pattern at each iteration against a negative control set of sequences. If alternate choices exist as to which pattern elements to include or in which alphabet to represent them, the decision is made only on some optimization function over the defining set. Thus, a less "optimal" pattern, but a more specific one, can be missed. This is one of the problems addressed by the machine-learning methods outlined in the next section. However, there is no inherent reason why most of the methods could not incorporate alternate element choices or weights by testing at each iteration against a negative control set (see under the heading, "Pattern Evaluation").

Machine Learning Methods

Machine learning is a somewhat ill-defined category, particularly in the area of protein pattern recognition. Whereas the use of neural network training algorithms is clearly considered to be machine learning, the automation of some of the preceding methods, when coupled with a knowledge or rule base and/or negative control set feedback, might equally be considered machine learning. We have arbitrarily separated the machine learning methods into those requiring a direct or iterative multisequence alignment and those that do not. This recognizes that there is additional information in a multisequence alignment that the "machine" does not have to learn.

Smith, Lathrop, and Cohen

In machine learning methods involving neural nets, the pattern is encoded as a set of nodal thresholds and transmission weights and the net connectivity. Within limits, the machine learning involves only the learning of the optimal weights. The maximum connectivity is specified in advance. We note that a node-to-node transmission weight of zero, or a firing threshold of infinity, effectively reduces the net's connectivity. Typically a neural net is alternately "shown" the defining and negative control sets, and the weights are updated continually. This can be a very computationally intensive process. The assignment of weights normally involves a backpropagation strategy (Rumelhart, Hinton, and Williams, 1986). Weights are adjusted to improve the network's performance on a set of training sequences containing both positive and negative examples. Once trained, the net can be used to ascertain whether another sequence contains the pattern encoded by the weights.

The inclusion of hidden layers (intermediate nodes between the input and output) can allow a neural net to incorporate nonlinear information such as correlations between widely separated pattern positions. In addition, the network's pattern representation can be independent of our understanding of how the different alternatives are encoded in the sequence. This is both a strength and a major disadvantage, for it is very difficult to determine what features compose the encoded pattern. Thus, it is difficult to relate the various aspects of a pattern to structure or function. There are neural net analysis tools available that allow various correlations and other encoded features to be identified, but this remains a complex problem. The backpropagation (Rumelhart et al., 1986; Cherkassky and Vassilas, 1989) algorithm is used to adjust the weights for a neural net trained to recognize or classify a sequence or subsequence (Holley and Karplus, 1989; Bengio and Pouliot, 1990; Kneller, Cohen, and Langridge, 1990). This algorithm corresponds to a steepest-gradient descent in the weight-space to determine the minimization of the overall error. The error being minimized can be any of a number of measures on the success of the classification of correct "matches" in the defining and negative control sets. A number of modifications in the basic training paradigm may be made in any particular implementation. Finding methods that converge most rapidly to the desired weights, or that do not oscillate between weight sets, still constitutes an active area of research. In some applications, it is necessary to know an approximate solution so that the gradient minimization algorithm can avoid local minima.

The input of the neural net need not be the sequence or even some linear string of correlated properties. The input can be some set of overall statistical properties, such as composition, lists of common words or subpatterns, and even information from multisequence alignments. However, in cases where an entire sequence is the input, the regions of interest must be identifiable (in order to be classified into positive and negative categories for the error function calculation). Any long sequence can contain both negative and positive domains. Thus, the use of neural networks has been largely restricted to predicting local features such as secondary structure and intron and exon boundaries, given some sequence "window" in which classification can be defined.

It has been assumed by many researchers that the neural net learning surface is smooth and well-conditioned. This is not necessarily true. Work by Kneller, Cohen, and Langridge (1990) and McGregor and Cohen (personal communication) suggests that the complexity of protein secondary structure may resemble the conformational energy surface of a polypeptide. Thus, to avoid local minima, some attention must be paid to the choice of the starting weights for the training computation. Convergence to a global minimum is difficult to demonstrate. This is particularly relevant for the "jackknife" validation of a network where, once the network has converged to a final set of weights, each member of the learning set is removed individually, and the weights are reoptimized. It is assumed that the network retains no memory of the excised example, and the predictive accuracy is quantified in these cases. Because retained memory seems likely, the jackknife strategy cannot be invoked reliably unless the network is retrained, starting from a random set of weights.

The idea of using the computer to learn a set of properties to discriminate one set of sequences from another is not limited to neural networks. As early as 1985, attempts were made to using machine learning and the concepts of expert system analysis (Haiech and Sallatin, 1985). Although such attempts were hampered by a lack of both sufficient data and experience in protein pattern analyses, the motivation was correct and has been continued (Gascuel and Danchin, 1986; Fishleigh, Robson, Garnier, and Finn, 1987; Lathrop, Webster, Smith, and Winston, 1990).

Gascuel and Danchin (1986) employed an explicit machine learning approach to pattern discovery. The pattern representation is built on a series of feature extractors, each of which extracts a feature from a sequence. The system contains a language for describing a wide range of sequence feature extractors. For example, (NUMBER X) might yield the number of amino acids of type X, or (SPACING X Y) might yield the distribution of spacings between all amino acid pairs of type X and Y. There is a separate grammar for combining these extracted primitive features into a complex pattern. The interaction between these primitive features and the combining grammar defines the pattern space. A learned or discovered pattern is considered to be any combination of features that discriminate the set of interest from other sequences, a negative control set. The software, PLAGE (Gascuel and Danchin, 1986), executes its functions in three steps. First, it generates the pattern space based on the available features. Second, it searches for matches between versions of each pattern from the most general to the most specific. Third, it evaluates the extent to which any of the pattern versions discriminate between the two sets.

The ARIEL software (Lathrop et al., 1990) also adopts an explicit machine learning approach. The aim is to identify hierarchical patterns characteristic of a structural motif that correlate with a particular function. The search begins with an initial pattern. In practice, this could be a primary sequence, regular expression, or a more complex pattern. Next, a higher-level "vocabulary" of protein structural properties inferable from the primary sequence or from any lower-level inferred properties must be defined. These inferences can employ various external sources (e.g., alignments, x-ray structures, statistical correlations, empirical or heuristic prediction algorithms). Both the defining and negative control sets then are annotated with this vocabulary. Every occurrence of the lowest-level defined inferred properties is located. Given the initial pattern and the annotated control sets, a feedback loop is established between the current pattern matches in the two sets and a pattern refinement procedure. Each pattern is represented as a hierarchical structure, and the evaluation of matches is carried out within that context (Lathrop et al., 1987). The procedure is run in a computer user–assistant mode. The user suggests what vocabulary elements are to be considered, and the system identifies the effect of refining the previous patterns by addition, generalization, or specialization. This involves the calculation of both the sensitivity and the specificity of all combinations on the defining and negative control sets. The user then receives a refined pattern set that maximally improves these in accordance with some user-defined function. The very large number of next iteration patterns obtainable from all combinations of suggested element additions and generalizations has been handled by implementation on a massively parallel system (Lathrop et al., 1990).

Most of the machine learning methods that use a defining and negative control set associate weights with each pattern element. This allows pattern elements common to every member of the defining set to be weighed as a function of whether they are also relatively common within the control set. Whereas this is inherent in the neural net procedures, it is not in others. In the ARIEL software (Lathrop et al., 1990), for example, a weight can be associated with each pattern element that is inversely proportional to the decrease in specificity (see later) on deletion of that element. This is related to the profile or weight matrix when the log likelihood measures are used. There the values approach zero if the amino acid preferences at any pattern position approach those expected by chance.

PATTERN EVALUATION

Basic Statistical Considerations

The literature is replete with patterns believed to identify sequence-structure or sequence-function correlations, most relying on intuitive "model" construction (Dickerson, 1971; Rossman et al., 1974; Von Heijne, 1983; Wierenga, Terpstra, and Hol, 1986; Pavletich and Pabo, 1991; Neer, Schmidt, Nambudripad, and Smith, 1994). Occasionally, these patterns are the product of a series of experiments directed at probing the precise conformational and functional role of particular amino acids. Some have been derived from carefully selected defining sets, but most have not been fully evaluated as to their diagnostic utility. The diagnostic ability of patterns often is overestimated at best, or underreported at worst. Traditionally, such a deficiency could be addressed by constructing strict training sets and a test subset from the defining functional set and negative control set. The first would be used in pattern development, whereas the sequestered test set would be used for evaluation. The behavior of the pattern on these sets then would generate measures of the diagnostic capability of the pattern. If, for example, the negative control set cannot be distinguished from the positives in the test set, then one concludes that the pattern is not a useful discriminator. Although there is usually a discrepancy between the performance on the training and test set, if the discrepancy is large, it might be appropriate to conclude that the training set had been memorized. This can happen in patterns represented as either neural nets or sets of simple common words. For example, there is a reasonable likelihood that any small set of unrelated sequences can have some set of amino acid words in common that would, purely by chance, be minimally shared as a set with all other (nonhomologous) proteins. We will return to this problem in a discussion of the descriptive language's richness.

At a minimum, two quantities should be reported for any pattern: the sensitivity and the specificity (type I and type II errors):

Sensitivity $= TP/(TP + FN)$

Specificity (selectivity) $= TN/(TN + FP)$

where TP is true positives, FN is false negatives, TN is true negatives, and FP is false positives. Here TP is the number of defining (or positive) set sequence matches and FP is the number of control (or negative) set matches. The sensitivity of a pattern (e.g., diagnostic of zinc fingers) reflects the likelihood that all occurrences (e.g., all "real" zinc fingers) have been, and thus will be, found. It reflects how sensitive the pattern is as a detector (e.g., of zinc fingers). The specificity is a measure of the probability of correctly identifying the negative instances. It reflects how specific the pattern is as a detector: (For instance, does it find only zinc fingers?) The diagnostic ability of a pattern can thus be conveniently summarized in the normal two-by-two truth or contingency table, given a positive and negative control set.

There are other measures that have been defined from such a truth table. These include:

Accuracy $= (TP + TN)/(TP + TN + FP + FN)$

Positive predictive value $= TP/(TP + FP)$

The computer science and statistical pattern-recognition literature is replete with discussions of estimating class membership using the Bayesian approach (Pao, 1989). Hence, it may be useful to identify the relationship between these discussions and sensitivity and specificity. Let p_i be a pattern and f_j the protein structure or functional class. One can then define the relationship between the conditional probabilities from the definition of joint probability. This is known as Bayes's relation:

$$P(f_j|p_i) = \frac{P(p_i|f_j)P(f_j)}{P(p_i)}$$

Here $P(f_j)$ and $P(p_i)$ are the probabilities of a protein having the function (f_j) and matching the pattern (p_i) respectively. The conditional probability, $P(f_j|p_i)$, of having the function (j), given that you have matched the pattern (i), is referred to as the *posteriori probability*. The conditional probability, $P(p_i|f_j)$, of having the pattern given the function, is referred to as the *prior probability*. If these probabilities can be taken from the observed frequencies in the two-by-two truth table, then sensitivity is $P(f_j|p_i)$ and specificity is $P(\hat{f_j}|\hat{p_i})$. Here $\hat{f_j}$ and $\hat{p_i}$ indicate negation, not having the function and not matching the pattern. Next, we introduce Bayes's decision rule for posteriori (or joint) probabilities. The logic assumes that one must classify all cases, even if there are only two hypotheses (having or not having the function); that is there is no indeterminate classification. In our case, the decision rule is that a protein with pattern (i) is assumed to have the function (j) if and only if:

$$g_j(p_i) > g_k(p_i)$$

for all k not equal to j, where $g_j(p_i)$ is a monotonic function of $P(f_j|p_i)$. This famous decision rule is equivalent to requiring that:

$$TP/(TP + FP) > FP/(TP + FP)$$

Note that one can use this rule to set a value for any pattern-match score cutoff or threshold used to define a match, as one must in profile analyses. This rule is insensitive to the number of true negatives (TN), which is a strong function of database size. However, the number of false positives is also a function of database size. The functions $g_j(p_i)$, used in Bayes's decision rule, normally include "costs" or penalties associated with the different decision outcomes but, in protein pattern analysis, no relation has been proposed for such penalties.

The relevance of sensitivity, specificity, and/or predictive value for neural networks or other machine-learning methods that employ both a defining and a negative control set may be obvious. However, both these quantities should be reported for any pattern when possible. Patterns can be useful even if they lack near 100-percent sensitivity and specificity. One approach is to use an initial filter or sequential pass approach for patterns designed to identify the same function, but with different specificities. If two patterns have nearly 100-percent sensitivity, but only 70-percent specificity, their sequential application could have a greatly improved specificity if their false positives are uncorrelated. This is similar to combining the two pattern match sets by means of the Boolean AND. If the false positives are uncorrelated random matches, the combined specificity, $sp(a, b)$, will behave on a sufficiently large data set as:

$$1.0 - [1 - sp(a)][1 - sp(b)]$$

Such sequential application of patterns (of different representations) using different match algorithms has been employed by Presnell et al. (1992) to improve the prediction of α-helical structures in all helical proteins, for example.

Control Sets

The evaluation and discovery procedures presume that appropriate control sets can be constructed and that a scoring strategy exists for identifying the success or failure of any match. There are at least two problems. First, identical functions can be encoded by evolutionary and structurally unrelated proteins sharing some common elements, as seen in the work on the cAMP-dependent protein kinase (Knighton et al., 1991), where a nucleotide is bound to a GXXXXG loop but not as part of a Rossman fold (Rossman et al., 1974). Second, many of the available protein sequences are of unknown function. This suggests that the two-by-two table be extended to a three-by-two to include unknowns among the matched and unmatched elements. Sensitivity and specificity can then be defined as being restricted to the sets of known functions or one can subsume under the FP and FN all unknowns, maximizing the conservative nature of the measures.

The construction of proper control sets is one of the most difficult problems, not just for the evaluation of a pattern but particularly for those discovery methods that rely on them. The currently known protein sequences and structures are a biased set, based on everything from ease of isolation and crystallization to species and grant priority preferences. They cannot be assumed to be representative of all realizable or observable proteins. They belong to a complex of families of varying degrees of relatedness or non-independence (Sibbald and Argos, 1990). This is most obvious at the sequence level, where the evolutionary lineages are readily traceable through the sequence similarities.

The choice of positive defining sets is relatively straightforward but still difficult. The positive set is constructed by identifying the defining feature or function, such as the catalysis of a common biochemical function. Major problems involve nomenclature inconsistencies and the incompleteness of the database annotation (or literature). Depending on whether two or one positive control set is required, the family (homolog) relationships may be dealt with somewhat differently. If only one positive set is to be constructed from a set of functionally related sequences, one should incorporate homolog information, (e.g., as sequence-associated weights, small for close homologs and larger for more distantly related homologs). This will allow one to base sensitivity measurements on a set of weighted matches and mismatches. If two sets are required—one for testing and one for learning—a decision must be made to remove close homologs. The problem is that if there are shared close homologs between the two positive sets, the evaluation may indicate more about how the learning procedure memorized the data particulars rather than its generalizations. On the other hand, all homologs cannot generally be removed as there is a very high likelihood that all proteins carrying out the same function are evolutionary homologs! In addition, there are many cases for which there are very few positive or defining examples. Thus, even if one tried to use most of the available information to discover the pattern, little or none is available for validation. This problem arises in any method but may be particularly troublesome for machine learning. As with science in general, prediction is a most convincing test. The match of a pattern to a sequence of unknown function, followed by experimental verification, is still the best test.

For patterns including secondary structural components, even the assignment of known examples in the defining or positive set is complicated. Different assignment strategies produce different definitions of alpha-helices and beta-strands, even in structures determined from x-ray analysis (Kabsch and Sander, 1983; Richards and Kundrot, 1988). Frequently, these differences are small and confined to the termini of the helices and strands. Therefore, it is not appropriate to assign equal weight to the success of predicting or matching the secondary conformation of each residue. Presumably, it is most meaningful to match the center of a helix or strand. If solvent exposure is part of the helical or strand definition, the degree of match between the hydrophobicity of the amino acid and the position's exposure may be most relevant (Presta and Rose, 1988).

The real statistical challenge comes in the construction of the negative control set(s). As noted earlier, the nonindependence or evolutionary relationships of the known proteins is a potential problem, particularly if the entire database is being used (Claverie and Sauvaget, 1985). The emphasis must be placed on carefully defining the statistical question being asked. Traditionally, this means stating the null hypothesis under test. In our case, this is related to testing a pattern's ability "correctly" *not* to match those sequences known not to have the function or structure presumably represented by the pattern.

To define a pattern's specificity precisely, and for computational reasons, a representative negative control set often is constructed. Such a set allows one to define the basis of a pattern's specificity or the null hypothesis in a manner that can be reproduced. Through careful construction of representative negative control sets, one can prevent the overestimation of specificity likely from the use of the total or a random database. One can ask how well a particular pattern (and its representation) can identify a given function from among a set of proteins of known function. The negative control might be a simple uniform sampling of the space of known functions and/or structures, one example from each homologous family. This is preferred over random sampling as the set of known proteins is a highly biased and nonuniform sampling. A representative set can also be constructed to ask particular questions about the specificity of various components of a pattern. For example, if one is trying to evaluate the addition of secondary structure annotation to a primary sequence pattern for the nucleotide-binding motif, one should include in the negative control set all non-nucleotide-binding proteins containing the primary sequence part of the pattern. This is an extension of the filtering logic discussed previously. Conversely, one could construct a negative control set composed of all internal beta-strands from alpha- and beta-proteins. (Most nucleotide-binding sites are contained within such structures.) Finally, the selection of membership in a negative control set can be used to reflect a non-sequence-specific pattern component, such as cell compartment (e.g., mitochondrial or nuclear).

Other Considerations

There are limitations on the interpretive value of even highly diagnostic patterns. One is inherent in the evolutionary process and the second is inherent in the methodology. Among a set of evolutionarily related proteins, there will be sequence elements conserved by chance as well as by selection. Chance here is a result of more than the random sharing of two elements; it includes those not yet having had time to change or not yet having been observed to change. Thus, most patterns contain some elements common only to the limited defining set and not essential to the encoded function or structure represented.

The second limitation, inherent in all the pattern discovery methods, is more serious. It is possible to construct highly diagnostic patterns for many sets of proteins including functionally and structurally unrelated proteins! The limitation is directly related to the power or "richness" of the representative languages used in the discovery methods. In the extreme, if the language allows a global OR, there is the pattern composed of the entire positive control set ORed together. This situation can be encountered in neural net training and is referred to as *memorizing the data*. The problem is less obvious in other cases, but is no less a problem. If, for example, one allows a pattern element to be any arbitrary combination of amino acids observed at a given aligned position, one can create a unique and highly specific pattern of such elements for nearly any small (fewer than 10 sequences) random positive control set. Fortunately, part of the solution is found in the use of meaningful vocabularies. When generating a profile or regular expression, one can restrict the allowed vocabulary to "meaningful" or naturally occurring elements, such as amino acid classes limited to shared biochemical properties. This is done in most alignment procedures used in pattern generation, either explicitly or by the use of an amino acid–to–amino acid similarity matrix. The higher-order pattern elements can also be restricted by understood principles. For example, one would not allow alpha-helices shorter than 4 or a periodicity other than 3.6. As our understanding of protein structure increases, we may be able to restrict not only the vocabulary but also the pattern combination rules or grammar to that which is meaningful. The "play-off" is the discovery power of a rich descriptive language versus the identification of arbitrary common features. One of the current problems is that we have no real sense of the likelihood of patterns equivalent to those discovered for biologically well-defined protein sets among arbitrary or mixed sets.

Other measures may exist that are useful to characterize a pattern and do not necessarily require the use of control sets. The most obvious is information content or density. Information content traditionally is defined using a Shannon measure (Shannon, 1949) based on the probabilities of observing the pattern elements. These may be a priori, expected, or measured probabilities. The simplest information measure used assumes an equally probable likelihood of $\frac{1}{20}$ for each amino acid. This generates a measure for primary sequence patterns in amino acid equivalencies (Smith and Smith, 1990). Such measures can provide information on whether a pattern is typical of verified biologically meaningful patterns.

DISCUSSION AND APPLICATIONS

The importance of comparative sequence analyses in functional identification has been obvious ever since the link between an oncoprotein and a growth factor was noted by Doolittle et al. in 1983. The wealth of new data from the various genome sequencing projects already demonstrates that approximately one in three new sequences matches a known sequence or pattern, allowing assignment of a probable function. The development of the various pattern approaches has grown to more fully exploit the information available among larger sets of functionally related sequences in such functional identifications. Successful structural motifs now range from the simple 3.6 periodicity of the amphipathic surface helices (Cornette et al., 1987; Eisenberg, Wilcox, and Eshita, 1987) to the complex linear patterns reflecting tertiary folding constraints (Cohen et al., 1983; Cohen, Abarbanel, Kuntz, and Fletterick, 1986; Webster, Lathrop, and Smith, 1987; Bowie et al., 1991). The utility of many patterns is in their ability to identify not only the likely function but also the key encoding sequence elements, thus narrowing the focus of the experimental studies to verify that function. Smith and coworkers (Bradley, Smith, Lathrop, Livingston, and Webster, 1987; Figge, Webster, Smith, and Paucha, 1988; Smith and Smith, 1989) have suggested functions for proteins on the basis of sequence patterns that have led to experimental tests and subsequent verification (Dyson, Howley, Münger, and Harlow, 1989; Dyson et al., 1990; Breese, Friedrich, Andersen, Smith, and Figge, 1991).

Neural networks have produced a more mixed outcome. Bohr et al. (1990) have investigated the utility of neural networks in tertiary structure prediction. A binary distance or contact map is used as a computational intermediate and, in this case, the network learns the structures of proteases. A homologous structure is recognized by the network, and a tertiary structure is constructed. Although this is intriguing, insertions and deletions in the aligned sequence pose significant problems for the network. There are extreme difficulties in trying to identify the correct distance matrix when no homologous protein is included in the training set (McGregor and Cohen, personal communication).

Recently, neural networks have been applied to other structure recognition problems for proteins. Qian and Sejnowski (1988) studied the problem of protein secondary structure prediction. They were able to achieve 64.3-percent accuracy using a perceptron network. More complex networks with hidden layers offered no performance advantage. The positional residue weights correlated well with the much older secondary structure propensities determined by Chou and Fasman (1978). These results were confirmed by Holley and Karplus (1989).

Without the addition of long-range interactions, neural networks could do no better than previous statistical methods (Chou and Fasman, 1978; Garnier, Osguthorpe, and Robson, 1978). Kneller et al. (1990) attempted to address this issue by dividing proteins into folding classes and adding input nodes for the Fourier transform of the hydrophobicity, module the periodicity of alpha- and beta-structure. For all alpha-proteins, the neural network was able to predict correctly approximately 80 percent of the helical residues. All beta-proteins were more problematic (70 percent accuracy), and alpha- and beta-proteins fared no better than in the original work by Qian and Sejnowski (1988). In a related study, McGregor, Flores, and Sternberg (1989) focused on beta-turns. They found that it was possible to predict 71 percent of the beta-turns correctly. Kim and coworkers (Holbrook, Muskal, and Kim, 1990; Muskal, Holbrook, and Kim, 1990) have explored the utility of neural networks to predict qualitative side-chain solvent accessibility and disulfide bridging. They succeed in correctly identifying 80-percent of these features.

The greatest success in this area has been the work of Lukashin, Anshelevich, Amirikyan, Gragerov, and Frank-Kamenetskii (1989) in recognizing promoter defining patterns. A two-block network with hidden layers was trained on a set of E. coli promoter sites. The network was able to identify successfully 94 to 99 percent of the promoters with only 2 to 6 percent false positive identifications. Presumably, this success relates to the importance of short-range interactions in promoter sites (approximately 25 base pairs in length) compared to the long-range interactions that are ubiquitous in protein structure.

Protein structures are amazingly robust. To a first approximation, nearly any amino acid can be substituted at any position, at least singly, and the structure will adjust, retaining its general fold. Even the function generally is unaffected by most substitutions outside of the active sites. This amazing robustness is, no doubt, a requirement of the evolutionary process. Therefore, it must be remembered that all comparative and pattern analyses are limited by the extent to which our databases underrepresent the natural or allowable variation within any particular structural or functional theme. This implies that the information derived from existing sequences and structures may, in many cases, be inadequate to identify all new sequences belonging to any particular functional or structural family. However, as our experience with the identification of functional patterns increases, we will learn how to generalize in more meaningful ways given our limited data sets.

NOTE

1. For example, the primary sequence elements, G followed by K or R, and then S or T, in the nucleotide binding fold can be linked with the N-terminus of a predicted alpha-helix, while the D (aspartic acid) is linked to the C-terminus of the next (predicted) beta-strand following the helix.

SUGGESTED READING

Lander, E. S., and Waterman, M. S. (Eds.). (1995). *Calculating the secrets of life: Applications of the mathematical sciences in molecular biology.* Washington, D.C.: National Academy Press.

Lim, H. A., Fickett, J. W., Cantor, C. R., and Robbins, R. J. (Eds.). (1993). *Proceedings of the Second International Conference on Bioinformatics, Supercomputing and Complex Genome Analysis.* Singapore: World Scientific.

Sankoff, D., and Kruskal, J. B. (1983). *Time warps, string edits, and macromolecules: The theory and practice of sequence comparison.* Reading, MA: Addison-Wesley.

Smith, D. W. (Ed). (1994). *Biocomputing: Informatics and genome projects.* San Diego: Academic Press.

Waterman, M. S. (Ed). (1989). *Mathematical methods for DNA sequences.* Boca Raton, FL: CRC Press.

4 Comparative Genomics: A New Integrative Biology

Robert J. Robbins

THE CHALLENGE

Although reductionism in molecular biology has led to tremendous insights, we must now emphasize integrative activities. This will require special skills, as the intuition of the experimentalist alone is not likely to be adequate. Formalization is needed in database design, although not perhaps for bench research.

Considerable reductionist data, such as metabolic pathways, protein structures, gene sequences, and genomic maps, are now available. Conceptual integration in molecular biology will depend on access to properly managed and well-integrated data resources, with databases providing the raw material for future analyses. For integrative efforts, databases do not merely provide summaries of previous findings and indices to the literature. Instead, they drive new scientific investigation, both helping to shape new bench research and also permitting a new kind of *in silico* studies (see chapter 5).

Privately (and sometimes publicly), biologists sometimes have claimed that database managers seem overly concerned with the niceties of data representation. Expert in separating signal from noise, many biologists suspect that computer scientists' concern with data modeling and data structures is just fussiness. However, if databases are to provide input for further analysis, then errors in data models and formats are not equivalent to simple untidiness, they are equivalent to the sloppy preparation of laboratory material.

A bench researcher would be appalled should a technician say, "There was a bit of mold on the plates, but we were in a hurry so I did the DNA extraction anyway." A database designer reacts similarly to, "We knew the data model wasn't quite right, but we were in a hurry so we built the system anyway." We must collectively take responsibility for maintaining databases with the same precision and rigor with which we maintain laboratory preparations and raw materials.

Conceptual integration in molecular biology will be facilitated by a technical integration of data and of analytical resources. At present, however, there are impediments that interfere with both technical and conceptual integration. We will consider these impediments later in this chapter.

The Human Genome Project

The international human genome project (HGP) illustrates the need for integrative approaches to molecular biology. The original goals of the project were (1) construction of a high-resolution genetic map of the human genome; (2) production of a variety of physical maps of all human chromosomes and of the DNA of selected model organisms; (3) determination of the complete sequence of human and selected model-organism DNA; (4) development of capabilities for collecting, storing, distributing, and analyzing the data produced; and (5) creation of appropriate technologies necessary to achieve these objectives (United States Department of Energy, 1990). The ultimate goal is the integration of these diverse findings into a coherent understanding of the human genome and that of several model organisms. Physical map data must be integrated with genetic map data; analyses of the mouse genome must be merged with those of the human genome; sequencing information must be connected to map information. Success of the HGP will depend, in large part, on integrative approaches to molecular biology.

In April 1993, a group of informatics experts met in Baltimore to consider the role of community databases in the support of genome research. A report from this meeting (Robbins, 1994b) noted:

The success of the genome project will increasingly depend on the ease with which accurate and timely answers to interesting questions about genomic data can be obtained. All extant community databases have serious deficiencies and fall short of meeting community needs.

An embarrassment to the Human Genome Project is our inability to answer simple questions such as, "How many genes on the long arm of chromosome 21 have been sequenced?"

Although relating genes and sequences is central to the HGP, and although much consideration has been given to sophisticated integrative approaches to molecular information, the simple fact remains that we cannot now ask an integrative question as simple as that relating map and sequence data.

This report also provided examples of integrative issues that will become crucial for continued success in genome research.

• Return a list of the distinct human genes that have been sequenced.

• Return all sequences that map "close" to marker M on human chromosome 19, are putative members of the olfactory receptor family, and have been mapped on a contig (see Glossary) map of the region; return also the contig descriptions.

• Return all genomic sequences for which alu elements are located internal to a gene domain.

• Return the map location, where known, of all alu elements having homology greater than h with the alu sequence S.

• Return all human gene sequences, with annotation information, for which a putative functional homolog has been identified in a nonvertebrate organism; return also the GenBank accession number of the homolog sequence where available.

• Return any annotation added to my sequence number # # # # since I last updated it.

• Return the genes for zinc-finger proteins on chromosome 19 that have been sequenced. (*Note:* Complying with this requires either query by sequence similarity or uniformity of nomenclature.)

• Return all sequences, for which at least two sequence variants are known, from regions of the genome within plus or minus one chromosome band of DS14#.

• Return all G1/S serine-threonine kinase genes (and their translated proteins) that are known (experimentally) or are believed (by similarity) also to exhibit tyrosine phosphorylation activity. Keep clear the distinction in the output.

As these examples show, there is a need to integrate data resources just within the HGP. The comparative studies that will arise as detailed genomic information becomes available for many species will depend on easy information integration.

Molecular biological data are accumulating exponentially. Sequence database growth (figure 4.1) now is such that 40 percent of the data in the database were provided in the last 5 months. Similar expansion is occurring in other genome databases as well, such as the Genome Data Base (GDB) and the Protein Data Bank (PDB). These information resources must be structured so that data can be extracted and used for further analytical techniques without the need for manual adjustments or interpretation. If it is not done right soon, it may never be done because the data volume will be too great. Although some may argue that *new* data are not accumulating very rapidly (many reported sequences are of proteins or genome regions previously sequenced), *all* the reported sequences must somehow be managed. Extracting and managing a nonredundant subset is yet another challenge.

Whatever one thinks of the HGP or of its goal of bulk-sequencing human and model-organism DNA, the technological advances that it spawns will have profound effects on biology. If the HGP even approximates meeting its goals, two things will happen: First, the amount of sequence data in the databases will increase more than a hundredfold over present levels (even if no one besides genome researchers ever sequences another nucleotide), and second, sequencing will become so inexpensive that the production of very large sequences from nongenome laboratories will become commonplace.

When sequencing is reduced to pennies per base, or less, it will be possible to begin investigating new organisms by collecting large amounts of sequences from them. As this happens, we will need analytical tools that allow

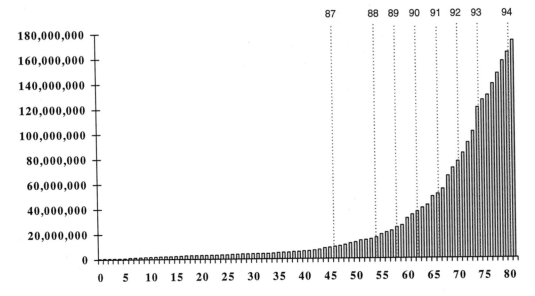

GenBank Release Numbers

Figure 4.1 The growth of base pairs of sequence data in GenBank,[1] from Release 1 through Release 82. The year boundaries beginning with 1987 are given at the top of the figure. The database is currently doubling in size every 2 years. (Data provided by Los Alamos National Laboratory and by the National Center for Biotechnology Information.)

integrative studies of organisms for which little else is known aside from large amounts of sequence information deduced from sequence.

Comparative Genomics: An Integrative Biology

Comparative genomics will emerge as a new scientific discipline as a result of the success of the genome project. To be sure, some journals already publish genomic articles with comparative content, but these usually involve relatively small genomic fragments. Here, *comparative genomics* refers to studies that involve data sets whose scope is entire genomes.

For example, imagine that complete human and mouse sequences were available. Whole-genome comparisons to look for chromosomal rearrangements, conserved linkage groups, and so forth could provide remarkable insights. In principle, one might do such comparisons with a giant dot plot that compared one 3-billion-base-pair sequence against the other.

Although this might not be computationally possible, similar results could be obtained through more practical means. For example, whole-genome *map* dot plots can be constructed now. The axes represent linear versions of the genetic maps from two species, and homologous genes are represented as points, with the *x* coordinate corresponding to the map position in species one and the *y* coordinate that for species two. Perfectly congruent maps

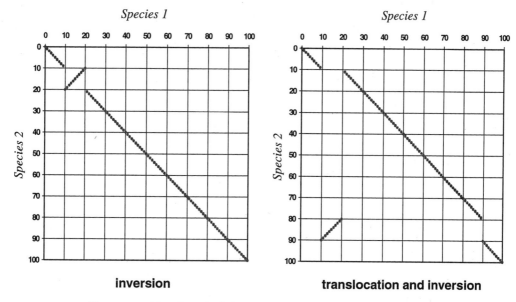

Figure 4.2 Hypothetical whole genome map dot plots. On the left is the pattern expected from two maps that differ only by a simple inversion. On the right is the plot for two maps that differ by a translocated inversion.

produce a single diagonal set of dots. Maps that differ by simple chromosomal rearrangements show recognizable patterns corresponding to those rearrangements (figure 4.2).

To construct such a plot, we need only have maps and homology documentation for the markers on the maps. Sufficient data are available for some bacteria. Figure 4.3 shows whole genome map dot plots comparing *Escherichia coli*, *Bacillus subtilis*, and *Salmonella typhimurium*. These plots were prepared to demonstrate that, in principle, interesting biological comparisons can be done using whole genome–sized data sets. To do such comparisons, however, we must have access to whole-genome data sets on which we can begin to compute right-away, without extensive manual adjustments. Even now when we have, at most, a few thousand markers per organism, the requirement of significant manual adjustments is a strong deterrent to comparative genomic analysis. Soon, with tens or hundreds of thousands of known markers per genome, the requirement of manual data adjustments would render the process essentially impossible. Present nomenclatural inconsistencies now makes the automatic production of such data sets difficult. This is discussed later under the heading, Conceptual Impediments.

TECHNICAL IMPEDIMENTS

Developing the necessary information infrastructure for integrative molecular biology will be hard work, with many technical difficulties to be overcome. First, basic design and implementation of information systems is always

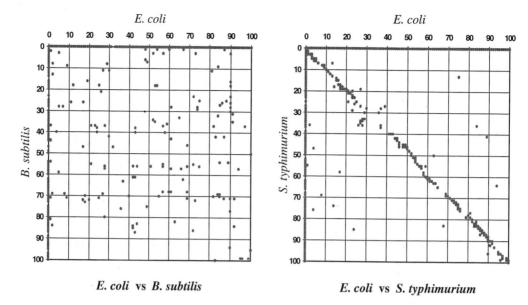

E. coli vs *B. subtilis*   *E. coli* vs *S. typhimurium*

Figure 4.3 Whole genome map dot plots show essentially random distributions of homologous genes when the maps of *E. coli* and *B. subtilis* are compared. However, a similar plot of *E. coli* and *S. typhimurium* shows an obvious diagonal and gives some evidence of the well-known inversion in the 30 to 40-minute region (Sanderson and Hall, 1970; Riley and Krawiec, 1987). Individual outliers may indicate small rearrangements or simply errors in the data. (Homology data from Abel and Cedergren, 1990.)

challenging. Second, achieving some semantic consistency across multiple, independently operated information resources will be a continuing challenge that has both technical and conceptual components. Finally, generating interoperability, getting the different components of an information infrastructure for molecular biology to work together smoothly, will be the biggest technical challenge.

Basic Design Challenges

Building a database is surprisingly difficult. Without adequate design and appropriate plans for integration of the components, efforts can fail spectacularly. In the 1980s, the General Services Administration tried to develop a database of the properties owned and occupied by the federal government. Although this may seem trivial (how difficult can it be to implement a database about buildings?), the project ultimately collapsed with no deliverables, despite having consumed more than $100 million (Levine, 1988):

The General Services Administration said last week it will discontinue development of Stride, a complex computer system to automate the Public Buildings Service. GSA has spent $100 million since 1983 trying to make the system work—a figure that does not include $78 million spent on the system Stride was intended to replace.

GSA officials said Stride fell apart largely because of a failure to create a workable systems-integration plan for it. The most glaring problem with Stride was that it lacked an integration design. Stride went directly from the functional design to work packages without the intermediate step of a detailed design to show how all of the packages would fit together.

To those unfamiliar with the risks of software development, the idea of spending this much money with no results seems inconceivable. To appreciate how this can happen, one must realize that some early decisions in building databases are so fundamental that if they prove to be incorrect, the entire effort is wasted. An equivalent problem in architecture might be the discovery, on completion of a new research facility, that another 12 inches of real clearance is required on each floor if the building is to house the equipment planned for it. How does one add 12 inches of real clearance to each floor of a completed building? The answer is that one does not. The only options are to abandon the original equipment or to abandon the building. Unfortunately, such disasters occur all too frequently in software projects.

Such abject failure rarely occurs in architecture, because much of architectural design and construction involves the assembly of previously designed and well-understood components into new combinations, with the entire enterprise bounded by the reality-check limitations of time and space. Software, like poetry, is constructed of pure thought. Individual programs and even complete software systems frequently are constructed entirely *de novo*. Architectural construction would be more like software development if architects and contractors themselves had to devise most of the components used during construction.

The occasional major failure in architectural construction (such as the 1981 walkway collapse in the Hyatt Regency Hotel in Kansas City that left 114 dead and more than 200 injured) often is associated with first attempts at new designs or with the need for last-minute, on-the-spot design changes that characterize most software development (Levy and Salvadori, 1992; Petrosky, 1992).

Solving the design challenge for molecular biology databases requires a thorough understanding of the scientific subject matter and an appreciation of the subtleties of database development and information modeling. Some conceptual challenges that affect design will be discussed under the heading, Conceptual Impediments. For a more extensive discussion of the requirements and design issues facing molecular biology databases, see Robbins (1993).

Semantic Consistency

Reasonable interoperability among molecular biology information resources cannot be achieved unless some minimum level of semantic consistency exists in the participating systems. No amount of syntactical connectivity can compensate for semantic mismatches.

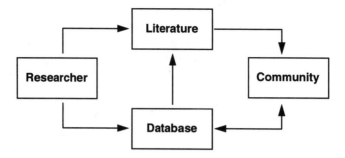

Figure 4.4 Mature electronic data publishing is an alternative form of publishing. Databases parallel, or even precede, the print literature. Authors are responsible for submitting their findings, authorship of submissions is recognized, and volunteer editors and reviewers help ensure the quality and trustworthiness of the resulting system.

Developing an information infrastructure to support integrative approaches to molecular biology will require increased effort to ensure semantic consistency. Controlled vocabularies and common-denominator semantics are important. The same unique identifiers must be used for the same biological objects in all interoperating databases. Participating databases must provide stable, arbitrary external identifiers (accession numbers) for data objects under their curation, and references to these objects in other databases should always be made via accession numbers, not via biological nomenclature. Linking data objects between databases requires that the other objects be unambiguously identifiable (accomplished via accession numbers) and relevant (accomplished via semantic consistency). Although perfect semantic consistency probably is unattainable, efforts to improve consistency are essential. In particular, community databases must document the *semantics* of their systems.

Interoperability Challenges

Databases have evolved from simple indices into a new kind of primary publication (figure 4.4). Electronic data publishing (Cinkosky, Fickett, Gilna, and Burks, 1991; Robbins, 1994a) has transformed some areas of science so that data submission has become a requisite part of sharing one's findings with the community.

A crisis occurred in the databases in the mid-1980s, when the data flow began to outstrip the ability of the database staff to keep up (Kabat, 1989; Lewin, 1986). A conceptual change in the relationship of databases to the scientific community, coupled with technical advances, solved the problem.

Now we face a data-integration crisis of the 1990s. Even if the various separate databases each keep up with the flow of data, there will still be a tremendous backlog in the integration of information in them. The implication is similar to that of the 1980s: Either a solution will soon emerge or biological databases collectively will experience a massive failure.

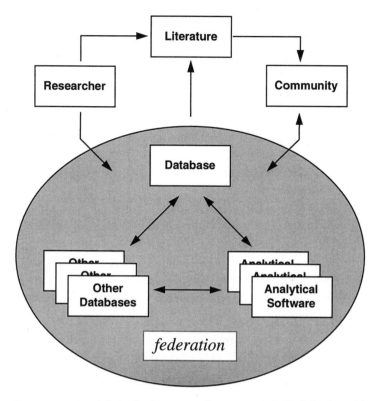

Figure 4.5 In a federated information infrastructure, individual databases interoperate with other databases and with on-line analytical tools. Individual researchers can interact with federated resources collectively, either in submitting or accessing information.

One possible solution for the data integration problem is the concept of *federation* (Robbins, 1994b, d). In a federated information infrastructure, individual scientists would interact simultaneously with multiple resources, both for submitting and for accessing information (figure 4.5). Although this is being widely discussed in biology, the means for implementing such a federation is not immediately at hand. True federation is still a research topic in computer science (Sheth and Larson, 1990; Bright, Hurson, and Pakzad, 1992; Hurson, Bright, and Pakzad, 1994). However, great success has attended the spread of loosely coupled federations of text-server systems such as gopher, WAIS, and World Wide Web (www). Whether such a loosely coupled federated approach can be extended to include structured data from complex databases and, more importantly, to support true semantic joins across different databases remains to be seen (cf. Robbins, 1995).

CONCEPTUAL IMPEDIMENTS

"Every database is a model of some real world system" (Hammer and McLeod, 1981). If the model is inadequate, the database will fail, sometimes totally. In scientific databases, the challenge is to represent the real world in a way that accommodates the subtlety of our present knowledge and that also can evolve gracefully with changing concepts.

Scientific knowledge does change over time, sometimes dramatically. Examples of assertions that seemed unassailably true when made, yet that seem wildly wrong now are:

If the genes are conceived as chemical substances, only one class of compounds need be given to which they can be reckoned as belonging, and that is the proteins in the wider sense, on account of the inexhaustible possibilities for variation which they offer... Such being the case, the most likely role for the nucleic acids seems to be that of the structure-determining supporting substance. (Caspersson, 1936)

Fifty years from now it seems very likely that the most significant development of genetics in the current decade (1945–1955) will stand out as being the discovery of pseudoallelism. (Glass, 1955)

Undoubtedly, similarly misguided beliefs are now current. Identifying and avoiding them must be a key goal for a designer of scientific databases. There exist for genomic databases several such beliefs to be avoided, among them the following:

Gene   A hereditary unit that, in the classical sense, occupies a specific position (locus) within the genome or chromosome; a unit that has one or more specific effects upon the phenotype of the organism; a unit that can mutate to various allelic forms; a unit that codes for a single protein or functional RNA molecule.

The ultimate...map [will be] the complete DNA sequence of the human genome.

A database built according to these concepts (excerpted from the US National Academy of Sciences study (1988) that helped launch the HGP) would fail.

Inadequate Data Models

A database that contains information about a particular class of objects must contain some sort of definition of those objects. For example, a database of people must define persons in terms of their attributes, or possible attributes. Should a person be defined as having one, two, or an unlimited number of different telephone numbers? Does a person have one unchanging address, or can a person have more than one address? Definitions of database objects limit the capabilities of the database, with bad definitions yielding bad databases

In genomics, definitional problems exist with our most fundamental concepts. "What is a gene?" and "What is a map?" are questions that still have no clear answers.

What Is a Gene?
According to the classical definition, the *gene* was the fundamental unit of heredity, mutation, and recombination. Classical genes were envisioned as

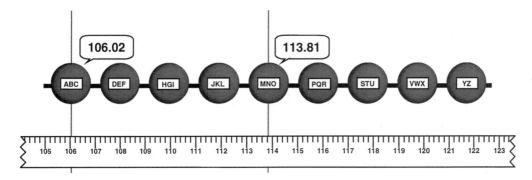

Figure 4.6 According to the "beads-on-a-string" model of genes, mapping just involved determining the correct order of the beads and the correct address, or locus, for each bead.

discrete, indivisible objects, each with its own unique location in the genome. Unchanging locations in the genome were believed to exist independently of the genes that occupied those locations: "The genes are arranged in a manner similar to beads strung on a loose string" (Sturtevant and Beadle, 1939). Mapping (figure 4.6) involved identifying the correct order of the beads and the proper address (i.e., position on the string) for each bead. Addresses (i.e., loci) were considered points, as the beads were believed to be small and indivisible.

Although no biologist still employs this model when designing bench research, all early map databases (and some present ones still) indirectly use it by employing data structures that assign simple, single-valued addresses as attributes to genes. This is clearly beads-on-a-string revisited, because it assumes that a coordinate space (the string) exists independently of the genes (the beads) and that the genes are discrete, nonoverlapping entities.

Later, physiological analyses led to the definition of genes through their products, first as "one gene, one enzyme," then "one gene, one polypeptide." Although we now know that very complex many-to-many relationships can exist between genes and their products, and between primary products and subsequent products (Riley, 1993), the notion that single genes can be unambiguously associated with single products still infects many genomic databases.

With the recognition of DNA as the hereditary substance, some began to define genes in terms of sequence—for example, "the smallest segment of the gene-string consistently associated with the occurrence of a specific genetic effect." Benzer's (1956) fine-structure analysis resulted in an operational definition, the *cis-trans* test, and in the concept of the *cistron*. This idea mapped nicely to DNA and, for many, provided the final definition of a gene: a transcribed region of DNA, flanked by upstream start regulatory sequences and downstream stop regulatory sequences (figure 4.7). This discrete coding-sequence model of a gene was extended to include mapping by asserting that a genome can be represented as a continuous linear string of nucleotides,

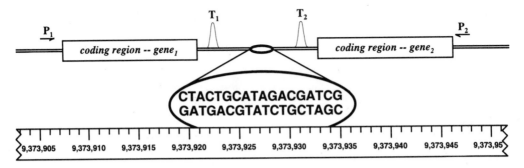

Figure 4.7 A simplistic view of the discrete coding-sequence model of the gene. Here, the fact that genes may be oriented in either direction in DNA is shown by reversing the placement of the promoter and terminator signals for the second gene. By this model, the locus of a gene corresponds to the address in base pairs of its start and stop signals.

with landmarks identified by the chromosome number followed by the offset number of the nucleotide at the beginning and end of the region of interest.

Although it ignores the fact that human chromosomes may vary in length by tens of millions of nucleotides and that some regions of the genome exhibit very complex patterns of expression, this simplistic concept continues to be widely espoused today. A major molecular-genetics textbook (Lewin, 1983ff.) has carried the following definitions, unchanged, through five editions spanning the years 1983 through 1994:

Gene (cistron) is the segment of DNA involved in producing a polypeptide chain; it includes regions preceding and following the coding region (leader and trailer) as well as intervening sequences (introns) between individual coding segments (exons).

Allele is one of several alternative forms of a gene occupying a given locus on a chromosome.

Locus is the position on a chromosome at which the gene for a particular trait resides; locus may be occupied by any one of the alleles for the gene.

This particular book has been a leader in emphasizing the molecular approach to genetics, yet these definitions carry much conceptual baggage from the classical era. By implying a single coding region for a single product, they derive from the "one gene, one polypeptide" model. By logically distinguishing the gene from its location, they even connect to the beads-on-a-string approach.

Whether or not the gene-as-sequence should include just the coding region or also the upstream and downstream regions such as promoters and terminators also is unresolved. Geneticists studying prokaryotes routinely restrict the concept of gene to the coding region and eukaryotic geneticists just as routinely extend it to include the promoter and terminator and everything in between. These differing views are so strongly held within their respective communities that they are considered to be virtually self-evident. Prokaryotic geneticists usually are astounded to hear that some would include flanking regions in the concept of gene, and eukaryotic geneticists are equally

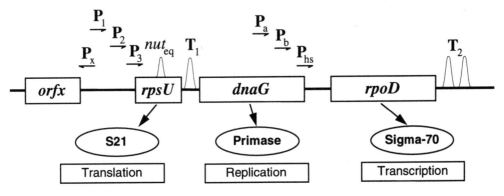

Figure 4.8 The MMS operon in *E. coli* contains a complex set of control and coding regions. Transcription of the *rpoD* coding region can begin from any of a half dozen promoters, some of which are embedded in the coding region of the *dnaG* gene. The subscripted P's and T's represent different promoters and terminators that function in the region. (Reprinted from Lupski and Godson, 1989.)

astounded at the notion they should be excluded. Despite these differences across different biological communities, databases such as GenBank are expected to include consistent concepts of the "gene" in their annotation of sequences from all possible taxa.

Many regions in prokaryotic and eukaryotic genomes are now known to possess a level of functional complexity that renders simplistic definitions wholly inadequate. Figure 4.8 illustrates the macromolecular synthesis (MMS) operon in *E. coli*, a complex region of overlapping control and coding regions. If control regions were considered part of the gene, there are no simple genes-as-nonoverlapping-sequences here.

Eukaryotic systems have provided us with examples of fragmented genes (exon and introns), alternative splicing (different protein products from the same transcript), and nested genes (complete genes carried on the introns of other genes). The concept of a gene as a *discrete* sequence of DNA is no longer viable.

In humans, the UDP-glucuronosyltransferase 1 (*UGT1*) locus actually is a nested gene family (figure 4.9). The first exon and promoter of an ancestral five-exon gene have apparently been replicated five times, with functional divergence occurring in the multiple first exons. Alternative splicing does not seem to be involved; rather, each promoter initiates the transcription of a transcript that is processed to yield a single mRNA. If the *cis-trans* test were applied to mutations of phenol UDP-glucuronosyltransferase and bilirubin UDP–glucuronosyltransferase, the determination of whether or not they were encoded by the same cistron would depend on which exon carried the mutations.

Must cistron now be redefined as equivalent to exon, not to gene? Given its original intent as the shortest stretch of contiguous DNA comprising a functional genetic unit, such a definition for *cistron* would seem appropriate.

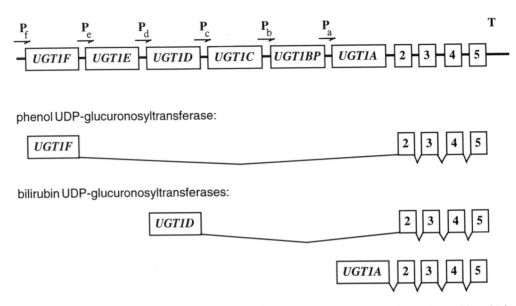

Figure 4.9 The *UGT1* locus in humans is actually a nested gene family that yields multiple transcripts through alternative promotion. Each promoter produces a transcript that is spliced so that the exon immediately adjacent to the promoter is joined with the four terminal exons shared by all of the transcripts. (Reprinted from Ritter et al., 1992.)

However, that interpretation is not common. Such complexities have led some authors to declare that a single definition for *gene* is no longer possible. For example. Singer and Berg (1991) say:

The unexpected features of eukaryotic genes have stimulated discussion about how a gene, a single unit of hereditary information, should be defined. Several different possible definitions are plausible, but no single one is entirely satisfactory or appropriate for every gene.

Most biologists readily accommodate new complexities into their mental models of genome function and proceed to interpret their experimental results accordingly, albeit without ever developing a formal modification of their linguistic definition of a gene. This is fine for guiding bench research, but a database cannot be built to represent genes if genes cannot be defined. What is needed is a new, more abstract definition that can accommodate complexity.

Some authors have begun to rethink the essence of gene, and some fairly radical concepts can be found casually presented in textbooks. For example, Watson, Hopkins, Roberts, Steitz, and Weiner (1992) offered these thoughts:

DNA molecules (chromosomes) should thus be functionally regarded as linear collections of discrete transcriptional units, each designed for the synthesis of a specific RNA molecule. Whether such "transcriptional units" should now be redefined as genes, or whether the term gene should be restricted to the smaller segments that directly code for individual mature rRNA or tRNA molecules or for individual peptide chains is now an open question.

Although this holds to the established notion of *discrete* transcriptional units, it also suggests a radical redefinition of *gene* to mean "unit of transcription." Restricting the concept of gene to functions that occur closest to the level of DNA is appealing but not widespread.

Despite denying that *gene* can be defined, Singer and Berg (1991) adopted a working definition:

For the purposes of this book, we have adopted a molecular definition. A eukaryotic gene is a combination of DNA segments that together constitute an expressible unit, expression leading to the formation of one or more specific functional gene products that may be either RNA molecules or polypeptides.

This approach has potential for database design, in that it abandons the concept of gene as discrete DNA sequence and explicitly embraces the potentially many-to-many relationship among genes and their products. In fact, an extension of this—"A map object consists of a set of not necessarily discrete and not necessarily contiguous regions of the genome"—may well prove to be the definition on which database designs converge.

Such a definition would include, essentially, any subset of the genome to which one might wish to attach attributes, with *gene* being just one subclass of such map objects. Dislodging the gene concept from a central position in a database of genes may be distasteful to geneticists, but it is proving essential to achieving working database designs.

That biologists assimilate complex new findings without necessarily reducing them to precise definitions has been known for some time. After attempting to formalize Mendelian genetics, Woodger (1952) noted that the *language* of geneticists usually is not as complex as their *thoughts*:

Geneticists, like all good scientists, proceed in the first instance intuitively and...their intuition has vastly outstripped the possibilities of expression in the ordinary usages of natural languages. They know what they mean, but the current linguistic apparatus makes it very difficult for them to say what they mean. This apparatus conceals the complexity of the intuitions. It is part of the business of [formalizing] genetical methodology first to discover what geneticists mean and then to devise the simplest method of saying what they mean. If the result proves to be more complex than one would expect from the current expositions, that is because these devices are succeeding in making apparent a real complexity in the subject matter which the natural language conceals.

In short, what biologists say about biology does not match what they understand about it. Anyone attempting to design biological databases must recognize that essential truth.

What Is a Map?

A *genetic map* is some ordered relationship among different genetic or physical markers in the genome. Devising appropriate data models for maps depends on the attributes of the objects being mapped (figure 4.10). Complete

Ordered List:

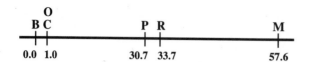

gene	locus
B	0.0
C	1.0
O	1.0
P	30.7
R	33.7
M	57.6

Directed Acyclic Graph:
(transitive reduction)

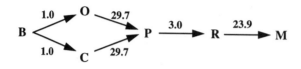

arc	length
B, O	1.0
B, C	1.0
O, P	29.7
C, P	29.7
P, R	3.0
R, M	23.9

Directed Acyclic Graph:
(transitive closure)

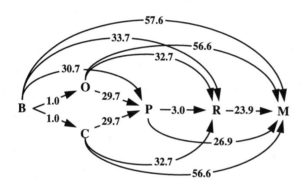

arc	length
B, O	1.0
B, C	1.0
B, P	30.7
B, R	33.7
B, M	57.6
O, P	29.7
O, R	32.7
O, M	56.6
C, P	29.7
C, R	32.7
C, M	56.6

Figure 4.10 Data structures for genetic maps reflect the underlying conceptual models for the maps. Many geneticists still think of maps as ordered lists, and ordered list representations are used in many genome databases. Directed acyclical graph (DAG) data structures can be represented pictorially (left) or tabularly (right). Depending on their use, DAGs may be represented as transitive reductions or transitive closures.

maps of indivisible beads can be simple ordered lists. Partial maps can be represented with directed graphs. Partial maps of complex overlapping objects might involve directed graphs of end points of atomic components of complex objects.

Even the notion of *map* itself is problematic, because generic descriptions of genome structure have much more in common with anatomies than with true maps. Maps describe the actual properties of individual objects, whereas anatomies describe the average properties of populations of objects (Robbins, 1994c).

Inadequate Semantics

Databases often exhibit semantic differences in their treatment of related material. For example, information about human beta-hemoglobin can be found in several databases, such as PIR-International, SwissProt, GenBank, GDB, On-line Mendelian Inheritance in Man (OMIM), and others. Although it would seem a simple matter to provide links that allow the easy traversal of these entries, these databases may have fundamental semantic differences that interfere with the development of sensible links. In the past, PIR-International data objects were proteins in the chemical sense so that any two proteins with the same structure were the *same* protein.

Thus, the PIR-International entry for human beta-hemoglobin was also the entry for chimpanzee and pygmy chimpanzee beta-hemoglobin. Although this policy has been discontinued by PIR-International, it remains evident in SwissProt release 28.0, where entry P02023 represents beta-hemoglobin for all three species, with cross-references to the three different entries in PIR-International.

In GenBank, objects are reported sequences, which may or may not correspond precisely with a gene or particular protein. GenBank may have hundreds or thousands of entries of genomic RNA, cDNA, DNA, or even individual exon sequences that relate in some way to human beta-hemoglobin.

In GDB, objects include genes, probes, and polymorphisms. There will be one GDB entry for the beta-hemoglobin *gene* but multiple entries for associated polymorphisms and probes.

In OMIM, objects are essays on inherited human traits, some of which are associated with one locus, some with multiple loci, and some whose genetic component (if any) is unknown.

Different community databases vary in the richness of their semantic concepts. GDB has more subtleties in its concept of a gene than does GenBank. GenBank's concept of nucleotide sequence is richer than that of other databases. To facilitate integration, participating databases should attempt to accommodate the semantics of other databases, especially when the other semantics are richer or subtler.

Inadequate Nomenclature

A future goal of the genome project is comparative analyses of results in human and selected model organisms. To test the feasibility of an integrative approach in which gene homologies were deduced from data other than sequence similarities, information was extracted from approximately 16,500 genes from nine different organismal information resources (table 4.1).

The data were reduced into a single comparable format that included taxon, gene symbol, gene name, map position, and gene-product function. Enzyme name and enzyme commission (EC) number also were included for genes with known enzymes as products. Then the data were examined to see

Table 4.1 Taxon distribution of the 16,500 gene records from different species used in the feasibility test of an integrative approach

Taxon	Genes
Escherichia coli	1,391
Salmonella typhimurium	754
Saccharomyces cerevisiae	1,427
Caenorhabditis elegans	229
Drosophila melanogaster	5,684
Zea mays	1,035
Arabidopsis thaliana	236
Mus musculus	1,361
Homo sapiens	4,383

Note: The data on *E. coli* were provided directly by Mary Berlyn, those on *S. typhimurium* by Ken Sanderson, those on yeast by Mike Cherry, those on corn from Stan Letovsky. The others were obtained from networked servers.

how readily one might detect potentially homologous genes. In some cases, homologies were already indicated in the databases, but these were ignored as the purpose of the test was to determine whether there was enough information present in the basic data to allow even the identification of candidate homologies. The examples in table 4.2 were chosen to illustrate a number of patterns found in the data.

In many cases (such as *acd*, *ag*, and *CHS1*), identical gene symbols were assigned to wholly unrelated genes. In others, symbols that differed only in case[2] (e.g., *cal*/*Cal* and *cat*/*Cat*/*CAT*) were assigned to unrelated, related, or similar genes, but with no consistency on the use of case. Some genes with phenomenological names had identical symbols and names (e.g., *dw*) but no homology at all, and others had identical symbols but completely different names (e.g., *ft*). Sometimes the names were similar, the symbols differed only in case, and the EC numbers suggested homology (e.g., *Gad1*/*GAD1*). Although a biologist should recognize easily the equivalence of glutamic acid decarboxylase 1 and glutamate acid decarboxylase 1, writing software to do so for every possible pair of equivalent enzyme names would be nearly impossible.

EC numbers proved to offer the best identifier of potential homologies, but they are far from wholly adequate for such purposes. Because EC numbers are actually attributes of chemical reactions that relate to enzymes only transitively through the catalysis of a particular reaction by a particular protein, a single protein can have multiple EC numbers and the same EC number may be assigned to multiple polypeptides.

The connection of EC numbers to genes is even less direct. EC numbers are unrelated to gene names and symbols. Of 3,886 pairs of genes (from different organisms) with the same EC number, only 105 used the same symbol for

Table 4.2 Examples of similar gene symbols as used in different organisms

Symbol	Taxon	Gene Name
acd	Ec	Acetaldehyde CoA deHase
acd	At	Accelerated cell death
acd	Mm	Adrenocortical dysplasia
ag	Dm	Agametic
ag	At	Agamous
CHS1	Hs	Cohen syndrome 1
CHS1	Sc	Chitin synthetase
cal	Dm	Coal
Cal	Dm	Calmodulin
cal	Ce	CALmodulin-related
cal	At	Cauliflower (defective inflourescence)
Cat	Dm	Catalase (EC: 1.11.1.6)
CAT	HS	Catalase (EC: 1.11.1.6)
cat	Ce	CATecholamine abnormality
Cat	Mm	Dominant cataract
cat1	Zm	Catalase 1 (EC: 1.11.1.6)
cat1	Sc	Catabolite repression
dw	Dm	Dwarf
dw	Mm	Dwarf
ft	Dm	Fat
ft	At	Late-flowering
ft	Mm	Flaky tail
Gad1	Dm	Glutamic acid decarboxylase 1 (EC:
GAD1	HS	4.1.1.15)

At = *Arabidopsis thaliana*; Ce = *Caenorhabditis elegans*; Dm = *Drosophila melanogaster*; Ec = *Eschorichia coli*; Hs = *Homo sapiens*; Mm = *Mus musculus*; Sc = *Saccharomyces cerevisiae*; St = *Salmonella typhimurium*; Zm = *Zea mays*.

both species (183, ignoring case). Also, numbers were poorly correlated with canonical enzyme names. Of 1,824 records accompanied by an EC number, only 250 had an associated enzyme name that matched the canonical name according to PIR-International. If different use of hyphens was ignored, that number rose to 314. If substring and superstring differences (i.e., if the canonical name was a substring of what appeared in the database) were also ignored, the number of matching names still was only 411 of 1,824.

Possible solutions for this dilemma are as technologically simple as they are sociologically impossible. What is needed is the development of a common genetic nomenclature that is independent of the taxon in which the gene is identified. Also needed is the recognition that a provisional nomenclature should be used until most genes are known. Although some efforts have been made to standardize genetic nomenclature within some taxonomic groups (e.g., for enteric bacteria, for mammals), no efforts have been seriously proposed to do so across all life.

TECHNICAL ADVANCES

Continuing technical improvements in data-handling capacities are needed, with scalable systems and processes especially important. A better technical approach to data resource integration is essential. A federated approach to information infrastructure may provide a solution.

Scalable Data Entry Systems

The data acquisition crisis of the 1980s was solved through direct data submission. However, as the data volume in molecular biology continues to grow, a new data acquisition crisis may occur, unless further changes are made in the procedures for entering data into the databases.

Direct data submission involved the research community in the process of loading the database by allowing researchers to prepare electronic files for submission. This method is similar to old batch-processing technology, in which computer users prepared files of punched cards for submission to the computer. What is needed is an advance in data submission equivalent to the development of direct, on-line computing. Researchers must shift from direct data *submission* to direct data *entry*.

The availability of direct, on-line data entry systems will allow improvements in the data-entry tools to propagate immediately. With direct-submission tools, each researcher must obtain his or her own copy of the software. If the software is upgraded, it takes some time before the upgrade propagates to all users. In contrast, with direct data entry, the entry software resides on one or more centralized systems. Improvements to the software become available immediately to all users.

The next generation of data-entry software should be tightly coupled with data analysis tools, so that researchers develop a submission at the same time that they analyze the sequence. Requiring that researchers use one set of software to analyze a sequence in their laboratory and then use a separate set to copy the results of that analysis into a format appropriate for submission is wasteful.

The Federation Concept

Some have addressed the interoperability challenge by calling for a federated approach to information resources (DOE Informatics Summit, reported in Robbins, 1994b):

We must think of the computational infrastructure of genome research as a federated information infrastructure of interlocking pieces.

Each database should be designed as a component of a larger information infrastructure for computational biology.

Adding a new database to the federation should be no more difficult than adding another computer to the Internet.

Any biologist should be able to submit research results to multiple appropriate databases with a single electronic transaction.

This vision is attractive, especially when one recognizes that different biological communities have overlapping information infrastructure needs. Access to databases is required for genome research, for molecular biology, for ecosystems work, and for a national biological survey. Many of these involve similar sorts of data—for example, nucleotide sequences.

The trend in information system development has been from the specific to the generic. Originally, each new information resource was developed as a stand-alone system. Anyone wanting access to it had to obtain a copy of the software and the data and install both on a local machine. The process was onerous, and few were willing to invest considerable time just to test the system. Often, the system was developed for a particular hardware and software platform, so even the interested user might not be able to acquire the system without making special purchases.

Next came a move toward networked, client-server systems. Developers produced data resources that were available over the networks, so local users needed only to obtain copies of the client software. Still, the client software was dedicated to accessing a particular data resource and often required specific hardware on which to run. In addition, client systems frequently involved embedded commercial software so that users were obliged to purchase appropriate licenses before accessing the system.

Now generic client-server systems are appearing, in which each data-resource developer "publishes" his or her resource onto the networks via some generic server system. Users need obtain only one copy of the generic client software, and then they can use this to access multiple, independent data resources.

The first big successes for this approach came with simple text and file retrieval using gopher and WAIS. Now, with Mosaic as the generic client and WWW as the generic server, the ability exists to produce interresource integration through the establishment of hypertext links. This has been used to develop data-resource accession systems with considerable integration, For example, the GenQuest server[3] now available through Johns Hopkins University allows users to carry out sequence homology searches using the algorithm of choice (FASTA, BLAST, Smith-Waterman) and then to receive in minutes the results of the analysis with live hot links to all other referenced databases. If the search identifies a potential protein homolog for which a structure is available, a simple mouse click retrieves the structure and launches an appropriate viewer so that the three-dimensional image may be viewed and manipulated.

Although many molecular biology databases are now publishing their data in WWW servers, complete integration is not yet possible, as WWW browsers do not process queries that involve real joins across multiple databases, nor do they process ad hoc queries to the underlying databases. However, many sites are at work developing middleware systems to meet these needs, and we can expect WWW clients to acquire increasing sophistication.

Almost certainly, a federated information infrastructure for molecular biology will not be achieved through a massive, centralized, single-site development project. Instead, we will see an increasing trend toward the integration of components developed at multiple sites. Although we are not yet at a point where such interoperability is truly "plug and play," we are approaching a "plug, tap, tweak, and play" situation in which the required tapping and tweaking is declining.

CONCEPTUAL SOLUTIONS

As technical solutions become available from a variety of sources, the limiting factors for data integration in molecular biology are likely to be conceptual and sociological. Better data models are needed, especially in the conceptual sense. A more consistent approach to genetic and genomic nomenclature is essential. Although there is some movement toward a more generic approach to nomenclature within some taxonomic groups, such as mammals, the overall approach to gene nomenclature is still phenomenological at base and independent in application. Indeed, the independence in genetic nomenclature approaches anarchy. For example, imagine that a new organism is discovered. A biologist who proposed a new name for the species would be called a taxonomist, whereas one who went so far as to propose new names for its anatomical components or for its enzymes would be pronounced a lunatic. Yet any biologist who offered new phenomenological names for all its genes would simply be deemed a classical geneticist.

Better Data Models

Most developers of biological databases now recognize the need to employ subtle and complex data representations to represent biological knowledge. To meet the various needs of many users, databases should employ internal data structures that most individual users consider too complex. Users of and advisors to biological databases must come to recognize this. User needs must guide the design of the system's external behavior, but user opinions should not be allowed to dictate internal data structures. The latter should be designed to accommodate diversity, to support local requirements for viewing data, and to facilitate complex integrative analyses.

The problems associated with the various definitions of *gene* and *genetic map* discussed earlier offer examples of where better data models are needed. Although no biologists still use the beads-on-a-string model to drive their experimental work, many biologists still fall back on that model when thinking about data models for genetic maps. One prominent database researcher was dissuaded from addressing the problems of genomic databases when an equally prominent biologist informed him that, "There are no more than 100,000 genes in the human genome, each with relatively few attributes. The final genetic map will just be a list of these genes in the correct order."

The Enzyme Nomenclature Solution

In the 1950s, enzyme nomenclature was spiraling out of control because of its phenomenological basis and its independent application (Webb,1993):

[W]orkers were inventing names for new enzymes, and there was a widespread view that the results were chaotic and unsatisfactory. In many cases the enzymes became known by several different names, while conversely the same name was sometimes given to different enzymes. Many of the names conveyed little or no idea of the nature of the reactions catalyzed, and similar names were sometimes given to enzymes of quite different types.

Responding to requests from leaders in the community, the International Union of Biochemistry in 1956 established an International Commission on Enzymes. The Commission produced an interim report in 1958 and a final report in 1961 that included a standardized approach to enzyme nomenclature and a hierarchical numbering system for identifying reactions catalyzed by enzymes. Continuing efforts have resulted in the publication of five editions of *Enzyme Nomenclature*, early of which provides summaries of the current classification and nomenclature of known enzymes.

The Commission was reasonably successful but the effort was not without some problems and controversy. Webb (1993) summarized the goals of the commission and its difficulties as follows:

A major part of this assignment was to see how the nomenclature of enzymes could best be brought into a satisfactory state and whether a code of systematic rules could be devised that would serve as a guide for the consistent naming of new enzymes in the future.... [T]he overriding consideration was to reduce the confusion and to prevent further confusion from arising. *This task could not have been accomplished without causing some inconvenience, for this was the inevitable result of not tackling the problem earlier.* [emphasis added]

Figure 4.11 plots the number of known enzymes and of known human genes over time. At the time that Webb claims was too late a start, there were only approximately 600 known enzymes. Now there are more than 4,000 known human genes and more than 10,000 known genes in other organisms. If efforts to standardize enzyme nomenclature began too late when there were only a few hundred enzymes, how would one describe the present situation with regard to genetic nomenclature?

In addition to standardizing the names of existing enzymes, the Commission on Enzymes and its successor have offered advice on the naming of new enzymes. In particular, the habit of applying phenomenological names has been strongly disapproved (Nomenclature Committee of the International Union of Biochemistry, 1984):

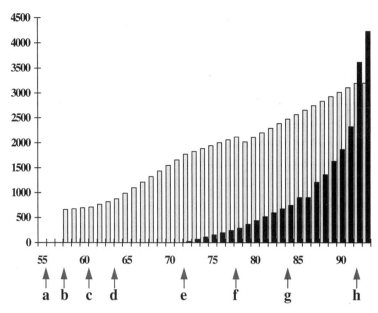

Figure 4.11 Increase in the number of known enzymes (*shaded bars*) and known human genes (*solid bars*). The arrows indicate significant events in the history of enzyme nomenclature: *a* = Enzyme Commission formed; *b* = Commission preliminary report; *c* = Commission final report; *d–h* = appearance of five editions of *Enzyme Nomenclature*.

In this context it is appropriate to express disapproval of a loose and misleading practice that is found in the biological literature. It consists in designation of a natural substance...that cannot be described in terms of a definite chemical reaction, by the name of the phenomenon in conjugation with the suffix -*ase*.... Some recent examples of such *phenomenase* nomenclature, which should be discouraged even if there are reasons to suppose that the particular agent may have enzymatic properties, are: *permease, translocase, replicase*, ... etc.

A review of titles and abstracts of papers in the MedLine database from 1986 through 1993 shows that this advice has been honored mainly in the breach. *Permease* appears in 520 citations, *translocase* in 160, and *replicase* in 362.

Though all the injunctions regarding enzyme nomenclature are not being followed, the efforts of the International Union of Biochemistry to standardize this nomenclature have been generally successful. Without the efforts of the Commission, enzyme nomenclature might have collapsed in chaos. Even where the nomenclature has not yet been fully standardized, the development of a structured numbering scheme allows a relatively stable approach to characterizing enzymes that catalyze particular reactions.

A Modest Proposal

A new, integrated approach to genetic nomenclature is needed. It makes little sense to have systematic rules and standards for the naming of organisms, for the naming of parts of organisms, and for the naming of proteins in organisms, yet to have no equivalent standards for the naming of genes in organisms. A more rigorous approach to the molecular notion of homology is also needed, because clear concepts of homology will be essential for a consistent genetic nomenclature.

Currently, some molecular workers use *homology* in the strict sense of meaning "similar by descent," whereas others use it only to mean "similar in sequence" (cf., Donoghue, 1992). Genomic databases should consider that providing sufficient data to allow the detection of probable homologies is part of their mission. One or more databases specifically dedicated to maintaining and publishing (electronically) asserted homologies and the underlying evidence would be useful.

Although full sequences are available for some viruses (see chapter 2), a complete sequence for a free-living organism has not yet been obtained. Significant efforts are being made to identify and characterize all the coding regions for some microbes (e.g., see chapter 7), but at present only a small percentage of genes have been identified and characterized in any species. Until many more are described, and until considerable comparative data become available, it will be impossible to develop a coherent plan for a global genetic nomenclature. Now that genomic research is discovering new genes rapidly, it might be wise to adopt an explicitly provisional approach to nomenclature while awaiting the development of a better scheme.

To be sure, some proposals have been made for new approaches to the provisional naming of genes. For example, in microbiology the following suggestions have been offered (Nierlich, 1992):

1. Where applicable, the new gene may be given the same name as a homologous gene already identified in another organism.

2. The gene may be given a provisional name based on its map location in the style *yaaA*, analogous to the style used for recording transposon insertion. (That is, *y* designates a provisional name, the next two letters represent the map position, and the final uppercase letter is a serial number indicating occurrence. For example, *ybbC* would be the third provisionally named reading frame in the 11 to 12-minute interval of the map.)

3. A unique, provisional name may be given in the Demerec style.

However, this plan addresses only the need for short-term provisional nomenclature to allow the temporary naming of an open reading frame until a full name can be assigned. Also, all these recommendations are flawed in one way or another, especially in long-term adequacy.

1. Adopting names from other taxa may work sometimes, but it would also allow, perhaps encourage, bad nomenclature to propagate from species to species.

2. Embedding semantic content into arbitrary identifiers is never a good idea. If it turns out that a provisionally identified gene was incorrectly located, either its name must change (thereby invalidating the whole point of assigning a name) or the embedded semantics of map position will no longer

be valid (thereby invalidating the whole point of embedding the semantics). In addition, this approach allows only 26 provisionally named genes per minute of map. Yura et al. (1992) found more than 26 open reading frames in one minute of the *E. coli* chromosome

3. The recommendation that provisional names be syntactically correct provides no specific guidance for provisional naming at all. If we use up a reasonable name space for provisional names, later to be revised, there will be no names left, if debilitating synonymies and homonymies are to be avoided.

An Example from Taxonomy

More than a century ago, de Candolle (1867) recognized the conflict between provisional and final naming in botanical taxonomy:

There will come a time when all the plant forms in existence will have been described; when herbaria will contain indubitable material of them; when botanists will have made, unmade, often remade, raised, or lowered, and above all modified several hundred thousand taxa ranging from classes to simple varieties, and when synonyms will have become much more numerous than accepted taxa. Then science will have need of some great renovation of its formulae. This nomenclature which we now strive to improve will then appear like an old scaffolding, laboriously patched together and surrounded and encumbered by the debris of rejected parts. The edifice of science will have been built, but the rubbish incident to its construction not cleared away. Then perhaps there will arise something wholly different from Linnaean nomenclature, something so designed as to give certain and definite names to certain and definite taxa.

That is the secret of the future, a future still very far off.

In the meantime, let us perfect the binomial system introduced by Linnaeus. Let us try to adapt it better to the continual, necessary changes in science ... drive out small abuses, the little negligences and, if possible, come to agreement on controversial points. Thus we shall prepare the way for the better progress of taxonomy.

Geneticists first need to achieve a Linnaean-like comprehensive, formal approach to genetic nomenclature. Then they could do far worse than to heed the advice of de Candolle and recognize that present nomenclature is mere scaffolding, sure to be supplanted when known genes number in the hundreds of thousands and full sequences are available for many species.

CONCLUSIONS

It is now time to emphasize more integrative approaches to molecular biology, and this will undoubtedly require the participation of those with special interests in integrative methods, because the intuition of bench researchers cannot be fully adequate across the range of materials to be brought together. Electronic information resources will play a crucial role in the integrative methods, provided that the data in them are of sufficient quality and in a format proper to be used as raw input for future analyses.

A new discipline of comparative genomics is emerging, facilitated by the technical advances accompanying the human genome project. This field will require new analytical methods that permit whole-genome data sets to be manipulated and analyzed. This, in turn, will require the availability of appropriate data. Several impediments, both technical and conceptual, currently block the development of appropriate information resources.

Having overcome the data acquisition crisis of the 1980s, we now face a data integration crisis of the 1990s. Proposals to build a federated information infrastructure for biology are promising, but the technical methods for implementing such a system have yet to be devised. Conceptual advances will be required before full semantic integration can be achieved. Some of our most basic biological concepts—for example, *gene* and *genomic map*—need to be considered.

Genetic nomenclature is unsystematic, especially across widely divergent taxa where the results are chaotic and wholly unsatisfactory. A new and comprehensive comparative approach to genomic nomenclature is needed, with special emphasis on the provisional naming of genes. Geneticists should recognize that present nomenclature is mere scaffolding, sure to be supplanted when our knowledge of the numbers of expands and we are aware of full sequences for many species.

NOTES

1. Looking at this figure, it is hard to believe that in the mid-1980s there was a crisis in the sequence databases, with the rate of data production apparently far outstripping the ability of the databases to keep up (Lewin, 1986). Then, data took more than a year to go from publication to entry into the databases. Now, data appear in the databases within a few days of submission. Technical and sociological changes were required to solve the problem. The development of software to allow scientists to prepare and submit data directly to the databases provided the technical fix. Sociologically, many journals began to require that scientists take personal responsibility for submitting data prior to publishing.

2. However, case cannot be safely ignored in all species. In *Drosophia melanogaster*, there are more than 100 pairs of genes whose symbols differ only in the use of case. Examples for loci with one-letter symbols include: a = arc, A = abnormal abdomen; b = black body, B = bar eye; d = dachs, D = dichaete; h = hairy, H = hairless; j = jaunty, J = jammed; p = pink, P = pale; r = rudimentary, R = roughened; s = sable, S = star; w = white, W = wrinkled; z = zeste, Z = Zerknittert.

3. Available as a choice on the main WWW page at URL http://www.gdb.org.

SUGGESTED READING

Cinkosky, M. J., Fickett, J. W., Gilna, P., and Burks, C. (1991). Electronic data publishing and GenBank. *Science, 252,* 1273–1277. Provides insights into the query: Are databases becoming a new scientific literature?

Frenkel, K. A. (1991). The human genome project and informatics. *Communications of the ACM, 34,* 41–51. Provides useful general background on computational implications of genomics.

Gilbert, W. (1991). Towards a paradigm shift in biology. *Nature, 349,* 99. Offers a compelling vision of how information infrastructure is affecting biology.

Hawksworth, D. L. (Ed.). (1988). *Prospects in systematics.* Oxford: Clarendon Press. Systematists have been managing large amounts of complex information for centuries. Useful insights from that community may be found in this text.

Lander, E. S., Langridge, R., and Saccocia, D. M. (1991). Mapping and interpreting biological information. *Communications of the ACM, 34,* 33–39. Addresses the computational implications of genomics.

Robbins, R. J. (1992). Database and computational challenges in the human genome project. *IEEE Engineering in Medicine and Biology Magazine, 11,* 25–34. Provides useful general background on computational implications of genomics.

Robbins, R. J. (1994). Biological databases: A new scientific literature. *Publishing Research Quarterly, 10,* 1–27. Addresses the issue of whether databases are becoming a new scientific literature.

[Special issue.] (1992). *Trends in Biotechnology, 10*(1). Includes many articles on biological information management.

[Special issue.] (1995). *IEEE Engineering in Biology and Medicine, 14*(6), Nov/Dec. Features a number of articles on computational issues relating to genome informatics.

5 On Genomes and Cosmologies

Antoine Danchin

Such a provocative title was selected for this chapter because the status of informatics in biology, and especially in genome studies, probably has not yet reached the place where it ought to be. It is this author's hope that readers will be stimulated not only to read what follows but to consider whether biology has changed recently since it has been accepted that many genomes have been, or soon will be, amenable to complete sequencing. We will presently possess the total chemical definition (including intermediary metabolism) of many living organisms. Nonetheless, perspective we shall soon have about living organisms will, in many ways, mirror our perspective of stars in cosmology, insofar as we shall be able to propose models to interpret the data we collect but shall forever lack the ability to construct *direct* experiments to test the models. As in cosmology, this does not mean that we cannot say pertinent things about genomes or that we can say nothing: Models can be tested by their internal consistency and by their predictive value, when they predict, for example, that a new object can (or will) be discovered (existential prediction rather than falsification).

Information, a generally poorly defined concept, will be central in our discussion. This concept goes back to the Aristotelian discussion of the Being as a relationship among substance, matter, and form: *Substance* is the being itself, in which one can distinguish a determinable *matter* and a determining *form*. The differentiation between individual entities usually was ascribed to specific features of form, and the study of causes asked for a special process *informing* matter. Thus, information is related not only to the specificity given by form—and DNA sequences are related to a very abstract conception of form—but also, implicitly, to causality. This may explain many of the wrong paths that have been followed when using the corresponding concept of information: In ordinary language, something very different is meant when the word *information* is used: *Information* means some kind of usefulness brought about by a given piece of knowledge on a system. Information is therefore also implicitly associated with the idea of *value*. This observation must be born in mind when considering the various theories of information that can be used to represent living organisms and, in particular, to represent DNA sequences.

Invention of techniques that permit DNA sequencing has brought about a true revolution in our way of considering genomes as a whole. Initially, most studies of DNA sequences dealt with a *local* analysis of the DNA information content. They took into account, however, the availability of homologous structures or functions from different organisms and thus provided some general view of global properties of genomes. This remains the dominant view of informatics applied to genetics. As evidenced in many different periodicals and other publications, and more recently, researchers have come to think of genomes as *global* entities. In the future, computers can be used as experimental tools, generating a new source of investigation of living organisms, their study *in silico* (in contrast to in vivo or in vitro).

A LOCAL VIEW OF GENOMES: ALIGNMENT AND CONSENSUS OF MOLECULAR SEQUENCES

Since protein and nucleotide sequences have been available (and this goes back to the early 1950s for protein sequences), scientists have attempted to compare sequences to one another and to align those sequences that were supposed to match one another after having diverged from a common ancestor. A paragon study of such comparison was the *Atlas of Protein Sequences and Structures* developed by Margaret Dayhoff and her coworkers (Dayhoff, Schwartz, and Orcutt, 1978). After thorough analyses of the phylogenetic divergence in families of globular proteins such as cytochromes or globins, this work permitted construction of a correspondence matrix between amino acids that reflected some of the evolutionary fate of each residue. Hundreds of articles have been written proposing general and less general approaches for creating alignment of sequences and generating the corresponding consensus (e.g., see Altschul and Erickson, 1985; Waterman and Eggert, 1987; Fischel-Ghodsian, Mathiowitz, and Smith, 1990; Gotoh, 1990a, b; Spouge, 1991).

Although it was relatively easy to align similar sequences, it often was impossible to align very divergent sequences unless some other related sequences also were known and were compared in parallel (multiple rather than pairwise alignment). This resulted in the hunt for intelligent algorithms permitting alignments and multialignments and providing the corresponding consensus of multiple sequences. However, it is clear that the production of alignments requires not only production of a matrix for correspondence between nucleotides (usually the identity matrix in this case) or similar amino acid residues (Altschul, 1991; Alexandrov 1992; Landès, Hénaut, and Risler, 1992), but also a choice for the penalties assigned to the gaps present in the sequences when they are aligned (Altschul, 1989; Barton and Sternberg, 1990; Gotoh, 1990b; Blaisdell, 1991; Brouillet, Risler, Hénaut, and Slonimski, 1992; Pearson and Miller, 1992; Rose and Eisenmenger, 1991; Saqi and Sternberg, 1991; Zhu, Sali, and Blundell, 1992).

The use of dynamic programming can provide a first set of aligned pairs (Smith, 1988). Many programs, derived from or created independently from the Needleman and Wunsch algorithm, including use of dedicated hardware, have been proposed (Waterman, 1983; Waterman and Byers, 1985; Gribskov et al., 1987; Taylor and Orengo, 1989; Fischel-Ghodsian et al., 1990; Smith and Smith, 1990; Rose and Eisenmenger, 1991; Saqi and Sternberg, 1991; Pearson and Miller, 1992; Fagin, Watt, and Gross, 1993; Subbiah, Laurents, and Levitt, 1993; Vogt and Argos, 1993). Unfortunately, these programs cannot be used for sets larger than three sequences because of combinatorial explosion.

In contrast, looking for common words in a composite sequence made of the concatenation of all sequences to align, followed by a manual fine alignment in which the unformalized knowledge of the biologist is used, can yield interesting results (see Schuler, Altschul, and Lipman, 1991; Zuker, 1991; and Pearson and Miller, 1992, for comparison of different approaches and inclusion of statistical error assessment). A typical example of such combination is the Smarties software[1] designed by Viari and coworkers for multialignment of proteins, using an adaptation of the algorithm of Karp, Miller, and Rosenberg (1972), in which one combines an algorithm searching for matching patterns with "anchor" sites that correspond to matching patterns chosen by the biologist and aligned forcefully. In this process, matrices meant to represent similarities between amino acids can be used for protein alignment. It is interesting to use matrices that take into account the constraints on amino acid residues that have been measured from comparison of known three-dimensional structures (Risler, Delorme, Delacroix, and Hénaut, 1988; Miyazawa and Jernigan, 1993); multiple alignments can also implement knowledge from such structures (Holm and Sander, 1993; Johnson, Overington, and Blundell, 1993; Orengo, Flores, Jones, Taylor, and Thornton, 1993; Pongor et al., 1993; Subbiah et al., 1993). Nonetheless, it must be stressed that a single matrix is used throughout the sequences, meaning that one deals with a composite view of the equivalence rules between amino acids at each position in the sequence (Saqi and Sternberg, 1991). This cannot represent what happens in reality because at some positions for example, it is the aromatic feature that is important (meaning that F [phenylalanine], Y [tyrosine], and W [arginine] are equivalent), whereas at other positions the fact that the residue is big could be critical (meaning that F, Y, R, H [histidine], and W [tryptophan] are equivalent), so that one should use an equivalence matrix for each position in the sequence. Clearly, new research program still need to be developed in this field, despite the large number of articles already published (cf. Day and McMorris, 1993).

Learning Techniques, Perceptron, and Neural Networks

When significant numbers of sequences are compared with one another, one may wish to extract a significant pattern not as a consensus sequence but rather as a matrix of probability reflecting the frequency of a given motif at a given place. This is reminiscent of pattern recognition problems involving, for example, learning techniques or techniques used in visual pattern recognition (Nussinov and Wolfson, 1991; Fischer et al., 1992; Bachar et al., 1993). It is not unexpected, therefore, that methods involving neural networks or the precursor of this new field of informatics, the Perceptron, have been used for generating pertinent patterns. Stormo and coworkers (1982b) were pioneers the field, using the Perceptron, a classical learning method in pattern recognition research, for creating matrices that permitted identification of the ribosome binding sites in *Escherichia coli* messenger RNAs, and they have been followed by many other authors studying both DNA and protein sequences (Schneider, Stormo, Gold, and Ehrenfeucht, 1986; Alexandrov and Mironov, 1987; Bohr et al., 1988; Nakata, Kanehisa, and Maizel, 1988; Stormo, 1988; Stormo and Hartzell, 1989; Hirst and Sternberg, 1991; Farber, Lapedes, and Sirotkin, 1992; Prestridge and Stormo, 1993; Snyder and Stormo, 1993). The main drawback of the Perceptron approaches, which unfortunately is present also in many neural network learning techniques, is that they involve formal "synapses" that have a transmitting strength which evolves in a *quantitative* fashion as a function of their actual use (Stormo et al., 1982b; Geourjon and Deleage, 1993; Gracy, Chiche, and Sallantin, 1993; Han and Kim, 1993; Hirosawa, Hoshida, Ishikawa, and Toya, 1993; Jones, Orengo, Taylor, and Thornton, 1993; Landès, Hénaut, and Risler, 1993; Miyazawa and Jernigan, 1993; Orengo et al., 1993; Orengo and Taylor, 1993; Orengo and Thornton, 1993; Panjukov, 1993; Pongor et al., 1993; Lawrence et al., 1993). This means that, except in the cases where there is obvious and strong similarity between objects in the training set, the synapse strength fluctuates. As a consequence, when the set size increases, the strength of each synapse generally goes through an optimum value and then slowly recedes toward a more or less average value as more exceptions invade the training set, thus losing discriminative power. Accordingly, as the training set of examples increases in size, the actual efficiency of the consensus matrix created by the learning procedure loses accuracy. Ancestors of neural networks have been proposed that evolve in such a way that the effective transmitting capacity of a synapse goes irreversibly to zero when its value falls below a certain threshold value, thereby freezing the learning state at an optimal value (Danchin, 1979). This feature has recently been implemented in neural networks by Horton and Kanehisa (1992). It would therefore be important to compare and develop new approaches involving neural networks, with emphasis on the *stability* of their learning capacity as a function of the training sets.

Other learning techniques rest on the explicit construction of grammars—that is, sets of structures and rules permitting the complete description of a sequence or a network of genes (see chapter 9). They can be divided roughly into two classes, learning by assimilation (i.e., by measuring *conformity* to features of a training set) and learning by discrimination (i.e., by constructing descriptors of the *differences* between the sets of examples supposed to differ from one another). An example of the first type is the Calm software, used for identifying the calcium-binding sites in calmodulin, for instance (Haiech and Sallantin, 1985), whereas an example of the second type is the Plage software used in identifying the relevant features of signal peptides in *E. coli* (Gascuel and Danchin, 1986; Gascuel, 1993). Both types depend on a grammar that is employed to generate descriptors used on the training sets. In general, grammars are essential to describe biological phenomena, and their efficiency rests heavily on the actual biological knowledge of their conceptors, providing therefore an interesting way to incorporate biological knowledge into automated procedures. For this reason, one probably should consider always as the most pertinent ones the grammars devoted to a specific question. A case in point is the work of Collado-Vides (1991a, b, 1992), who constructed a grammar for the identification of regulated promoters in *E. coli* (see also chapter 9).

Sequence acquisition has been exponentially growing for 10 years. In parallel, the power of computing has also increased in an exponential fashion, and it has been possible to foresee treatment of biological data that would have been impossible just a few years ago (see Barton, 1992). Because DNA and protein sequences can be seen, to first approximation, as linear texts of an alphabetical writing, they pertain to the theories of information that have precisely the study of such texts as privileged objects. In what follows, we attempt to develop ideas showing that the alphabetical metaphor, and the corresponding treatment of information, is likely to have a very deep meaning in terms of genomes. This will require a progression in the conceptualization of information and, because biologists are not always familiar with the underlying mathematics, rough principles are provided so that such researchers can follow the reasons for working to improve our concepts of information.

Shannon and Weaver's Model for the Information Seen During DNA Replication

When considering living organisms, one often casually refers to the *information* they carry, namely in their genomes' DNA sequence. Information is a "prospective character" (Myhill, 1952), which refers to the identification or localization of control processes involving small amounts of energy that result in large changes in processes involving large quantities of matter or energy. A very elementary, but useful and often-used, concept of information has been derived by Shannon and, independently, by Weaver, who published their findings together (Shannon and Weaver, 1949). These authors, by comparing the sequence before and after transmission, defined the information of a collection of digit sequences transmitted through an electromagnetic wave channel submitted to some noise (Shannon, 1949). The information concept they invented is a very crude view of a message, explicitly considering not its meaning but the local accuracy of its transmission. Genomes can be analyzed using this view, especially if one considers only the information carried out during the process of replication, leaving aside the information involved in gene expression. This concept of information can shed light on interesting properties, but one should always bear in mind the fact that the actual information content of a sequence must be much richer (it has a meaning), unless one considers only the information seen by the replication machinery. Later we shall see other ways of considering the information content of sequences, cases in which the global signification is better taken into account.

Shannon and Weaver's theory of information derives from a subset of the general theory of measurable spaces (Benzécri, 1984). Given a measurable space Ω (space of events) and a a part of Ω, if ω is an event, chosen at random, one would like to say that if one knows that ω belongs to a, this means that we have some information on ω. Thus, information brought about by asserting $\omega \in a$, is noted $H(a)$. H (Shannon's information) is defined in probabilistic terms as follows: $H(a) = f(p_a)$, where p_a is the probability that the information on a is true (the smaller a, the easier the location of ω). First, f is supposed to be positive and decreasing and such that a random assertion having equal probabilities to be true or false gives a unit of information: $f(1/2) = 1$ (this is a natural condition for normalization). Second, f is such that if A and B are two partitions of W and if the assertions $\omega \in A$ and $\omega \in B$ are independent, then $H(A \otimes B) = H(A) + H(B)$. Finally, one adds the natural condition that $f(1) = 0$, meaning that if an event is known to be true for certain (ω belongs to Ω is trivial), this is equivalent to give no information on it. A consequence of this definition of information is that if Ω is split into a number of n of equivalent separate subsets Ω_n, one has, after the final property of H:

$$H(W_n) = n \times (1/n) \times f(1/n) = f(1/n)$$

This can be used as a starting point to show that function f is:

$$\forall p \in (0, 1) \qquad f(p) = -\log_2(p)$$

More generally, let $i \in I$ be a random variable of probability law p_I (where $p_I = \{p_i | i \in I\}$; then Shannon's information becomes:

$$H(p_i) = -\sum \{p_i \log_2 p_i | i \in I\}$$

It can be seen that such definition of information is highly schematic and far from being able to describe the properties of what one's common sense would consider information. Indeed, a more elaborate view would indicate that to consider an event in isolation is meaningless in most actual cases: One needs, for instance, to possess a means to recover the considered information. Also, when one considers life, the present approach of information does not take semantics into account, as would be necessary if one considers gene expression.

Another way to consider the information carried in a message, when it corresponds to an instance of a family of messages considered as similar, leads to the same equation. If p_r is the probability of a message when received, and p_s the probability of this message when it is sent, then the information corresponding to the quality of the message is: $H = \log_2(p_r/p_s)$. If one does not take into account the loss of information during transmission (message transmitted without error), then p_r is equal to unity. If there is a large number of similar messages, the corresponding information is $H = -\sum p_s \log_2 p_s$. One notes here that the equation obtained has still the same form as in the preceding description.

In the case of DNA, what is taken into account is the probability of a message that, in the usual description of genomes, reduces to the actualization of a succession of letters (chosen among A, T, C, or G) at a given position of the genomic text. The specific information analyzed corresponds to the difference between an actual collection of DNA sequences and a random sequence having the same base composition. An evident improvement would be to consider not the one-letter successions of the program but the relevant words (sequences of letters of definite length). However, this assumes that one knows the ones that are relevant! It would appear, therefore, that analysis of the genomic text using classification methods is a very important prerequisite for an in-depth analysis of its information content.

The mathematical aspect of Shannon's information is developed to display the logarithmic law to which it leads. Indeed, the form of this law is similar to that of another law discovered in the domain of thermophysics. And this coincidence has had unfortunate conceptual consequences, when information was related to life, because the ideas of the latter scientific domain contaminated the ideas of the former one, as we shall now see (see also Yockey, 1992).

Information and Shannon's Entropy

It is quite usual to read that there is *identity* between a form of thermodynamic entropy and the information carried by a given process (but see the recent analysis by Zurek, 1989). In fact, Shannon's information often is named *Shannon's entropy* (but this was not a choice made by Claude Shannon!). It is a consequence of the remark that there is identity between the form of Shannon's law H and a form of entropy described in terms of statistical mechanics, particularly when describing the specific case of the perfect gas. However, identity in the form of a mathematical description of a process is by no means a sufficient reason to identify the underlying principles or the very processes with one another. Moreover several words such as *information, entropy,* and *order* have been used in many different contexts, meaning something very precise in each but usually meaning something very different from the word's significance in another context. Because the ideas of information and order are used often in biology and may influence the exercise of the corresponding knowledge, it seems useful to explore some of the underlying reasons that have permitted authors sometimes, to identify the negative value of information and entropy.

The frequent identification between information and entropy is ideological in nature and corresponds to a mistaken identification of disorder and entropy coupled with an unsaid view of the world as spontaneously evolving toward a "bad" ordering. The underlying assumption is that the world is constantly moving toward an unavoidable disorder, a fact against which we should fight (see Schrödinger's *What Is Life?* and my foreword to its French translation [Danchin, 1986]). The starting point of such pessimistic views of the world is the mechanostatistical model of thermodynamics, founded on the very specific and restricted case of the perfect gas. In this model, gas atoms are hard, impenetrable spherical points that interact by pure elastic collisions. The popular description then is that of a container made of two independent chambers containing different gases and separated by a rigid wall. At the start of the process, a small hole is drilled through the wall. It is easy to observe subsequently that the container evolves in such a way that in time there is an equivalent mixture of both gases in both chambers, each gas having a tendency to occupy the volume from which it was absent, until a final equilibrium is achieved, wherein the partial pressure of the gas is equal in both chambers, an equal number of gas atoms coming from each chamber to the other. If one wanted to prevent homogenization, it would be necessary to place a little demon at the hole who had control of a door that could be closed when the wrong gas was going toward the chamber in which it did not belong. One then asserts that the action of the demon corresponds to the knowledge of information on the gases' nature.

This description is considered adequate to justify the identification between information and entropy. Many technical objections to such a primitive description could be given, but one is compelling: In this way of accounting for a system, its entropy depends on its level of description (see also Zurek, 1989). However, *entropy* can also be defined as a macroscopical physical entity linked to the total energy of a system and its temperature. It therefore is independent of the model. Thus, the identification between entropy and disorder is disputable. Let us consider a two-dimensional representation: A blue gas in a square chamber is separated from a yellow gas in a similar chamber. Drilling a hole in wall between them will yield a green rectangle created by mixture of the blue and yellow gases. Is the green rectangle more disordered than the juxtaposition of a blue square with a yellow one? Identification between entropy and information also is misleading, because one can immediately notice here not only that information is a qualification of a state of a given system but that it requires some knowledge of both the system and of the observer. Like order, information is a concept that is *relative* and must take into account several systems.

Finally, when considering the entropy of a system made of many components, one should take into account not only the distribution of the components in space and according to their energy states but also the various scales at which they occur: There is some entropy contribution in the building up of a hierarchy, for instance. If one wished to have information on a system, it might be important to consider the structure of the hierarchy. There is also a most important phenomenon: If one considers processes that make a large number of components evolve, there are at least two well-defined evolutions that should be taken into account. One corresponds to systems in which the overall structure of individual levels is preserved, even if the individual components of each level seem to evolve in a stochastic way. This is generally the case of ergodic processes. However, there is another case in which the overall tendency is the mixing up of everything, the overall tendency being that of homogenization. In this case, calculated entropy has a special form, which tells us something about the tendency of the system to lose memory of past events.

Without qualifications about the general structure of the underlying dynamics (ergodic or mixing) or about the observer, the description of information in statistical terms is highly misleading. Shannon's measure of entropy reflects only a mathematical property of statistical analysis, as discussed later (see Benzécri, 1984). For the moment, however, limiting Shannon's information to what its mathematical form assumes it is, this information can be used for a first description of DNA sequences' information content. This first level of local information can be used to identify coding regions in DNA (Shepherd, 1981; Smith, Waterman, and Sadler, 1983; Almagor, 1985; Graniero-Porati and Porati, 1988; Luo, Tsai, and Zhou, 1988; Gusein-Zade and Borodovsky, 1990; States and Botstein, 1991; Farber, Lapedes, and Sirotkin, 1992; Robson and Greaney, 1992). A more evolved use of Shannon's information is also present in the analysis of gene sequences using periodical Markov chain analysis, as demonstrated by Borodovsky and McIninch (1992). A consequence of this type of work is obtaining a large collection of gene sequences from a given organism, which brings us to the use of informatics in the analysis of whole genomes.

LANDSCAPES: A GLOBAL VIEW OF GENOMES

A genome is the result of a historical process that has driven a variety of genes to function collectively. Many biologists using computers have had the following experience: When one scans through a sequence library for similarity with a query sequence, one often ends up with a list of sequences that have in common only the organism to which the initial sequence belongs. This is as if there existed a *style* specific for each organism, independent of the function (an image proposed by P. Słonimski at a public conference is that of columns in a temple, which can be of Corinthian, Doric, or Ionian style but having always the same function). Is this impression true? Several ways of investigating the overall structure of genomes have confirmed and extended this observation (Fickett, Torney, and Wolf, 1992; Karlin and Brendel, 1992).

The indication is that there must exist some sort of mutual interaction which permits the cell to discriminate between self and nonself. It therefore seems interesting to investigate the meaning, at the lowest possible level, of what could be mutual information.

Mutual Information and Standard Statistical Analysis

In Shannon's terms, it is natural to define mutual information between pairs of events. This is a function that is always less than or equal to the information that one possesses on individual corresponding events:

$$H(p_{IJ}; p_I \cdot p_J) = H(p_I) + H(p_{IJ}) - H(p_{IJ})$$
$$= \sum \{ p_{ij} \log_2(p_{ij}/p_i \cdot p_j) | i \in I, j \in J \}$$
$$= \sum \{ p_i \cdot p_j (p_{ij}/p_i \cdot p_j) \log_2(p_{ij}/p_i \cdot p_j) | i \in I, j \in J \}$$
$$= \sum \{ p_i \cdot p_j f(p_{ij}/p_i \cdot p_j) | i \in I, j \in J \}$$

Now, in usual statistics, there is another means to evaluate correlation between events, using a specific metric known as the χ^2 *test*. This metric is meant to evaluate the mutual distance (this can be interpreted as some mutual information) between two laws. It is designed in such a way that the mutual distance p_{IJ} is equal to $p_I \cdot p_J$ when all the knowledge about the association of the two laws is the result of the knowledge about each one:

$$\| p_{IJ} - p_I \cdot p_J \|^2 = \sum \{ (p_{ij} - p_i \cdot p_j)^2/p_i \cdot p_j | i \in I, j \in J \}$$
$$= \sum \{ (p_i \cdot p_j [(p_{ij}/p_i \cdot p_j)^2 - 1] | i \in I, j \in J \}$$
$$= \sum \{ p_i \cdot p_j f(p_{ij}/p_i \cdot p_j) | i \in I, j \in J \}$$

knowing by definition of the laws p_I (normalization) that the sum of the product $p_i p_j$ as well as that of the mutual law p_{ij} is equal to 1.

One can then remark that the final expression has a general form similar to the form of the Shannon's mutual information law: One has simply to exchange the form of the distance function f involved in the χ^2 paradigm, where it is $x^2 - 1$, for the form in Shannon's law where it is $x \log_2 x$. Both functions have a very similar behavior around $x_0 = 1$: Their value, their first derivative, and their second derivative are equal. In fact, if one proposes to measure mutual information by various plausible laws (mutual "distances"), one is led to the conclusion that the envelope of all such laws is Shannon's information function. This has the consequence that the function $x \log_2 x$ is the natural function that should be used in most cases. However, this conclusion can also be extended, implying that it is natural to find this same function in many situations otherwise unrelated: *When finding that the representation of a phenomenon yields a function of the form* x log$_2$ x, *one should not infer that the analogy extends to the nature of the phenomena described by this same function.*

Methods using χ^2 are numerous (they can even be used in learning techniques for the partition of training samples), but because it is very powerful yet rarely used in biology, we shall refer to only one such method, factorial correspondence analysis (FCA), which makes use of χ^2 distances for classification of objects without a priori knowledge of the classes (Hill, 1974). The underlying idea is that one can construct pertinent classes where objects do not obligatorily share simultaneously *all* the characters defining the objects in a given class and nevertheless can be said to belong to the same class because they share most of the characters defining a class (Benzécri, 1984). Such an analysis permits one not only to build up classes but also to build up trees describing the relationships between classes. This permits one, under some conditions, to build up hierarchical clusters, reminiscent of phylogenies.

FCA has been used to study codon usage in individual genomes. In the absence of any other knowledge, coding sequences (CDSs) can be described by the usage of codons specifying each amino acid residue of the polypeptide they encode. Accordingly, each CDS is represented as a point in a 61-dimensional space, each dimension corresponding to the relative frequency of each of the 61 codons (Blake and Hinds, 1984; Alff-Steinberger, 1987; Hanai and Wada, 1989; Bulmer, 1990; Lawrence and Hartl, 1991; Sharp, 1991). The set of CDSs is displayed as a cloud of points in the space of codon frequencies. Using the χ^2 distance between each CDS pair, FCA allows calculation of the two-dimensional projection of the cloud of points yielding maximum scattering (Hill, 1974; Lebart, Morineau, and Warwick, 1984). On such projection, genes that have a similar codon usage will appear as neighbor (but the converse is not necessarily true).

To analyze this graphical representation in terms of individual classes, it is necessary to use a second method that automatically clusters the objects (here, the CDSs) that are close to one another. For example, in a first step, one splits the collection of objects into k groups by a dynamic clustering method; then, in a second step, objects that are always clustered together in the course of the different partition processes are selected (Delorme and Hénaut, 1988). Analysis of *E. coli* CDSs revealed that the hypothesis requiring the minimum assumptions is a clustering into three well-separated classes (Médigue, Rouxel, Vigier, Hénaut, and Danchin, 1991a). The codon bias is very strong in class II, intermediate in class I, and weak in class III. For example, in class II, CTA triplet is used in fewer than 1 percent of all leucine codons, whereas CTG is used in 76 percent of all cases (this corresponds to a major leucine tRNA, $tRNA_1^{leu}$).

The proof that this partition is significant comes from the observation that these three classes of *E. coli* genes can also be clearly distinguished by their biological properties. Class I contains almost all genes of intermediary metabolism, with the noticeable exception of genes involved in the core of carbon assimilation (glycolysis, tricarboxylic acid cycle [TCA] cycle, and fatty acids synthesis). It also contains genes specifying gene regulation (activators and repressors) and genes responsible for DNA metabolism. Thus, class I comprises those genes that maintain a low or intermediary level of expression (Bennetzen and Hall, 1982; Gribskov, Devereux, and Burgess, 1984; McLachlan, Staden, and Boswell, 1984; Sharp and Li, 1986, 1987) but can be potentially expressed at a very high level (e.g., the lactose operon). In contrast, class II contains genes that are constitutively expressed at a high level during exponential growth. Most of these genes are involved in the translation and transcription machinery, as well as in the folding of proteins. Up to this point, the clustering of *E. coli* genes fits the codon usage clusters described by Gouy and Gautier (1982). It also fits the general gene clustering described by Sharp and coworkers (Sharp and Li, 1986, 1987). In contrast, in class III, the "frequent" codon CTG is used in only 30 percent of all leucine codons. In the same way, the second and first classes show significant bias against a few codons (mainly ATA, AGA, and AGG) that are not discriminated against in class III. In general, the codon usage characterizing the third class is different from that of classes I and II, for the distribution of codons is quite even. It therefore is significant that most of the corresponding genes have been shown to be involved in functions required for horizontal transfer of genes (Médigue, Rouxel, Vigier, Hénaut, and Danchin, 1991a).

With these examples, we observe that investigation at a first level of the information content of a genome is rewarding. In particular, it tells us that the actual selection pressure on the fine evolution of sequences inside genes is actually visible when one possesses a large collection of genes for analysis. Because this corresponds to a fine variation, such as a general bias in nucleotide distribution in genes, and comprises many genes at the same time, it becomes difficult to propose experiments to test whether the interpretation of the observation adequately reflects reality. Indeed, it must be difficult to submit a modified organism to a "normal" life cycle (i.e., a life cycle that differs strongly from laboratory conditions, including chemostats) and to test whether its DNA has evolved according to the predicted consequences of a model. In some cases, however, experiments have been performed that can answer some of the predictions: Cox and Yanofsky (1967), for example, have submitted a culture of *E. coli* lacking a functional *mutT* gene (the product of which prevents AT to GC transitions) for evolution during a large number of generations, and they have observed that, after some time, the overall content of the DNA was significantly more GC-rich than was the parental strain.

In the examples just given, the information content of a sequence was considered mostly as a function of its *local* value. A more appropriate way to consider the same information would be to evaluate it as a function of its context, as we shall now do.

Eukaryotes and Prokaryotes: Algorithmic Complexity

A genome sequence is certainly far from being random. This is due not only to the fact that it actually contains functional information (i.e., information used for the definition of molecules that operate in metabolism and cell construction), but also because the starting DNA molecules were certainly not random. For this reason, given a sequence, it is very difficult to estimate its actual probability: One has to compare it to an unknown baseline. When studying the frequency of words (such as *TA*, *CCTTGG*, for instance), one has to estimate the probability of occurrence for them if the sequence were random. However, this does not mean that the real sequence can compare to a purely stochastic sequence of letters of same overall probability of occurrence: A way to overcome this difficulty is to choose regions of the genome, such as coding sequences, where it is known that the major selection pressure has been operating only indirectly on DNA, using this as a means to calculate a baseline. For example, as proposed by Hanai and Wada (1989) or Hénaut, Limaiem, and Vigier (1985), one can study the frequency of words overlapping codons (four or more letters) by comparing the level of occurrence of a given word with its level of occurrence if all the codons used in every CDS were taken into account but their actual succession were random. In this case, most of the bias comes from the style of the genome (which can, in part, be reflected by a nonrandom frequency of dipeptides in proteins [Petrilli, 1993]), resulting in some contribution to the bias in the usage of words). To illustrate, it can be observed that among four-letter palindromic sequences, CTAG is counterselected in all organisms (except those in which C is modified, such as bacteriophage T_4 where C is replaced by hydroxymethyl cytosine) (Médigue et al., 1991a). This seems to be due to a special conformation of the double-stranded helix when it carries such sequences of motifs, resulting in some weakness (perhaps inducing spontaneous breaks) (Hunter, Langlois d'Estainot, and Kennard, 1989). In all other cases, the word's frequency reflects the specific style of each genome.

Is it possible to go farther and extract significant global properties of genomes? Relationships between objects, such as those that are of fundamental importance in biology, should be taken into consideration. In particular, the very nature of the information carried from generation to generation in the DNA molecule, which constitute genes, is not at all present in the analogy between entropy and Shannon's information. For example "context," "meaning," or "semantics" are, by construction, absent from this extremely primitive model. Biology provides us with a metaphor that displaces the idea of information toward a new field, that of programming and informatics. Is there more insight in these new fields than in the "natural" way of considering information ? At least since 1965, Kolmogorov and Chaitin, following Solomonoff in the USA and the Russian school of electronic engineers, have formulated the problem in details in the following way (for a general description, see Yockey, 1992). Let us consider the simple case of a chain of characters such as those found in computer sciences. What can be said about that chain's information content? A way to consider the chain is to try to reduce its length so that it can be accommodated in a memory—for example, using the minimum space—without altering its performance when used in a program, for example (at least in terms of accuracy, if not in terms of time)— in short, without losing its information content. This is called *compressing* the data. This problem is of very broad and general interest. Because a chain of characters can be identified with an integer, the universal Turing's computation rules apply. It is particularly possible, given a chain, to define the shortest formal program (in terms of Turing's universal computation algorithms—that is, algorithms that can be implemented on any machine operating on integers) that can compress an original chain, or restore it given its compression state. The information value of a chain S is therefore defined in this model as the minimal length of the universal program that can represent S in a compressed form.

With this definition, it appears that a completely random chain S cannot be compressed, implying that the minimal program required to compress S is identical to S. It should be noticed here that an *apparently* random chain can be compressible: This is the case for instance, with the chain formed by the digits of the decimal writing of the number π, which can be generated by a short program. In contrast, a chain made of repetitive elements can be summarized as a very short program: the sequence of the element and the instruction "repeat." In this context, the information of a sequence is defined as the measure of its compressibility (one often uses the concept of complexity rather than information in this context). It is, however, apparent here that this is not all of what can be said, should we wish to define information in a finer sense. Indeed, the program defining π is highly compressible, as is the program generating the sequence 01010101..., but we would like to say that the information embedded in π is richer than the one embedded in 01010101....

Represented as sequences of letters, genes and chromosomes have an intermediate information (complexity) content: One finds local repetitions or, in contrast, sequences that are impossible to predict locally (Lebbe and Vignes, 1993). Their complexity is intermediary between randomness and repetition. The apparent complexity of sequences originating from higher eukaryotes or prokaryotes is very different, and this links genomic sequences to both sides of the seemingly uninteresting fraction of information (because it corresponds on the eukaryotic side to repetition and on the prokaryotic side to randomness). The complexity of the former is more repetitive and usually looks much lower than that of the latter, which appears more random (Graniero-Porati and Porati, 1988). This is quite understandable if one remembers that bacterial or viral genomes are submitted to stringent economy constraints, implying that they must remain very compact. In contrast, genomes of higher eukaryotes are extremely loosely constrained, and they contain, for instance, many repetitive sequences.

Nonetheless, would it not be more natural to say that eukaryotes are more complex than prokaryotes? We can propose here that this difference in the form of algorithmic complexity is perhaps the actual *cause* of the main morphological differences between prokaryotes and differentiated organisms. The former, following their strategy for occupying biotopes owing to a rapid and versatile adaptive power, are forced to keep up with a small genome.

This implies that several levels of meaning must often be superimposed in a given sequence. The apparent result of this constraint is that their actual algorithmic complexity looks more like that of a random sequence. However, this constraint has an important physical consequence (because, after all, DNA is not only an abstract sequence of symbols but also a molecule that must occupy space): *Superposing signals on the same DNA stretch generally precludes combining synchronous recognition processes by specific proteins, for it is not possible to place two different objects at the same time at the same location.* To exclude such combinations (which alone can fulfil the logical principle of third-party exclusion), however, prevents investigating the refined principles of a delicate modulation of gene expression. The exactly opposite situation is found in differentiated organisms such as animals (and perhaps in the regions of prokaryotic genomes displaying a low complexity). *The lack of limitation in the length of the DNA text permits juxtaposition of recognition signals by regulatory proteins* and allows exploration of the properties of their association according to the rules of combinatorial logic. This is probably what permits differentiated expression as observed in the various cell types. Indeed, this is already well understood for the regions controlling transcription (promoters, enhancers, etc.), but this is probably also the reason for the multiplication of introns in higher eukaryotes, as they behave as inserts without apparent signification, in the very coding regions of genes. Thus, analysis of information in terms of algorithmic complexity permits us to propose explanations to account for the differences among major organism types, whereas Shannon's information gives a meaning only to conservation of a DNA sequence, with a low level of errors, from generation to generation, through the replication machinery.

Many analyses of genomes indirectly pertain to analysis of their algorithmic complexity, thus providing an interesting view of the general organization of base sequences in chromosomes, but such analyses consider genomes as fixed, synchronous entities. In general terms, higher eukaryotes seem to be much more complex than bacteria, be it only because they are able to exhibit time-dependent differentiation patterns. It appears, therefore, that something more should be said about their information content: Information cannot be identified simply by algorithmic complexity, so that Chaitin and Kolmogorov's information (see Yockey, 1992) represents only part of what interests us. In the examples just provided, the concept of information did not imply the notion of *value* (depending both on the object considered and on the nature of the problem that has to involve the object). In fact, if one considers a particular piece of information, its actual availability must have some *value*. Information that is within everyone's reach must differ from that which is difficult or slow to obtain. In the case of living organisms, knowledge of the paths that have been followed through evolution and have led to organisms as we know them today is very valuable and difficult to obtain. Indeed, having access to this type of information may require the introduction of *time*, as we shall now consider.

Evolution: Bennett's Logical Depth

We have seen that we would like to have an idea not only of the complexity of a sequence (as evaluated using the previously described approaches) but also of the availability of the information contained in the sequence. This is particularly relevant in the case of DNA sequences, because all that can be inferred from the knowledge of a sequence derives from the accessibility of its actual information content. To see more repetitions in eukaryotes and more apparent randomness in prokaryotes does not account for the paths that have led to such a large overall difference in genomes. We wish to know the *usefulness* of the information in the sequence. A very short recursive program can be written to express the digits of π and it is possible to write a program of the same length but generating a repetitive sequence. However, the availability of the information content of the corresponding programs' outcome is very different. Although it is easy to predict, after a very short computation, the value of the nth digit of the repetitive sequence, the same is not true for π. In the latter case, if n is large enough, a long computation time is required until the digit is generated, using the program permitting calculation of π.

This observation is essential when one considers sequences generated by recursive programs. In many cases, the writing of the program can be short because the procedure that is reiterated is nested inside the program. Therefore, the information content of the sequence, in terms of algorithmic complexity, is small, because it can be summarized as a short program. Nonetheless, the information on a given digit could be extremely tedious to obtain and might be obtained only after a very long computation time. In this respect, the knowledge of the precise value of this digit—the *information* on it—would seem to be very high, had we another quick means to obtain it. Bennett (1988) has formalized this aspect of sequences in terms of logical depth. The *logical depth* of a sequence S is the minimum time required for generating S using a universal Turing machine (Bennett, 1988). Obviously, knowing a sequence, it is not possible to know its algorithmic complexity. It is even less obvious to have an idea of its logical depth. For this reason, the conceptual analysis made earlier might appear to be merely philosophical (and philosophical is a bad word in English, though not in Latin or Greek civilizations!). We can propose efficient means to analyze both the algorithmic complexity and the logical depth, by proposing algorithms meant to generate the sequence being analyzed. For this we must propose educated guesses, and we contend that, starting from an analysis of the underlying biological processes, we can indeed propose constructive hypotheses. For example, trying to represent how RNA polymerase recognizes its promoter will permit us to generate an algorithm that, in turn, allows us to identify promoters from sequences. Cases in point are the algorithms proposed by d'Aubenton, Carafa, Brody, and Thermes (1990) for the recognition of transcription terminators or by Lisacek, Dinz, and Michel (1994) for the recognition of group I introns. The same is true in the case of logical depth: If this aspect of information carried by DNA is relevant, then the algorithms needed for analysis should require significant computing time to produce interesting outputs. This is particularly true in the case of phylogenetic analyses and indicates that algorithms that perform too fast are necessarily flawed with respect to the biological relevance of their output. A positive side of this very negative observation is that time adds a new dimension to sequence analyses: It could be interesting to add to the features permitting comparisons between sequences the time required for the computation algorithms performed on them, because similar sequences ought to perform with similar computation times under a given hardware and software environment.

Logical depth relates to the Aristotelian distinction between potentiality and reality. It rightly suggests that one should not be allowed to speak in terms of potentiality as if in terms of reality. *There are cases when to speak of potentiality is nonsense, because the actualization of the potential, in true reality, is impossible in a reasonable period of time.* Only those facts that are mechanistically generated could permit such a gross identification between potentiality and future reality. It must be stressed here that reflection on this aspect of information does not completely exhaust the nonclarified (natural) concept of information. In particular, it rests on the existence of an abstract, but perfect, universal Turing machine. In the case of living organisms—provided that one can identify their behavior as that of Turing machines (see Danchin, in press, for justification of this hypothesis)—the building up of the machine is the result of the actualization of a finite program present in the machine. Here one should take time into account once again and consider only those algorithms the length of which is smaller than the lifetime of the machine as a given, unaltered structure. Indeed, a time limit defining the algorithms permitting construction of the machine must also be taken into account: This limit defines a *critical depth*, likely to be of major importance in the case of life. We shall limit ourselves to these descriptions, however, to consider now the specific case of living organisms.

GENOMES *IN SILICO*: INFORMATION AND HEREDITY

The argument that has just been developed is meant to be a substratum for reflection on the metaphor of *program* that currently is used in molecular genetics. It is known that a giant molecule, DNA, made of the sequential arrangement of four—and only four—types of related molecular species, specifies entirely the structure and dynamics of any living organism (adding its own constraints to a world where physics and chemistry operate, permitting in particular the generation of compartmentalization and metabolism). We shall not consider here the problem of origins of such organisms (Danchin 1990), but shall try only to investigate the nature of the relationship between the DNA sequence and the building up of an organism.

As in the case of the reflection of Turing in number theory (Hofestadter, 1979), one can surmise that there must exist a machine to interpret the content of a DNA sequence. This machine is a preexisting living cell, made of a large but finite number of individual components organized in a highly ordered way. Given such a machine, a fragment of DNA, provided it contains a well-formed sequence, is necessary and sufficient to produce a specific behavior representing part or all of the structure and dynamics of an organism. The appropriateness of the alphabetical metaphor used to describe the DNA sequence is further demonstrated by the fact that it is possible to manipulate on paper four symbols—*A, T, G, C*—and to organize them in such a way that, when interpreted into chemical terms (i.e., linked into the proper sequence arrangement into an artificial DNA molecule), they produce a behavior having at least some of the properties of real natural biological entities. This is the basis of survival and multiplication of viruses as well as of a success story in biotechnology: It is possible to make bacteria synthesize a human growth hormone that has been demonstrated on children to be as active as the natural product. That the machine is, in a way, independent of DNA has also been further indicated by the fact that it is possible in some organisms (amphibians) to replace the nucleus of a cell by a new nucleus, resulting in the construction of a new organism (i.e., both a new DNA molecule and a new machine).

However, DNA alone is absolutely unable to develop into a living organism: This is why the question of origin must be considered if one wishes to resolve this chicken-and-egg paradox, through the coevolution of the machine and of the program. It must be stressed here that to apply the consequences of Turing's arguments, it is, in principle, necessary that the program and data be well-identified and separate entities. As stated earlier, biotechnology uses this fact to permit synthesis of human proteins in bacteria, but this corresponds to small genome fragments. Viruses, which infect cells, use bacteria in a similar way to reproduce. Finally, a recent discovery seems to indicate that, at least as a first approximation, bacteria behave as if separating data (the cell machinery, excluding DNA) and program (the chromosome): We have found that one-fifth of the *E. coli* chromosome is made of DNA on which selection operates in such a way that it favors and sustains horizontal exchanges not only between bacteria of the same species but also between distantly related organisms (class III genes) (Médigue et al., 1991a).

Now, if one considers the role of a DNA molecule in a cell, it is possible to describe the various functions derived from its presence as calculable in an algorithmic way. Many routines in the process are recursive (i.e., call themselves but must terminate). It is a generative process that permits generation of both a DNA molecule and of the machine and data necessary to interpret it. Although this has often been called *self-reproduction*, it is production of an organism that is similar if not identical to the starting organism. One should speak, therefore, of *self-reference*. A living organism is, in this context, the algorithmic actualization of a program, the initiation of which has been obtained through a selective process, keeping only actualizations that are stable in time. It should be noticed here that this means that selection does not operate on the phenotype (data) nor on the genotype (program) but on the unfolding of the algorithm (i.e., on the organism itself), thus placing in a new light the usual paradox facing all selective theories and standard Darwinism in particular.

In this respect, it is possible to discuss the information content of a DNA sequence or its complexity using the various definitions previously given. At first sight, most, if not all, of the sequence must be specified to generate the daughter sequence. Things would be simple, and the corresponding information—that which is required for reproduction—would even be rather poor, if there were an absolute fidelity in replicating the original. However, the rule of replication (A → T, T → A, G → C, C → G) is not absolutely faithful, even if it can be considered to result from an algorithmic process. Variations are embedded in the replication process as a function of the general environment of the organism as well as a function of the local sequence of the DNA segment that is replicated. This corresponds to information that should be taken into account, and that is very important indeed, as we shall see later.

However, it must be emphasized at this point that the existence of error management results in specific features of the DNA sequence. As already stressed, CTAG sequences are generally rare, and in E. coli this is compensated by an excess in CTGG words (Médigue et al., 1991a). Spontaneous deamination of cytosines is likely to be compensated by a bias in the fidelity of replication, tending to create more GC pairs. In turn, this mechanism can result in an overshoot, as seen in E. coli strains that lack the mutT gene. Many other processes are certainly modeling the overall GC content of genomes, resulting in a genome style (in this case G + C content) that has been taken into account by taxonomists for preliminary classification of organisms (see Bergey, 1984, for bacteria). Still, such evolution can be much subtler. Chi sites are oriented along the replication fork direction in E. coli (Médigue, Viari, Hénaut, and Danchin, 1993), and it is likely that, because replication is semi-conservative, there exists some bias in regions where Okasaki fragments are initiated (this should result in biases having a periodicity of the order of a few kilobases). Another bias should be introduced by the restriction modification systems present in bacteria, as well as by the existence of methylated sites (as have been observed in a few cases). Finally, there should exist biases in the DNA composition along genes corresponding to the fine tuning of gene expression as a function of environmental conditions or of location of the transcribed or translated regions in the genes. Limaiem and Hénaut (1984) have observed such biases as the enrichment in C + G of the third codon letter as a function of the position of the codon in the genes' sequence. Despite their evidence, these observations are difficult to submit to experimental falsification because this would require, for example, construction of artificial genes and study of their evolution for a very long period of time in conditions that may be very different from natural conditions (chemostats). Therefore, it becomes important to make predictive tests using computers, but to what extent can we make predictions?

Gödel has demonstrated that recursive processes have the very interesting property of being open, in such a way that there exist propositions in natural number theory that can be generated from a corpus of axioms and definitions and that are not decidable in the original corpus (Hofestadter, 1979). Gödel's demonstration is ingenious; it uses a coding process that is certainly reminiscent of the coding process known as the *genetic code*. However, this demonstration has a specific feature: The aim of the demonstration, permitting Gödel to generate his famous "sentences," is known beforehand. The situation faced by living organisms is different: They cannot predict the future; they can have no aim except for (re)producing themselves. Our hypothesis here will be that *the variation provided in the replication and recombination machinery, permitting the generation of new DNA sequences, is the very process that permits living organisms to generate the equivalent of Gödel's sentences in terms of new, deterministic but utterly unpredictable organisms*. The process involves both a procedure for generating variants and a procedure that permits accumulation of "adapted" variants. The latter is so-called natural selection, which must be understood as a process screening stability (and *not* fitness) of the surviving organisms. According to this view, the Spencerian "survival of the fittest" is therefore but a surviving variant of the "instructive" Lamarckian thought.

If this is accepted, one is faced with the puzzling observation that to evaluate the actual information content of a DNA molecule, its history must be traced, which requires an extremely slow procedure: The information content of a DNA sequence is so deep (in Bennett's terms) that it would require the age of life on Earth to evaluate its true complexity. Bearing this in mind, we should be very modest when trying to predict the future of living organisms. In addition, one observes that the process of natural selection associated with generation of mutations provides a blind procedure for generating the equivalent of Gödel's sentences. (In fact, Gödel was cheating somewhat, as we intimated earlier, because he knew what he wanted to obtain, which living organisms cannot do.) This process also generates many more sentences of a type the existence of which we are unable to predict (but the existence of which we can conjecture, simply because at least one type—Gödel's type—does indeed exist). This corresponds to the truly *creative* phenomena that are derived from living organisms.

CONCLUSION

Genome analysis will yield major insights into the chemical definition of the nucleic acids and proteins involved in the construction of a living organism. Further insight comes from chemical definition of the small molecules that are the building blocks of organisms, through the generation of intermediary metabolism. Because life requires also in its definition the processes of metabolism and compartmentalization, it is important to relate intermediary metabolism to genome structure, function and evolution. This requires elaboration of systems for constructing actual metabolic pathways (see chapter 11) and, when possible, dynamic modeling of metabolism (see chapter 6) and for correlating pathways to genes and gene expression (as discussed at the experimental level by Neidhardt [see chapter 9] and theoretically by Thomas [see chapter 8]). In fact, most of the corresponding work cannot be of much use because the data on which it rests have been collected from extremely heterogeneous sources and most often are obtained by in vitro studies.

The initial need is to collect, organize, and actualize the existing data. For the data to be effective and lasting, data collection should proceed through the creation of specialized databases (see Karp and Riley, 1994). To manage the flood of data issued from the programs aiming at sequencing whole genomes, specialized databases have been developed (Médigue et al., 1993; Rudd, 1993), that make it possible not only to bypass meticulous and time-consuming literature searches but also to organize data into consistent patterns through the use of appropriate procedures aimed at illustrating collective properties of genes or sequences. In addition to sequence databases, it has become important to create databases wherein the knowledge progressively acquired on intermediary metabolism could be symbolized, organized, and made available for interrogation according to multiple criteria (Rouxel, Danchin, and Hénaut, 1993).

Using organized data, it will become possible to make *in silico* assays of plausible pathways or regulation before making the actual test in vivo. Well-constructed databases will also permit investigation of properties of life that go back to life's origin, placing biology in a situation not unlike that of cosmology.

NOTES

1. This software, though still unpublished, is available through Internet; mail to viari@radium.jussieu.fr.

SUGGESTED READING

Benzécri, J. P. (Ed.). (1984). *L'analyse des données: I. La taxinomie; II. Les Correspondances* [in French; there is no equivalent in English]. Paris: Dunod.

Cover, T. M., and Thomas, J. A. (1991). *Elements of information theory.* New York: Wiley.

Hénaut, A., and Danchin, A. (in press). Analysis and predictions for *Escherichia coli* sequences, or *E. coli in silico.* In F. C. Neidhardt, J. L. Ingraham, K. Brooks Low, B. Magasanik, M. Schaechter, and H. E. Umbarger (Eds.), Escherichia coli and Salmonella typhimurium (Vol. 2) (2nd ed.). Cellular and molecular biology: Washington, DC: American Society for Microbiology.

Yockey, H. P. (1992). *Information theory and molecular biology.* Cambridge: Cambridge University Press.

II Regulation, Metabolism, and Differentiation: Experimental and Theoretical Integration

What can we expect when attempting to integrate the biology of protein-protein interactions, physiology, metabolism, gene regulation and even differentiation? The second part of this book provides a sketch of what is currently being investigated in the direction of answering this question.

First, Michael Savageau, in chapter 6, offers a critique of one of the basic formal tools in enzymology, the Michaelis-Menten paradigm. Biochemical systems theory provides a theoretical perspective with successful predictions in bacterial systems of regulation. This methodology is, in fact, based on an evolutionary assumption centered on the distinction between positive and negative systems of regulation. Subsequent chapters illustrate different extensions on gene regulation. The contribution of Frederick Neidhardt (chapter 7) shows how different genes are coordinated and integrated in the cell and how new techniques of molecular biology permit a sensitive monitoring of the activities of extensive regulatory networks of genes. The work of René Thomas is classic among theoretical approaches to the study of gene regulation; chapter 8 summarizes the developments of the logical formalization of networks of regulatory positive and negative interacting loops.

These two theoretical approaches—from an evolutionary (Savageau) and a logical (Thomas) perspective—emphasize both the distinction between positive and negative regulation and the dynamic properties of networks of regulators and regulated genes and their associated products. The next two chapters offer a view of gene regulation from a more structural perspective. The attempt to integrate knowledge on the regulation of gene expression using generative grammars (Julio Collado-Vides, chapter 9) is so far devoted mostly to description of the anatomy of regulatory elements as they are located close to the promoter sites. Here again, as in the case of Savageau and Thomas, the formal approach has been developed centered on bacterial systems. The subsequent steps in these methodologies, some already in process, will deal with regulatory systems of higher organisms. Thomas Oehler and Leonard Guarente offer in chapter 10 a summary, from a strictly experimental approach, of the regulation of gene expression in yeast. Interactions of this type are, at least conceptually, connected to the biology of differentiation as addressed by Reinitz and Sharp in a later chapter.

Keeping the order imposed by the levels of biological organization, the next two chapters address metabolic integration. The computational analysis of metabolic pathways demonstrated by Michael Mavrovouniotis (chapter 11) provides the third example of how to understand the behavior of an entire biological network via the properties of its building blocks. These building blocks are regulatory loops for Thomas, individual genes for Neidhardt, and individual biochemical reactions for Mavrovouniotis. In chapter 12, Jack Cohen and Sean Rice then highlight important limitations in descriptions of metabolism and physiology. The question here is how to implement new methods whereby these problems can be at least partially solved. A connection with chaotic behavior is suggested as another avenue by which to investigate biology, a connection also made by Savageau.

The common concern for describing biological structures raises common conceptual problems across several of the approaches here presented. The importance of *finding the adequate representation* and its consequences in the methods being developed is explicitly made in the work of Savageau, Mavrovouniotis, and Collado-Vides, as well as in the type of work described by Robbins, and Smith and collaborators in the previous section. It also is clearly fundamental to the work of Thomas (who has emphasized this in his previous work) and in the approach of John Reinitz and David Sharp in chapter 13. This interesting chapter by Reinitz and Sharp offers a combined theoretical and experimental work about how gene regulation can generate three-dimensional patterns in higher organisms, specifically formation of eve stripes 2 and 5 in early development of *Drosophila*.

Science is an activity not merely devoted to finding new methods; it also has room for raising new questions. Boris Magasanik, collecting various experimental observations on nitrogen regulation in bacteria, in chapter 14 raises an intriguing evolutionary perspective on the origin of a complex mechanism of regulation. We emphasized that structure and its description, as illustrated by the previous experimental and formal chapters, is becoming dominant in this area of molecular biology. Closing the loop, Magasanik brings us back again to the question of the evolutionary origin of such complex structures.

Finally, Robert Berwick, drawing on lessons from computational studies of natural language, addresses some issues related to the many computational approaches to molecular biology.

6

A Kinetic Formalism for Integrative Molecular Biology: Manifestation in Biochemical Systems Theory and Use in Elucidating Design Principles for Gene Circuits

Michael A. Savageau

It is becoming increasingly clear that integrative approaches to molecular biology are critical if we are not to be overwhelmed with information about the constituent elements of living organisms. The systems strategy provides one approach to this problem. In the first phase, *divide and conquer*, emphasis is on reduction of the complexity to basic elements that can be rigorously characterized; in the second phase, *reconstruct and predict*, emphasis is on integration of the constituent information into a testable working model of the intact system. In principle, the systems strategy is fairly clear; in practice, there are a number of difficulties that must be addressed. In the reconstruction phase, these include nonlinear dynamics, physical context, and formal representation. From this effort we have learned a number of valuable lessons. First, we must reassert the fundamental importance of dynamics, which has been relegated to a secondary role, if not abandoned, during the ascent of modern molecular biology. Second, we must become more knowledgeable about the role of intracellular organization. Molecular crowding in this context is a powerful force that has been ignored until recently. Third, we must pay more attention to the choice of representation, which determines, in large part, the scope of phenomena that can be understood, predicted, and controlled. The choice of an inappropriate representation can invalidate the integrative process from the onset, despite a wealth of detailed information concerning the constituent elements. Recent results concerning kinetics in situ have added support for a new kinetic formalism, one that promises to be more appropriate for integrative molecular biology than the formalisms used traditionally. This formalism provides the basis for biochemical systems theory and other recently developed theoretical approaches aimed at understanding intact metabolic systems. Application of this new formalism will be illustrated by an analysis of alternative gene circuits; it will be shown that the systems strategy provides a deeper, more integrative understanding of these basic cellular mechanisms.

BASIC ASSUMPTIONS

Discussions of integrative and reductionist approaches in science often founder because of unarticulated differences in philosophical assumptions.

Hence, we begin with a few of the working assumptions, which are concerned with the scope of integration, the relation between reduction and integration, complexity, and general strategy.

Scope of Integration

One aspect of integration in molecular biology is collecting, cataloging, and cross-referencing vast amounts of molecular data. This is the initial focus of most attempts to generate molecular databases. Another aspect is conceptual synthesis of understanding at different levels of organization, which provides the context for this chapter. Many different levels of conceptual synthesis are necessary for the development of an integrative molecular biology. These can be represented in terms of the conventional view of "structural" information flowing from its origin in DNA. First, one would like to know how the DNA sequence determines higher levels of genome organization (e.g., see (chapters 4 and 5). Second, one would like to know how the RNA sequence, which mimics that in the DNA, determines the three-dimensional structure of the folded RNA. Third, one would like to know how the sequence of amino acids in the protein, which has been translated from the RNA, determines the three-dimensional structure of the folded protein (e.g., see chapter 3). Next, for some RNAs and proteins one would like to know how the three-dimensional structure determines catalytic type, specificity, and rate. Given the array of catalysts, it would be useful to know how these determine cellular composition and structure in space and time. The last level in this hierarchy will be the focus of this chapter. We will address strategies for determining the integrated dynamics of molecular networks from the catalytic specification of their elements.

This is, of necessity, an oversimplification that portrays a linear progression emanating from the nucleotide sequence in the DNA. In fact, the processes at each level of this hierarchy are influenced by events at other levels through a rich network of interactions. Nevertheless, each level of the hierarchy presents fundamental problems of integration that can be pursued independently at this point.

Reduction and Integration

This discussion of hierarchical levels and integration *across* levels should make it clear that understanding cannot be attributed exclusively to either reduction *or* integration. Rather, understanding—at least deep understanding—comes only as a result of both reduction *and* integration. Nevertheless, the theme of this text is integrative approaches, so there will be an emphasis on this aspect of the problem of understanding in molecular biology. Reduction may not seem to get its fair share of attention, but this should not be construed as antireduction.

Organizational Complexity

Biological organisms have long been considered the *sine qua non* of complex systems. This is due in part to their size, number of constituents, and rich patterns of form and behavior. The size of a self-replicating cell is typically small (e.g., 10 μm). Smallness alone would not be so remarkable if it were not for the enormous number of cellular constituents. If we take the case of *Escherichia coli*, a particularly well-studied example, we know from various estimates that the number of different types of molecular elements is on the order of 10,000. Although one might not be interested in constructing a detailed model of 10,000 variables, these numbers allow us to appreciate the enormous variety of constituents that are contained within, and carry out the functions of, a typical cell.

It is variations in these cellular variables, and particularly specific constellations of their values, that determine the basic patterns in space and time that we associate with the living world. It is notable that the basic spatial patterns in developmental biology appear to be laid down on the millimeter scale. These basic patterns emerge between the single-cell stage and the attainment of this millimeter size (e.g., see chapter 13). Growth thereafter does little to change the basic pattern, although the form continues to evolve as a result of differential or relative growth among the parts of the developing organism during the process of morphogenesis.

Systems with large numbers of elements are found elsewhere in nature. Indeed, many of these systems also can be considered complex in some sense. However, they often are assemblages of large numbers of relatively similar elements, and their behavior often is amenable to averaging over a large ensemble of these constituent elements. The behavior of a gas comes to mind. The complexity of biological systems is distinguished by their organization. In studying a biological system, one cannot simply average over all the elements. The elements are very different in kind, and one must develop methods of study that respect the character of the elements and their specific organization. Indeed, it is the organization that one hopes to understand. We call this *organizational complexity* and distinguish it from other forms of complexity.

If we are to make sense of organizationally complex systems, we will have to go beyond documenting and cataloging the many instances that are presented to us in nature. We will have to discover underlying principles and, we hope, manageable rules that will allow one to predict the occurrence of specific forms of organization. Clearly, if each type of organization has its own rule, then we will have achieved little. The adjective *manageable* implies an appropriate simplification, one that abstracts the essential elements of the system without violating its organizational integrity. This is no easy task. There is a constant temptation either to oversimplify or to complicate excessively the systems under study. If the system is oversimplified, it may yield more readily to analysis, but the relevance of the result for the original system becomes doubtful. On the other hand, if the system is made excessively complex in the hope of omitting nothing of potential importance, then it becomes unlikely that any basis for rulelike behavior will be discovered. When an investigation is in progress, it often is difficult to know what is needed—more abstraction or more detail. For this reason, it is important that the investigation be carried out within a framework that allows for systematic development and evaluation of our understanding.

Systems Strategy

The systems strategy is one method for dealing in a comprehensive way with the problem of understanding integrated behavior of organizationally complex systems in relation to their underlying molecular determinants. Although the strategy may go by different names, the basic approach is the same: Define a system for study, separate and characterize underlying mechanisms, and reconstruct and verify integrated behavior.

The necessity of defining the system for study immediately brings to the fore the issues of limits and choices. One cannot study an unlimited amorphous collection of entities; one must draw boundaries and make choices about what to include and what to exclude in the construction of a model system. The most common choices involve spatial, temporal, and functional limitations.

Having defined a system for study does not diminish the task that confronts us. If it is at all complex, then one can obtain only limited information by restricting oneself to observations of the intact system. Further insight requires that the underlying determinants be identified and characterized. This is the reductionist phase of scientific investigation that has been the preoccupation of molecular biology from its inception.

Detailed characterization of the underlying mechanisms alone does not bring about deep understanding of the integrated system. Such understanding requires integration of molecular data into a formal model that is capable of interrelating the molecular and systemic levels of organization. The behavior inherent in the model then must be made manifest through systematic analysis. If the model is indeed an appropriate representation of the actual system, then this process of analysis will lead to verification of the integrated behavior exhibited by the actual system.

THE GOAL

Bearing in mind this prefatory material, we set out, in this chapter, to understand the integrated behavior of organizationally complex biological systems in relation to their underlying molecular determinants. In a sense, all biologists would claim this as a reasonable goal. However, there are different approaches that tend to complement one another. To illustrate these, let us consider the metaphor of a television set.

First, there is understanding that derives from *phenomenological description*. We know that it is roughly cubical in shape. It has a glass front and an assortment of knobs, one of which determines the level of sound, another of which selects the program, and so on. We can make the description quantitative by measuring the diagonal of the screen and measuring the decibel level of sound. All this deals rather directly with the integrated system and provides very useful information. However, we are all aware that this is not a very profound level of understanding.

Second, there is understanding that derives from *component characterization*. We can open the TV set, remove components one by one, and completely characterize their properties in isolation. In the end, we will have all the components isolated and characterized and, if we were careful to note the original location of each component, we could reconstruct the wiring diagram of the TV set. This seems to be understanding of a more fundamental sort but, in the process, we have lost sight of the integrated behavior. We have all had the experience of an appliance failing. You can take out the owner's manual and look at the wiring diagrams in the back but, if you are like this author, you quickly realize that they are of little help in identifying the problem.

Third, an approach that attempts to bridge the gap between the system level and component level is *selective mutilation*. If one mutilates one component, the sound becomes garbled. Mutilate another and the picture fills with snow. Mutilate yet another and the snow disappears. With this approach, one begins to learn which components are associated with which systemic functions. However, this is often a rather blunt instrument, and it is difficult to distinguish direct from indirect associations.

In each of these complementary approaches, one can see a rather direct analogy to the ways in which we attempt to understand cellular systems. There is another kind of understanding for the TV set that currently has little analogy in the biological world. This is the kind of understanding that derives from knowledge of the *design principles*; it is this that the physicist or engineer had or acquired when designing the TV set in the first place.[1] With this kind of understanding, one knows why the item in question is designed the way it is and not some other. We believe this kind of integrative understanding will become central to biology in the near future.

THE CURRENT PARADIGM: PITFALLS AND LESSONS

The Michaelis-Menten formalism,[2] which has provided the kinetic framework for the reductionist phase of molecular biology, was introduced at the turn of the century. It so thoroughly permeates the discipline that for many investigators it is considered reality itself.[3] This is true despite the fact that the fundamental character of the molecular elements was not made clear until the rise of modern molecular biology in the second half of this century. It is surprising that there has been so little critical examination of the Michaelis-Menten formalism in light of what is now known.

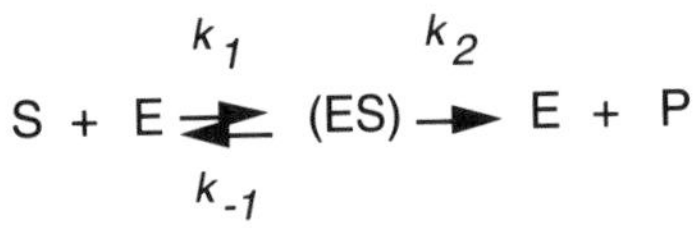

$$S + E \underset{k_{-1}}{\overset{k_1}{\rightleftharpoons}} (ES) \xrightarrow{k_2} E + P$$

Figure 6.1 Traditional Michaelis-Menten mechanism. S, free substrate; P, free product; E, free enzyme; ES, enzyme-substrate complex. The monomolecular rate constants are given by k_2 and k_{-1}, and the bimolecular rate constant is given by k_1.

Michaelis-Menten Paradigm

The textbook derivation of the Michaelis-Menten rate law exhibits the key assumptions of this formalism. The starting point is a mechanism (figure 6.1) in which a single molecule of substrate and a single molecule of enzyme associate to form a specific enzyme-substrate complex (ES) that dissociates into free enzyme (E) and either free substrate (S) or free product (P). The kinetic equations representing this mechanism are a composite description arrived at by treating the elementary steps according to traditional chemical kinetics.[4]

$$d(ES)/dt = k_1 S(E) - (k_{-1} + k_2)(ES)$$

$$dP/dt = v_P = k_2(ES)$$

where $d(ES)/dt$ is the rate of change of the concentration of enzyme-substrate complex, dP/dt (or v_P) is the rate of change of the concentration of free product, E is the concentration of free enzyme, S is the concentration free substrate, k_{-1} and k_2 are the rate constants for the monomolecular reactions, and k_1 is the rate constant for the bimolecular reaction.

These equations are nonlinear, and the inability to solve them readily at the turn of the century led to the quasi-steady-state assumption whereby one considers the dynamics of the enzyme-substrate complex to be so fast that its concentration can be treated as if it were in steady state. This is equivalent to setting the time derivative of the enzyme-substrate concentration to zero and reducing the differential equation to an algebraic one. This algebraic equation then is combined with the enzyme constraint equation, which states that the concentration of total enzyme (E_t) must equal the concentration of free enzyme plus the concentration of enzyme-substrate complex.

$$0 = k_1 S(E) - (k_{-1} + k_2)(ES)$$

$$E_t = (E) + (ES)$$

The result is a set of two algebraic equations that are linear in the two variables [(E) and (ES)]. This set of linear equations is readily solved to yield the concentration of the enzyme-substrate complex as a function of the rate constants for the elementary steps and of the concentrations of free substrate and total enzyme. This expression then is rearranged to give the rate of the reaction in its conventional form as a nonlinear (rational) function of substrate concentration and two aggregate parameters, the maximal velocity V_m and the Michaelis constant K_M.

$$v_P = \frac{V_m S}{S + K_M}$$

The result is an algebraic expression that provides a static representation of the mechanism in the absence of any dynamical context.

The Michaelis-Menten formalism has been phenomenally successful in guiding the experimental study of enzyme kinetics in vitro. There is no shortage of testimony to this fact, and it undoubtedly will continue to be so. Reasons for this success are discussed in some detail elsewhere (Savageau, 1992). However, our focus here is on integrative molecular biology. In what follows, we shall consider three major obstacles to integration that are inherent in the Michaelis-Menten formalism—namely, the banishment of dynamics, the elimination of the natural physical context, and the use of a limited formal representation. For a more complete critique, see also Savageau (1992).

Dynamics Banished

The near-absence of nonlinear dynamics from molecular biology is striking. This author believes that its banishment to the periphery can be traced to the introduction of the quasi-equilibrium (and subsequently the quasi-steady-state) assumption early in the century. In any case, it is now recognized that nonlinear dynamics is central, not peripheral, to understanding integrated molecular networks. It is the nonlinear dynamics that give rise to the most characteristic behaviors associated with living systems. Furthermore, complex behavior is inherent in very simple nonlinear mechanisms, which raises the possibility of discovering simple nonlinear mechanisms that underlie complex phenomena.

A mere increase in value for a parameter of a simple nonlinear mechanism can cause a wide range of dynamic behaviors to be exhibited (figure 6.2). Starting from a simple periodic motion, the behavior can "bifurcate" into a modulated oscillation involving two maxima and two minima. A further increase in the parameter causes the modulation to undergo "period doubling" to yield an oscillatory pattern with four maxima and four minima. Another doubling yields a pattern involving eight maxima and eight minima. Eventually, this cascade of period doubling leads to "chaos" in which the temporal behavior no longer repeats itself as it did in the periodic cases previously exhibited.

The recognition that chaotic behavior can be the result of an underlying nonlinear mechanism that is completely deterministic has had a profound influence on our most basic assumptions about nature (Ruelle, 1991). What is now the relationship between chance and determinism? What are the implications? In medicine, this has led to new views of what constitutes health and disease (Mackey and Milton, 1987). Certain disease states are now seen to be

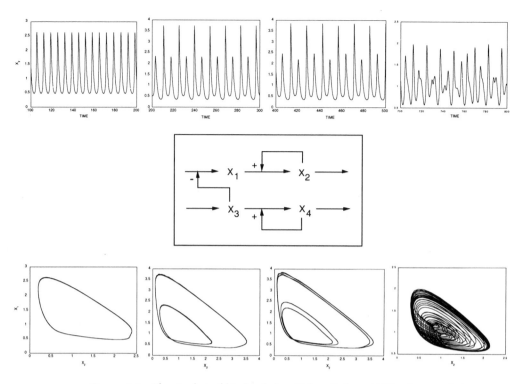

Figure 6.2 The simplest of biochemical oscillators (Selkov, 1968), when coupled, generate complex periodic behavior. One oscillator forcing another of the same intrinsic frequency causes the second to oscillate with a constant amplitude. An increase in the intrinsic frequency of the first oscillator causes the amplitude of the second to bifurcate and become modulated with a single period. As the intrinsic frequency is increased further, the period of modulation doubles to two, and so on, until the modulation becomes chaotic. The temporal responses are shown on the top, with corresponding phase portraits on the bottom.

the result of mistakes in timing (dynamics) rather than loss of an enzymatic function (gene deletion). Homeostasis has given way to more diverse concepts of health. In some cases, a change from steady behavior (e.g., smooth pursuit eye movement) to chaotic behavior (e.g., chaotic pursuit eye movement in schizophrenia) is considered pathology. However, in other cases, a change from chaotic behavior (e.g., visual gaze) to steady behavior (e.g., visual fixation) also is considered pathology. (See also chapter 12.)

Any attempt to understand integrated behavior in terms of underlying molecular mechanisms must be based on a formalism that includes nonlinear dynamics from the start. However, this is not the only criterion on which the Michaelis-Menten formalism falls short.

Physical Context Eliminated

The Michaelis-Menten formalism assumes that there are no interactions between different enzymes or between different forms of the same enzyme within a particular mechanism. The necessity for this assumption is clear from the derivation given earlier, for if there were interactions between different enzymes or between different forms of the same enzyme, then the kinetic equations would have terms containing products of the concentrations of these different enzyme forms. The resulting quasi-steady-state equations would no longer be linear in the enzyme forms; they would be nonlinear and they would no longer have a simple solution.

For example, a mechanism involving dimerization of an enzyme that catalyzes a reversible Michaelis-Menten reaction in both the monomer and diner state is characterized in quasi–steady state by the following set of equations consisting of five kinetic equations plus the enzyme constraint equation:

$$d(ES)/dt = 0 = k_1 S(E) + k_{-2}(EP) + k_{-8}(EES) - (k_{-1} + k_2)(ES) - \boxed{k_8(E)(ES)}$$

$$d(EP)/dt = 0 = k_{-3} P(E) + k_2(ES) + k_{-9}(EEP) - (k_{-2} + k_3)(EP) - \boxed{k_9(E)(EP)}$$

$$d(EE)/dt = 0 = \boxed{k_7(E)(E)} + k_{-4}(EES) + k_6(EEP) - (k_4 S + k_{-6}P)(EE) - k_{-7}(EE)$$

$$d(EES)/dt = 0 = \boxed{k_8(E)(ES)} + k_4 S(EE) + k_{-5}(EEP) - (k_{-4} + k_{-8} + k_5)(EES)$$

$$d(EEP)/dt = 0 = \boxed{k_9(E)(EP)} + k_{-6}P(EE) + k_5(EES) - (k_{-5} + k_{-9} + k_6)(EEP)$$

$$E_t = (E) + 2(EE) + (ES) + (EP) + 2(EES) + 2(EEP)$$

where $d(\cdot)/dt$ is the rate of change of the concentration of the molecular species in parenthesis, and the rate constants with a positive subscript are associated with elementary reactions in the forward direction, whereas those with a negative subscript are associated with the corresponding elementary reaction in the reverse direction. Note that this is a set of nonlinear algebraic equations because of the highlighted terms that involve products of the variables.

Elimination of the physical context by the study of purified enzymes in dilute solution promotes conformity with the assumptions of the Michaelis-Menten formalism.[5] This is one of the primary reasons that this formalism has served so well for the characterization of isolated enzymes in the test tube (Savageau, 1992). However, in the living cell, enzymes do not function in dilute conditions isolated from one another; recent studies of enzyme organization within cells (Clegg, 1984; Srere, Jones, and Matthews, 1989) have shown that essentially all enzymes, even the so-called soluble enzymes, are found in highly organized states. For example, glycolytic enzymes previously believed to be soluble have been shown to exist in a complex with actin filaments in muscle cells (Masters, Reid, and Don, 1987). Molecular crowding within cells is the rule, and there are important thermodynamic (Minton, 1992) and kinetic (Savageau, 1993b) consequences.

Thus, the physical context within cells dictates that enzymes interact with other enzymes and with various structural elements; any formalism that eliminates these interactions at the start cannot be the basis for a valid integrative approach. The physical context also includes molecular crowding, which requires a more general representation of kinetic processes than is provided for in the Michaelis-Menten formalism.

Formal Representation Limited

Traditional chemical kinetics provides the formal representation for the elementary steps in the Michaelis-Menten formalism. In this representation, the rate of an elemental process is expressed as a constant, the rate constant, times a product of concentration factors. The concentration in each factor is raised to a positive integer value, the kinetic order of the reaction with respect to the corresponding metabolite. These integers also represent molecularity, which is the number of molecules of each type that enter into the reaction. For example, in a bimolecular reaction with two molecules of a single substrate reacting to form a molecule of product, the rate will be proportional to substrate concentration squared.

Implicit in traditional chemical kinetics is the assumption that the reaction occurs in dilute solutions that are spatially homogeneous. However, it is now clear that elementary chemical kinetics are very different when reactions are diffusion-limited, are dimensionally restricted, or occur on fractal surfaces (Ovchinnikov and Zeldovich, 1978; Toussaint and Wilczek, 1983; Kang and Redner, 1984; Zumofen, Blumen, and Klafter, 1985; Kopelman, 1986; Bramson and Lebowitz, 1988; Galfi and Racz, 1988; Newhouse and Kopelman, 1988; Jiang and Ebner, 1990; Koo and Kopelman, 1991). For example, the bimolecular reaction considered previously has a rate law in which the exponent is no longer 2 (the molecularity of the reaction) but 2.46 when the reaction is confined to a two-dimensional surface, or 3.0 when restricted to a one-dimensional channel (Kopelman, 1986). In other words, the kinetic order no longer is equivalent to the molecularity of the reaction. The increase in kinetic order results in kinetic behavior that has a higher effective cooperativity. The fractional kinetic orders, in some cases, are a direct reflection of the fractal dimension of the surface on which the reaction occurs. For this reason, these more general kinetics sometimes are referred to as *fractal kinetics*.

The behavior of an enzyme-catalyzed reaction in which the elementary steps are represented in fractal kinetic terms but that is Michaelis-Menten-like in all other respects has been analyzed to show how the Michaelis-Menten formalism might be modified to overcome some of its limitations. The results show that the behavior is decidedly non-Michaelis-Menten (Savageau, 1995). The differences are summarized in table 6.1. First, the effective K_M decreases as the concentration of enzyme increases, whereas it is traditionally independent of enzyme concentration. Second, a Hill plot of rate versus substrate concentration shows sigmoid kinetics and not the traditional hyperbolic kinetics. It is as if there were cooperativity among different binding sites for substrate, when in fact there is but a single binding site. Third, the kinetic order of the overall reaction with respect to total enzyme is greater than

Table 6.1 Comparison of Michaelis-Mentex mechanisms with elementary steps that involve either traditional or fractal kinetics

Enzyme Kinetic Property	Elementary Kinetics	
	Traditional	Fractal
Effective K_M	Independent of enzyme level	Inverse function of enzyme level
Rate as a function of substrate concentration	Hyperbolic	Sigmoid
Kinetic order of rate with respect to enzyme level	Unity	Greater than unity

unity, whereas reaction rate is traditionally proportional to total enzyme (i.e., the kinetic order is unity). In this case, it is as if there were cooperativity of interaction among enzyme molecules, when in fact there is only a single molecule of enzyme participating in the reaction.

Thus, a formal representation must be sufficiently general if it is to serve as the basis for a valid integrative approach. The traditional Michaelis-Menten formalism fails because it uses a limited formal representation that is unable to describe accurately reactions that are diffusion-limited or dimensionally restricted in some manner. Far from being the exception, many (if not most) reactions in a living cell occur on two-dimensional membranes or one-dimensional filaments.

A NEW KINETIC FORMALISM

Recognition of the inadequacies of the Michaelis-Menten formalism has led to the examination of alternatives that might provide a more appropriate formalism for the representation of complex biochemical systems in situ. Any worthy successor to the Michaelis-Menten formalism must be capable of representing the kinetic behavior of complex biochemical systems at any level of organization from elemental chemistry to the physiology of the intact organism. The most regular features of these phenomena are now clear.

Fractal Kinetics and Allometric Morphogenesis

Power-law expressions are found at all hierarchical levels of organization from the molecular level of elementary chemical reactions to the organismal level of growth and allometric morphogenesis (figure 6.3). This recurrence of the power law at different levels of organization is reminiscent of fractal phenomena, which exhibit the same behavior regardless of scale (Mandelbrot, 1983). In the case of fractal phenomena, it has been shown that this self-similar property is intimately associated with the power-law expression (Schroeder, 1991). Hence, it is not surprising that one of the most promising alternatives to the Michaelis-Menten formalism is provided by the power-law

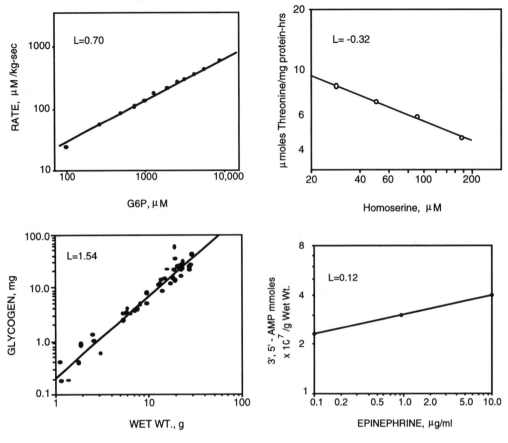

Figure 6.3 Allometric morphogenesis. Examples exhibited at different levels of organization. (Clockwise from the upper left) Enzyme-catalyzed reaction (glucose-6-phosphate dehydrogenase); repressible biosynthetic operon (threonine synthesis); hormonal control system (epinephrine control of fat pad); and whole organism composition (glycogen deposition). (After Savageau, 1992.) For additional discussion, see Bertalanffy (1960), Richards (1969), and Reiss (1989).

formalism (Savageau, 1996), which avoids the criticisms mentioned in the previous section and reduces to the Michaelis-Menten formalism as a special case under the appropriate restrictive conditions.

Power-Law Formalism

The power-law formalism is a mathematical language or representation with a structure consisting of ordinary nonlinear differential equations whose elements are products of power-law functions. The generalized mass-action representation within the power-law formalism can be written in conventional notation as follows:

$$\frac{dX_i}{dt} = \sum_{k=1}^{p} \alpha_{ik} \prod_{j=1}^{n+m} X_j^{g_{ijk}} - \sum_{k=1}^{p} \beta_{ik} \prod_{j=1}^{n+m} X_j^{h_{ijk}}$$

where $i = 1, 2, \ldots, n$; X_i is the concentration of molecular species X_i; dX_i/dt is the rate of change in the concentration of X_i; the exponential parameters g_{ijk} (and h_{ijk}) are the kinetic orders for elemental processes contributing to the production (and consumption) of X_i; and the corresponding multiplicative parameters α_{ik} (and β_{ik}) are the rate constants for these elemental processes. In words, the rate of each elemental process is represented by a product of power-law functions, one for each reactant or modifier that has an influence on the process, and a proportionality constant called the *rate constant*. Each power-law function consists of a concentration variable for some system component together with an exponent, called a *kinetic order*, which represents the influence of that particular component on the rate of the process. The concentration variables and multiplicative parameters are nonnegative real numbers, whereas the exponential parameters are real numbers that can be positive, negative, or zero. The basic definitions, fundamental concepts, methods of analysis, and applications to various classes of systems can be found elaborated elsewhere (e.g., see Voit, 1991).[6]

The first criterion that might be used to judge the appropriateness of the power-law formalism is the degree to which it is systematically structured. A formalism with a systematic structure is like a language with a well-defined grammar (see chapter 9); one can study the systematic structure and deduce general consequences that hold for all phenomena represented within such a formalism. A systematically structured formalism allows one to write a general expression within which all particular expressions are contained. One can then represent any particular case simply by selecting appropriate values for the parameters in the general expression. For example, if two metabolites are known to affect a given process, then within a structured formalism one can write a general expression for the rate of the process v_i as a function of their concentrations X_1 and X_2. The linear formalism provides the paradigm. The general expression in this case is a sum of variables. Each variable that influences the process is multiplied by an appropriate coefficient and included in the sum. In our two-variable example, the general expression is as follows:

$$v_i = b_i + a_{i1} X_1 + a_{i2} X_2$$

Two different mechanisms might be represented with different values for the parameters, but the *form* of the equation would be the same. The linear formalism implies linear relationships among the constituents of a system in quasi-steady state. This is an example of the kind of general information one can obtain by studying the structure of the formalism itself. In this case, however, the general information allows us to conclude that the linear formalism, despite its desirable systematic structure, is inappropriate for most biological systems because the relationships among their constituents are typically nonlinear (see figure 6.3). The power-law formalism, like the linear formalism, is systematically structured. The general expression, in this case, is given by a product of power-law functions. There is one power-law function

A Kinetic Formalism for Integrative Molecular Biology

for each variable that influences the process. Thus, with two variables the general expression becomes this:

$$v_i = \alpha_i X_1^{g_{i1}} X_2^{g_{i2}}$$

Again, two different mechanisms might be represented with different values for the parameters, but the *form* would be the same. Given this representation for the individual rate processes, all the well-known growth laws and allometric properties follow by deduction, which suggests that the power-law formalism is an appropriate representation for intact biological systems.

The second criterion is the degree to which actual systems in nature conform to the power-law formalism. This issue has been examined from the perspectives of three different representations (Savageau, 1996). The power-law formalism was first developed as a *local* representation for the kinetics of reactions exhibiting small variations about a nominal operating point in vivo. The theoretical basis for this development guarantees the accuracy of the local representation for a limited range of variations. It was subsequently recognized that essentially any nonlinear function could be systematically recast into an exact equivalent in the power-law formalism, which as a *recast* representation is accurate over global rather than merely local variations. Finally, it was shown that the power-law formalism can be considered a *fundamental* representation that includes as more limited special cases other representations—such as Michaelis-Menten, mass-action, and linear—that are considered fundamental in various disciplines.

Based on these criteria, as well as others, one can conclude that the power-law formalism indeed provides an appropriate representation for complex biological systems.

A Canonical Nonlinear Representation

The power-law formalism has a very systematic structure, whereas the Michaelis-Menten formalism has an ad hoc structure (Savageau, 1996). As noted earlier, a formalism with a systematic structure is like a language with a well-defined grammar, and there are important theoretical and practical consequences that follow. As a theoretical matter, such a formalism allows one to study its systematic structure and deduce general consequences that hold for all phenomena represented within that formalism. As a practical matter, any analytical or numerical method that is developed to extract information from the general expressions associated with systematically structured formalisms will automatically be applicable to an enormous number of particular cases. This provides a powerful stimulus to search for such methods. This is especially true of the power-law formalism, which can be considered a canonical nonlinear representation that is general enough to represent the kinetics in situ for any reaction of interest. Examples of advances in general methodology that were made possible by the power-law formalism are given next.

Advances in General Methodology

An algorithm has been developed that allows any suitably differentiable function to be recast systematically into the power-law formalism (Savageau and Voit, 1987). Recasting produces equations with nonlinear forms that are simpler than those of the original equations. Furthermore, the simpler form is canonical, which implies that methods developed to solve efficiently equations having this form will be applicable to a wide class of phenomena.

An example of what can be done along these lines is the efficient algorithm developed for solving differential equations represented in the canonical power-law form (Irvine and Savageau, 1990). This algorithm, when combined with recasting, can be used to obtain solutions for rather arbitrary nonlinear differential equations, and this canonical approach yields solutions in shorter time, with greater accuracy and with greater reliability than is typically possible with other methods. Such an approach has been used to solve problems readily that would be very difficult to solve otherwise (Voit and Rust, 1992). Another example is provided by new algorithms for finding the roots of nonlinear algebraic equations (Burns and Locascio, 1991; Savageau, 1993a).

A convenient software package based on this canonical power-law formalism is available (Voit, Irvine, and Savageau, 1990). This package includes the new methods for solving differential equations, tools for a complete network and sensitivity analysis in steady state, methods for the presentation and analysis of graphical data, and other data management utilities. For a detailed application that uses these methods, see Shiraishi and Savageau (1992a–d, 1993).

BIOCHEMICAL SYSTEMS THEORY

Of the three different forms of representation provided within the power-law formalism, the local representation was the first to be used in the development of a theory specifically intended for complex biochemical systems. The proposal for a biochemical systems theory based on local representation within the power-law formalism (Savageau, 1969, 1971) grew out of concepts from classical network and system-sensitivity theories (Bode, 1945). The principal difference is the extension from a linear to a logarithmic coordinate system that is more appropriate for the nonlinearities found in biological systems.

The goals of any such theory include (1) a strategy for constructing a quantitative model of the integrated biochemical system, (2) methods for assessing the quality of a model, (3) methods for predicting the changes in integrated system behavior that result from specific changes in the inputs to the system, (4) a strategy for elucidating the design principles of specific systems that have evolved through natural selection, and (5) a rational approach to the redesign of biological systems for technological or therapeutic purposes. To a large extent, these goals have been achieved through the subsequent development of biochemical systems theory (for recent reviews, see Savageau, 1991, and Voit, 1991).

Constructing Quantitative Models

Construction of a model within biochemical systems theory is a simple four-step exercise. First, relevant pools of metabolites (concentration variables) are identified. Second, the processes that convert one metabolite to another are specified along with the variables that act as modifiers of these processes. Third, Kirchhoff's node equations, which represent the conservation of mass for each metabolite, are written in terms of aggregate influxes and aggregate effluxes for each pool. Finally, the rate law for each aggregate flux is represented by a product of power-law functions, one for each variable that influences the aggregate rate. Thus, quantitative representation at the level of individual rate laws involves two types of fundamental parameters—kinetic orders and rate constants—each with a well-established mathematical definition and a straightforward graphical interpretation. This representation is based on Taylor's theorem and thus is guaranteed to be accurate over some range of variation, which empirical evidence suggests is reasonably broad (Sorribas and Savageau, 1989a).

Assessing Model Quality

Biochemical systems theory provides methods for assessing the logical consistency and robustness of a model. It is important to examine these measures of quality for the model so that one can gauge the reliability of predictions that arise from subsequent analysis. For systems that exhibit a nominal steady state, the most basic manifestations of logical inconsistency are the failure of the model to exhibit a steady state, or to yield a steady state that is in agreement with the actual steady state of the integrated system, or to yield a steady state that is dynamically stable. There are simple tests for the existence and stability of a steady state and, if satisfied, there is an explicit solution for the steady-state behavior in terms of the fundamental parameters and independent variables. Models that are consistent may nonetheless be lacking in robustness, which is manifested as a pathological sensitivity to small changes in the values of their parameters. The degree to which system behavior is sensitive to such parameter variation is measured by a conventional sensitivity coefficient. Hence, the robustness of a model can be assessed by determining the profile of its parameter sensitivities. The full set of parameter sensitivities, which are systemic properties, in turn can be predicted from the set of kinetic orders and rate constants, which are the fundamental component properties (Savageau, 1991). The dynamic behavior of a system cannot be predicted analytically, but it can be obtained by efficient numerical algorithms that have been developed specifically for this purpose. These methods have been incorporated into a convenient menu-driven program

(Voit et al., 1990) with which one can explore dynamic behavior in a systematic fashion. Model quality from a dynamic perspective can be assessed by comparing experimental measurements of such features as pool sizes, peak values, and response times (turnover times) with corresponding predictions from the model.

Predicting Integrated Behavior

The explicit steady-state solution provided by biochemical systems theory allows one to predict the nature of signal propagation through the intact system. A change in any independent concentration (input signal) propagates through the system, causing changes in dependent concentrations and fluxes (output signals). The degree of signal amplification in going from any input to any output is given by a gain factor in conventional network theory. Because changes in the logarithms of the variables are the more natural way to express these relationships in biochemical systems theory, the gain factor actually is defined as a logarithmic gain, which has a well-established mathematical definition and a straightforward graphical interpretation. Logarithmic gains with magnitude greater than one indicate amplification of the signal, whereas those with magnitudes less than one indicate attenuation. Logarithmic gains with a positive sign indicate that input and output signals change in the same direction, whereas those with a negative sign indicate that these signals change in opposite directions. The full set of logarithmic gains, which are systemic properties, can be predicted from the set of kinetic orders, which are fundamental component properties (Savageau, 1991). From the magnitudes of the logarithmic gains for a given systemic output (concentration or flux), one can determine the total influence and its distribution among all input signals. Conversely, from the magnitudes of the logarithmic gains for a given input signal, one can determine its total influence and how this influence is partitioned among all systemic outputs. The dynamic response of the intact system to changes in the input signals cannot be predicted analytically but, again, one can explore the dynamic behavior systematically by computer-assisted methods (Voit et al., 1990).

Elucidating Design Principles

Elucidation of design principles is a more challenging task than the analysis of a specific system because one is searching for regularities that apply to a general class of systems. In any two manifestations of the same design principle, one is likely to find different values for the parameters involved. Thus, the search requires a generic type of analysis that makes no assumptions about the specific values for the parameters in the design. This is possible within biochemical systems theory because it provides an explicit steady-state solution in symbolic form. Such solutions are rare for complex nonlinear systems but, when they exist, important consequences often follow. The existence of symbolic solutions for different systems that are being compared allows one to equate the expression for specific systemic responses while exploring the implications of alternative values for their component parameters. This provides the mathematical equivalent of a well-controlled experiment (Irvine, 1991). The strategy generally involves (1) representing each of the alternatives within the power-law formalism, (2) making these representations equivalent in every respect except for the difference in design under consideration, (3) establishing quantitative criteria for functional effectiveness against which the alternatives can be objectively compared, (4) rigorously analyzing the alternatives to uncover the functional implications of each design, and (5) making predictions that can be tested against experimental findings.

This strategy has been successful in a variety of metabolic contexts (Voit, 1991). For example, a qualitative analysis of feedforward inhibition mechanisms in amino acid biosynthetic pathways led us to predict two fundamental features of these systems (Savageau and Jacknow, 1979). First, there is an optimal position in the pathway for the feedforward inhibitor; namely, it should be the first intermediate in the pathway. Second, the first, or feedback-inhibited, enzyme should be found in a macromolecular complex with the last enzyme, the cognate aminoacyl-tRNA synthetase. These features of the design were demonstrated to enhance system stability and temporal responsiveness. Singer, Levinthal, and Williams (1984) tested these predictions in the case of the isoleucine-valine pathway of E. coli, and their results provided detailed confirmation. α-Ketobutyrate, the product of the first enzyme, was found to be an allosteric inhibitor of the last enzyme, isoleucyl-tRNA synthetase, and the first and last enzymes were found to be associated in a macromolecular complex. Another example involving alternative circuit designs for the regulation of gene expression is presented in the next section.

Redesigning Biological Systems

If one knew the basic design principles for the major classes of biochemical systems, then one could develop rational strategies to redirect normal expression for technological purposes or to correct pathological expression for therapeutic purposes. At the moment, these principles remain largely unknown. Nevertheless, because of the enormous practical implications, this is one of the more attractive applications of biochemical systems theory. With an adequate model, one is in a position to ask any number of "what if" questions and, by exhausting the possibilities, one might hope to uncover the most effective targets for an experimental intervention. This is, in effect, no different from the direct empirical approach that has characterized the field of metabolic engineering in the past. Although it might be considered more economical to do the search with a computer model rather than with living organisms, this assumes the existence of an adequate model. If one must first develop such a model by extensive experimental study, then the weighing of relative costs becomes more ambiguous. The cost and difficulty in developing detailed numerical models will be a limiting factor for some time. In this regard, the ability to carry out qualitative analyses that are independent of specific parameter values is a significant advantage provided by biochemical systems theory. Although such analyses may not yield specific numerical results (e.g., they do not predict specific numerical values for the inhibition constant in the example cited earlier), they are capable of identifying important differences that are qualitatively meaningful. The thermodynamic approach of Mavrovouniotis (chapter 11) and the logical approach of Thomas (chapter 8) focus on different types of questions, but they also yield meaningful qualitative information.

A case study that illustrates most of these points in detail can be found in a series of articles by Shiraishi and Savageau (1992a–d, 1993). It should be noted that the power-law formalism provides the basis for a number of other local theories that have followed some of the same developments as biochemical systems theory. One of these is metabolic control analysis (Kacser and Burns, 1973; Heinrich and Rapoport, 1974. The main difference between them is that the power-law formalism is explicit in biochemical systems theory, whereas it is implicit in metabolic control analysis. The consequences of this and other differences have been documented elsewhere (Sorribas and Savageau, 1989b; Savageau, 1991; Shiraishi and Savageau, 1992d).

DESIGN PRINCIPLES FOR GENE CIRCUITS

Before considering alternative designs for gene circuits, it will be helpful to review some basic concepts of gene regulation, which are treated in detail elsewhere (Savageau, 1989). This section will deal with inducible systems and focus on features shared by many of these systems; the more particular features that may be exhibited by one or the other system also could be treated, but this would take us well beyond the scope of this chapter.

Background

At the molecular level, there are two modes of gene control, positive and negative. This duality of modes is general, although there may be many realizations of each that differ in mechanistic detail (e.g., see chapters 10 and 14). Positive realizations include activators of transcription initiation, antiterminators of transcription termination, and modulators acting at a distance through transcriptional enhancers. Negative realizations include repressors of transcription initiation, proterminators of transcription termination, and modulators acting at a distance through transcriptional silencers. There are well-established rules that allow one to predict the mode of control that will evolve for the regulation of a given set of genes in a particular environment.

At the physiological level, specific regulation of gene expression is manifested in the steady-state induction characteristic. A basal level of expression exists at low concentrations of the specific inducer, and a maximal level of expression exists at high concentrations. The level of expression varies with concentration of inducer at intermediate levels.[7] The *threshold* concentration of inducer is given by the intersection of the basal plateau with the inclined portion of the induction characteristic. The *capacity* for regulated expression is defined as the ratio of maximal-to-basal levels of expression. The *logarithmic gain* of the system is defined as the slope of the inclined portion of the characteristic. It describes the degree to which the input signal (a percentage change in inducer concentration) is amplified (or attenuated) as it propagates through the system to produce the output response (a percentage change in gene expression).

Coupling of Gene Circuits

Regulation of gene expression is at the heart of modern molecular biology. Intense investigation of many systems has revealed a great variety of circuitry. Yet our ability to predict when an organism will evolve a particular circuit for a given set of genes is almost nonexistent. My colleagues and I have had some success in understanding circuits with extreme forms of coupling between regulator and effector genes. The extremes of complete uncoupling and perfect coupling are shown schematically in figure 6.4.

Our approach has involved (1) representing each of the alternatives within the power-law formalism, (2) making these representations equivalent in every respect except for the difference in coupling, (3) establishing quantitative criteria for effectiveness against which the alternatives can be objectively compared, (4) rigorously analyzing the alternatives to uncover optimal designs for each, and (5) making predictions that can be tested against the experimental findings.

Criteria for Functional Effectiveness

The criteria that we have adopted are listed in table 6.2. Each of these has been appropriately quantified and used as the basis for subsequent comparisons. More detailed discussion of these criteria can be found in Hlavacek and Savageau (1995).

Dynamic Equations

The simplest form of the dynamic equations representing the regulatable region of the induction characteristic in the power-law formalism is the following:

Completely uncoupled circuits:

$$dX_1/dt = \alpha_1 X_4^{g_{14}} X_3^{g_{13}} - \beta_1 X_1^{h_{11}}$$

$$dX_2/dt = \alpha_2 X_5^{g_{25}} X_1^{g_{21}} - \beta_2 X_1^{h_{22}}$$

$$dX_3/dt = \alpha_3 X_6^{g_{36}} X_2^{g_{32}} - \beta_3 X_2^{h_{32}} X_3^{h_{33}}$$

Completely Uncoupled Circuit

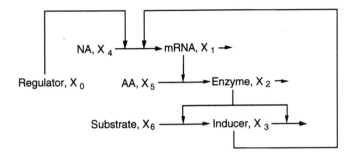

Perfectly Coupled Circuit

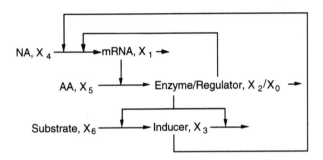

Figure 6.4 Extreme forms of gene coupling. (Top) Expression of regulator and effector proteins is completely uncoupled. The concentration of regulator protein is invariant, whereas the concentration of effector protein varies several fold during induction. (Bottom) Expression of regulator and effector proteins is perfectly coupled. The concentration of regulator protein varies coordinately with that of effector protein during induction.

Table 6.2 Criteria for functional effectiveness of an inducible catabolic system

Stability	The state of the system must be dynamically stable in an appropriate sense.
Robustness	The system should be insensitive to perturbations in the system's component parts.
Decisiveness	There should be a sharp threshold in concentration of substrate for induction.
Efficiency	The gain in product from a given suprathreshold increment in substrate should more than offset the cost of induction.
Responsiveness	The system should respond rapidly to changes in the environment.
Selectivity	There should be limited variation of regulator to prevent indiscriminate interactions.

A Kinetic Formalism for Integrative Molecular Biology

Perfectly coupled circuits:

$$dX_1/dt = \alpha_1 X_4^{g_{14}} X_2^{g_{12}} X_3^{g_{13}} - \beta_1 X_1^{h_{11}}$$

$$dX_2/dt = \alpha_2 X_5^{g_{25}} X_1^{g_{21}} - \beta_2 X_2^{h_{22}}$$

$$dX_3/dt = \alpha_3 X_6^{g_{36}} X_2^{g_{32}} - \beta_3 X_2^{h_{32}} X_3^{h_{33}}$$

where dX_i/dt is the rate of change of X_i, α_i (or β_i) is the rate constant for the production (or consumption) of X_i, g_{ij} (or h_{ij}) is the kinetic order describing the influence of X_j on the production (or consumption) of X_i, X_1 is the concentration of mRNA, X_2 is the concentration of enzyme, X_3 is the concentration of inducer, X_4 is the concentration of precursors for mRNA synthesis, X_5 is the concentration of precursors for protein synthesis, and X_6 is the concentration of substrate. The concentration of regulator, X_0, is represented implicitly in these equations: It is a constant that has been absorbed into the rate constant α_1 in the completely uncoupled case; it is a variable, which is proportional to the concentration of enzyme, that has been replaced by the variable X_2 in the perfectly coupled case.

This representation of the circuits in figure 6.4 involves variables (X_4, X_5, X_6) that are subject to independent experimental variation and variables (X_1, X_2, X_3) that depend on the values of the independent variables and parameters that define the system. Each of the dependent variables is governed by a differential equation representing the conservation of mass. It states that the change in dependent concentration is equal to the difference between increase due to synthesis and decrease due to degradation (and dilution by growth). Thus, for each circuit there are six rate processes to consider, one for synthesis and one for degradation of each dependent variable. The rate law for each process is represented by a product of power-law functions, one for each reactant and modifier that influences the process.

For example, the first term on the right in each set of equations represents the rate law for transcription of mRNA. In the completely uncoupled circuit, this process is influenced by the concentration of substrate (X_4), the concentration of inducer (X_3), and a number of variables (gene copy number, concentration of polymerase, concentration of regulatory protein, etc.) that may be considered constant and treated implicitly. Thus, the rate law is represented as a product of two power-law functions. In the perfectly coupled circuit, the concentration of regulator varies with the level of gene expression and so must be made an explicit dependent variable. The rate of transcription in this case is influenced by the concentration of substrate (X_4), the concentration of inducer (X_3), the concentration of regulator (in the simplest case equal to X_2), and a number of implicit variables. Thus, the rate law in this case is represented as a product of three power-law functions.

Each parameter, except for those representing the process of transcription, is assumed to have the same value in the alternative circuits and, when actual numerical values are of interest, these have been determined from experiments on the lactose operon of *E. coli*. This is one aspect of assuring that well-controlled comparisons are being performed.

Only parameters that reflect the differences in gene circuitry are allowed to differ in value. These are the regulatory parameters g_{12}, which is the kinetic order of transcription with respect to the regulator and represents the regulator's kinetic contribution to the regulation, and g_{13}, which is the kinetic order of transcription with respect to the inducer and represents the inducer's kinetic contribution to the regulation. The other parameters representing transcription, α_1 and g_{14}, may differ in value between the two circuits, but their values are fixed by the requirement that the alternative circuits be as nearly equivalent as possible (Irvine, 1991). For further discussion of the dynamic equations and the selection of parameter values, see Hlavacek and Savageau (1995).

Given these considerations, one can represent the design of the alternative circuits by a point within the two-dimensional space of regulatory parameters. This abstract perspective also provides a convenient means for graphically summarizing the results for a multitude of comparisons.

Space of Regulatory Parameters

The horizontal axis of the space of regulatory parameters represents the regulator's kinetic contribution to the regulation, whereas the vertical axis represents the inducer's kinetic contribution (figure 6.5). Each point in this plane represents a different design. Only those points that lie above the horizontal axis are of interest because they represent inducible systems; points that lie below the horizontal axis represent repressible systems. Points on the vertical axis represent completely uncoupled circuits in which expression

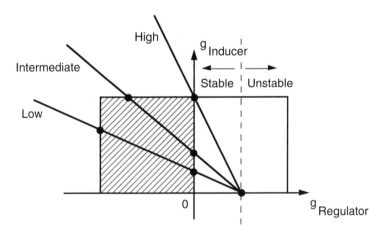

Figure 6.5 Space of regulatory parameters for circuits with the negative mode of gene control. The vertical axis represents the kinetic order of transcription with respect to the concentration of inducer (g_{13} is represented as $g_{Inducer}$); the horizontal axis represents the kinetic order of transcription with respect to the concentration of regulator (g_{12} is represented as $g_{Regulator}$). The shaded region represents the physically realizable region. The inclined lines radiating from a common point represent lines of equivalent logarithmic gain for systems with low, intermediate, and high logarithmic gain. See text for further discussion.

A Kinetic Formalism for Integrative Molecular Biology

of the regulator is invariant during induction, and hence its kinetic contribution to the regulation is zero. Points in the left half of the plane represent systems in which the regulator has a negative mode of action, whereas those in the right half of the plane represent systems in which the regulator has a positive mode of action.

The maximum allowable values for the regulatory parameters are fixed by the number of binding sites on the multimeric regulator (Wyman, 1964; Monod, Wyman, and Changeux, 1965), which is typically a dimer (Matthews, 1992; Kim and Little, 1992). As a consequence, realizable designs will be found only within the rectangular region about the origin. The dashed vertical line in the figure is the *boundary of instability*. Designs represented by points to the right of this line are unstable and therefore physically unrealizable. Thus, only designs represented by points lying to the left of the dashed line and within the rectangular region need be considered in our comparisons.

The inclined line in figure 6.5 is a *line of equivalent logarithmic gain*. It is the locus of points representing designs that possess the same induction characteristic. The comparison of designs that are represented along such a line of equivalent gain constitutes a well-controlled comparison.

Negative Mode with Low Gain

When the mode of gene control is negative (left half of plane, figure 6.5) and the logarithmic gain is low (line of equivalent gain with a shallow slope), the results of comparisons based on the criteria in table 6.2 can be summarized as follows: The completely uncoupled and perfectly coupled circuits are equivalent with respect to decisiveness and efficiency. This is a consequence of making well-controlled comparisons. The completely uncoupled circuits are superior with respect to selectivity. Their regulators are invariant during induction and therefore less likely to cause indiscriminate interaction with other genes. However, the difference in this case is not considered significant because low gain implies small variation in regulator levels even with perfect coupling. The perfectly coupled circuits are superior with respect to all the other criteria. Perfectly coupled circuits are represented further from the boundary of instability, and the greater the regulator's kinetic contribution, the greater is the degree of stability. This is a qualitative result that is clear from the geometry in figure 6.5. The perfectly coupled circuits are more robust, and they become more so as the regulator's kinetic contribution increases. This is a quantitative result that is independent of the particular values for the other parameters, although this is not intuitively obvious. The perfectly coupled circuits are more temporally responsive. To establish this result, it was necessary to explore a wide range of specific numerical values. Again, response time was found to decrease as the regulator's kinetic contribution increased and, again, this is a quantitative result that is not intuitively obvious. Thus, circuits with perfect coupling are equal or superior to those with complete uncoupling on the basis of all six criteria.

Negative Mode with High Gain

When the mode of gene control is negative (left half of plane, figure 6.5) and the logarithmic gain is high (line of equivalent gain with a steep slope), the spectrum of physically realizable designs becomes greatly restricted. In effect, only circuits that are completely uncoupled are possible; circuits that are perfectly coupled but otherwise equivalent would require the inducer's kinetic contribution to be greater than that allowed by the structure of the multimeric regulator. Thus, completely uncoupled circuits are the only designs expected for systems with the highest gain and the negative mode of gene control.

Positive Mode with Low Gain

When the mode of gene control is positive (right half of plane, figure 6.6) and the logarithmic gain is low (line of equivalent gain with a shallow slope), the results of comparison are, for the most part, just the reverse of those when the mode is negative. The completely uncoupled and perfectly coupled circuits are equivalent with respect to decisiveness and efficiency. The completely uncoupled circuits are superior with respect to all the other criteria. Their regulators are invariant during induction, they are represented further from the boundary of instability, they are more robust, and they have shorter

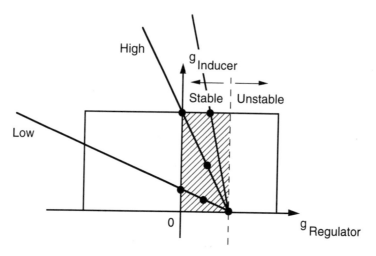

Figure 6.6 Space of regulatory parameters for circuits with the positive mode of gene control. The vertical axis represents the kinetic order of transcription with respect to the concentration of inducer (g_{13} is represented as $g_{Inducer}$); the horizontal axis represents the kinetic order of transcription with respect to the concentration of regulator (g_{12} is represented as $g_{Regulator}$). The shaded region represents the physically realizable region. The inclined lines radiating from a common point represent lines of equivalent logarithmic gain for systems with low, high, and very high logarithmic gain. See text for further discussion.

Table 6.3 Predictions regarding gene circuitry based on our analysis

Mode	Gain	Circuit	Example	Reference
Positive	Low	Complete uncoupling	Maltose	Schwartz, 1987
	High	Perfect coupling	Ethanolamine	Roof and Roth, 1992
Negative	Low	Perfect coupling	Histidine	Smith and Magasanik, 1971
	High	Complete uncoupling	Lactose	Miller and Reznikoff, 1980

response times. In this case, circuits with complete uncoupling are equal or superior to those with perfect coupling on the basis of all six criteria.

Positive Mode with High Gain

When the mode of gene control is positive (right half of plane, figure 6.6) and the logarithmic gain is high (line of equivalent gain with a steep slope), the spectrum of physically realizable designs is again truncated by the maximum allowable value for the inducer's kinetic contribution. In this case, the completely uncoupled circuit is not realizable. The optimal design among the perfectly coupled circuits is the one for which the inducer's kinetic contribution to the regulation is maximal. The perfectly coupled circuits represented along such a line of equivalent gain all have the same decisiveness, efficiency, and selectivity. The one for which the inducer's kinetic contribution is maximal is superior to all the others with respect to degree of stability, robustness, and temporal responsiveness. Thus, perfectly coupled circuits are the only design expected for systems with the highest gain and a positive mode of gene control. Furthermore, the geometry of the relationships displayed in figures 6.5 and 6.6 suggests that systems with a positive mode should, in general, be capable of higher logarithmic gains than otherwise equivalent systems with a negative mode.

Predictions

The results just reported are very general. They depend only on the differences in topology of the circuits depicted in figure 6.4, and they are independent of the particular values for the parameters in the power-law representation. The generality of these results suggests that extreme forms of gene coupling are governed by relatively simple rules that provide a deeper understanding of these designs. On this basis, we make the following predictions (table 6.3): Systems with the negative mode of gene control are expected to evolve with the perfectly coupled circuit when they have a low logarithmic gain, or with the completely uncoupled circuit when they have a high logarithmic gain. Systems with the positive mode of gene control are expected to evolve with the completely uncoupled circuit when they have a low logarithmic gain, or with the perfectly coupled circuit when they have a high logarithmic gain.

CONCLUSIONS

The outpouring of data concerned with the molecular elements of biological systems has outpaced our ability to synthesize these data into meaningful models of the intact system. This is particularly apparent when we try to understand more complex biological phenomena typical of higher levels of organization. We are in need of an integrative perspective that will complement the reductionist view that has guided the development of molecular biology to date.

An appropriate integrative approach can be expected to exhibit three characteristics. First, the focus will be systemic, integrated behavior. Second, understanding will not reflect mere superficial descriptions of behavior but will be based on the underlying molecular determinants. Third, the understanding will be relational; it must be able to relate knowledge at the systemic and the molecular levels. It will require development of rigorous means for determining the properties at one level of organization in terms of properties characteristic of the other. This type of understanding—quantitative understanding of design principles that relate molecular and systemic properties—will be at the very heart of integrative molecular biology in the future.

The development of an integrative perspective will require experimental methodologies that allow one to quantify many biochemical variables simultaneously (e.g., see chapter 7). It also will become increasingly important to make prescribed quantitative changes in specific molecular constituents and to make rapid nondestructive measurements of biochemical variables in situ. Understanding the integrated temporal, spatial, and functional behavior of complex biological systems also will require a systematic method of description that allows one to relate the molecular and systemic properties in a deep quantitative manner.

Mathematics, like it or not, is the only language available that is sufficiently systematic and precise to deal with the distinctive features of integrated biological systems. It offers three essential advantages. First, the large numbers of variables characteristic of organizationally complex systems pose an enormous bookkeeping problem, and mathematics provides the only systematic way of tracking all their values. Second, the richness of interactions and their nonlinear character lead to critical quantitative relationships that characterize systemic behavior. Only the quantitatively precise language of mathematics has the power to elucidate these system design principles, to compare alternative designs rigorously, and to represent them in an efficient manner. Finally, when systems are even moderately large, it becomes necessary to implement the methods for their analysis on a computer, and this is possible only when the problem can be formulated with mathematical precision. Although insights have been achieved in the past without the explicit use of any mathematical formalism, such instances do not make the preceding views any less compelling. There are numerous examples of relatively simple well-defined systems that defy intuitive attempts to predict their integrated behavior, yet the behavior of these same systems is readily predicted and understood in elementary mathematical terms. We can expect that this also will be true of organizationally complex systems, which, by definition, have long been resistant to intuitive understanding.

In the search for a formalism that is appropriate for the analysis of organizationally complex systems, one naturally looks to kinetic approaches. The Michaelis-Menten formalism has served admirably as the kinetic representation for the reductionist phase of molecular biology. As this author has argued elsewhere (Savageau, 1992), the canons of good enzymological practice have evolved in large part to ensure that the characterization of reactions in vitro fits the postulates of the Michaelis-Menten formalism. Though these canons are appropriate for the characterization of isolated reactions in vitro, they often are highly inappropriate for the characterization of reactions as they occur in situ within the integrated living system. Thus, far from being the main route that leads to a proper integrative approach, the Michaelis-Menten formalism can be seen as a cul-de-sac from which further progress toward this goal is blocked.[8] Other kinetic formalisms must provide the framework for the tasks ahead.

The power-law formalism currently is the most promising candidate for this role. As we have seen, it avoids the major pitfalls associated with the Michaelis-Menten formalism. First, it includes the full nonlinear dynamics. Second, it can incorporate all relevant interactions from the physical context in vivo. Third, it provides a sufficiently generic representation. It can accommodate the power functions with noninteger exponents that have recently been observed at the level of elementary chemical reactions; it also is consistent with the older observations on growth laws and allometric morphogenesis at the organismal level. Even if new forms of nonlinearity are discovered to be important for the description of biological systems, the power-law formalism will be capable of accommodating these as well, because it is in fact a canonical nonlinear representation.

Local representation within the power-law formalism provides the basis for biochemical systems theory. This theory has been shown to possess many of the qualities one expects of a rigorous integrative approach and to yield insights that would be difficult, if not impossible, to achieve by other means. As a concrete illustration of the utility of this approach, we have examined the design of alternative gene circuits.

Our analysis of the two extreme forms of coupling has shown that these designs can be understood in terms of simple rules. It is important to note that the conclusions do not depend on the choice of specific numerical values for the parameters in these models and, for this reason, the predictions can be considered quite general. In fact, the conclusions involve only qualitative properties: the topology of the circuitry (completely uncoupled or perfectly coupled), the signs of the interactions (positive or negative), and the capacity for regulated expression (high or low). Thus, the experimental work needed to test the predictions is greatly diminished. Many of the data already exist in the literature but in a form that makes the data difficult to find. If this qualitative information were integrated into a database such as that of Collado-Vides (see chapter 9), then one would immediately have numerous cases with which to make a rigorous test of the predictions.

Although these predictions now can be tested in those cases for which there is evidence of complete uncoupling or perfect coupling, such predictions should not be extended to other types of circuits that are neither perfectly coupled nor completely uncoupled. The design principles that govern circuits with more general forms of coupling have yet to be fully analyzed, but it is already clear that they exhibit a rich behavioral repertoire.

Finally, the importance of dynamics in integrative molecular biology must not be underestimated. Even in studies concerned only with steady-state analyses, it is necessary to validate these analyses by careful determination of the dynamics. This is especially true in the case of large and complex biochemical systems where these issues are not at all obvious (see Shiraishi and Savageau, 1992d). The importance of dynamics is illustrated clearly in our analysis of alternative gene circuits. The prediction of a completely uncoupled or perfectly coupled circuit depends critically on differences in their dynamic behavior (stability and temporal responsiveness) and would not have been revealed by approaches based only on steady-state considerations.

ACKNOWLEDGMENTS

This work was supported in part by grants from the US National Institutes of Health and the National Science Foundation.

NOTES

1. Although the use of an analogy between a biological organism and a technological device may tempt some to think of the "watchmaker" arguments of a century ago, this temptation should be avoided. The context today is very different. Then the objects were static, little was known about the organism, and by *design* one meant forethought in its creation. Today, we recognize that technological products are not static; they often have a short life span and evolve no less than biological organisms. The TV sets of today are very different from those of 40 years ago; they have evolved through variation and selection in the marketplace. Today, we know a great deal about organisms, particularly some of the simpler ones. Not only have we taken the organism apart and learned to identify its molecular components; we have learned a great deal about its functions or physiology. Today, by *design principles* we mean the laws of nature that govern the arrangement of parts, details, form and so on, especially so as to produce a complete and harmonious unit. Many of the design principles implicit in a TV set were unknown 40 years ago. Many variations in design were created and studied and, in the process, design principles were discovered. Now that they are known, one can guide future designs with greater assurance. Similarly, many of the design principles implicit in cells are unknown today. Yet investigators already are engineering cells for biotechnological and therapeutic purposes. We must study existing cells and their variants in an attempt to discover the design principles that could improve our ability to guide this powerful new technology.

2. I include within the term *Michaelis-Menten formalism* the underlying structure for essentially all of contemporary enzyme kinetics, which is based on a set of key assumptions shared with

the original development of Michaelis-Menten kinetics (Savageau, 1992). Thus, this formalism also provides the framework for treating various allosteric mechanisms, which often are referred to as *non-Michaelis-Menten*.

3. Data often are made to fit this perceived reality, as the important critique by Hill, Waight, and Bardsley (1977) suggests.

4. The rate laws of chemical kinetics are based on various probability considerations. Hence, there is explicit recognition of the stochastic nature of the discrete events that underlie kinetic observations. In most cases, one is interested in the mean behavior. When one is dealing with large numbers of events over long time scales, the means are well defined (the variance is small), and the continuous deterministic representation in terms of concentrations is justified. When the numbers are small and the time scale short, then the averages (both ensemble and temporal) may have large variances, and the mean behavior must be interpreted with caution.

To make this issue more concrete, consider the operator for a particular gene in *E. coli* and take the extreme case for which there is but a single copy in the cell. At any instant, one can consider gene expression to be either off or on depending on whether the operator is bound with its particular repressor. One cannot predict the state with certainty; it depends on the number of repressors in the cell, the number of other types of molecules in the cell, several physical parameters, and so forth. Binding is a discrete phenomenon that is inherently stochastic.

Nevertheless, given a certain number of repressors over a sufficiently long period of time in which the operator is unbound—say 60 percent of the time—we can consider the operator to be 60-percent unbound. In this statement, an inherently discrete stochastic description is equated with a continuous deterministic description. One clearly can get into trouble if the time period is very short, say microseconds. The number of repressors might indicate that the operator should be bound for a certain percentage of the period when it is, in fact, completely unbound throughout the period. However, the time scale that governs change of mRNA levels in a cell growing with a 60-minute doubling time can be on the order of 240 minutes. If one considers the percentage of time that the operator is unbound with repressor on this time scale, then one finds a reproducible and predictable percentage. In this context, it is more useful to consider the rate of gene expression as a continuous variable that can be deterministically related to concentration of repressor, than it is to think of the rate of gene expression as being either on or off depending on the state of the cell at a particular instant. Justification of the kinetic description, or indeed any other description, must come from the validity of the results obtained, and, in the end, this is determined by agreement of theoretical predictions and independent experimental tests.

We have found the kinetic description most useful for addressing questions of circuit design in cellular and molecular networks. An example involving enzyme circuitry is given in Savageau and Jacknow (1979), and an example involving genetic circuitry (Hlavacek and Savageau, 1995) will be described later in this chapter.

5. I thank John Campbell (personal communication) for pointing out that Monod, in his work on allosteric enzymes, found the classical emphasis on enzyme purification responsible for destroying or masking the allosteric properties of many enzymes and thus making their kinetics conform to the Michaelis-Menten expectation. Monod went on to suggest (personal communication to John Campbell) that enzymes be studied *before* any purification is carried out and that purification be carried out only with demonstration at each stage that an alteration in kinetics does *not* occur.

6. Spatial variations can be treated in the conventional fashion within the power-law formalism by adding diffusion terms, but the resulting partial differential equations have no simple analytical solution. In practice, one solves these equations numerically by discretizing the spatial dimensions of the problem; in effect, the set of partial differential equations is reduced to a set of ordinary differential equations such as those described in the text. Thus, this same power-law representation also characterizes heterogeneous spatial distributions of the molecu-

lar species when one considers the space to be divided into small discrete volumes within which chemical reaction occurs and between which diffusion occurs (Savageau, 1976). The subscript of X_i then indicates the name *and* location of each molecular species. The ordinary differential equations then are solved by conventional numerical methods, which also involve discretizing the temporal dimension. An example of this approach to dealing with heterogeneous spatial distributions is provided by the classical work of Meinhardt and Gierer (1974) on pattern formation during hydra development. Their equations consist of reaction terms involving power-law kinetics and traditional diffusion terms. These equations were reduced to ordinary differential equations, by discretizing the spatial dimension, and then solved numerically using conventional techniques. This same basic approach to dealing with spatial variation is used by Reinitz and Sharp (in chapter 13) in dealing with pattern formation during *Drosophila* development. As these examples show, the coupling of diffusion and chemical reaction provides rich possibilities for generating structure and organization in developing systems.

7. Gene expression can change from one state to another in response to small changes in the concentration of an inducer. On an appropriately small *concentration* scale, such switch behavior is seen to be a continuous process. When considered on a significantly larger concentration scale, such switch behavior can be approximated as an ideal step function that changes discontinuously from one state to another. One can easily realize either static or dynamic switch behavior using the continuous deterministic representation of generalized mass-action equations. Static switches exhibit no dynamics, hysteresis, or discontinuities. An allosteric rate law provides a well-known example (Monod et al., 1965). The greater the degree of cooperativity, the sharper is the switch from basal to full expression. Dynamic switches exhibit continuous dynamics that lead to hysteresis and discontinuous change when viewed from the perspective of small changes in the concentration of the critical variable. When starting from a history of basal expression, the concentration of the critical variable must be increased above an upper threshold, at which point the system switches to full expression; when starting from a history of full expression, the concentration of the critical variable must be decreased below a lower threshold, at which point the system switches to basal expression. An example is provided by lactose transport, which in *E. coli* is induced by an intracellular intermediate of lactose catabolism, responding to extracellular concentrations of lactose as the critical variable (Novick and Weiner, 1957; Cohn and Horibata, 1959).

8. The inclusion of Michaelis-Menten kinetics in a database for the purpose of generating kinetic models of intact systems would be inappropriate. Furthermore, since premature adoption of an inappropriate representation often makes it difficult later on to establish a more appropriate one, the use of Michaelis-Menten kinetics at this point would likely be counterproductive.

SUGGESTED READING

Goodwin, B. C. (1994). *How the leopard changed its spots: The evolution of complexity.* New York: Charles Scribner's Sons. The first part of this book provides a very readable discussion related to many of the general issues considered in this chapter.

Savageau, M. A. (1989). Are there rules governing patterns of gene regulation? In B. C. Goodwin and P. T. Saunders (Eds.), *Theoretical biology—epigenetic and evolutionary order*, Edinburgh: Edinburgh University Press, pp. 42–66. This review deals with the question, "Why are there both positive and negative regulators of gene expression?"

Savageau, M. A. (1992). A critique of the enzymologist's test tube. In E. E. Bittar (Ed.), *Fundamentals of medical cell biology*, Greenwich, CJ: JAI Press, vol. 3A, pp. 45–108. This chapter expands on the critique of the Michaelis-Menten formalism given here in abbreviated form.

Shiraishi, F., and Savageau, M. A. (1992). The tricarboxylic acid cycle in *Dictyostelium discoideum*: I. Formulation of alternative kinetic representations. *Journal of Biol. Chem. 267*, 22912–22918. The first part of this article provides a brief review of biochemical systems theory, emphasizing the basic foundations and recent developments.

Voit, E. O. (1991). *Canonical nonlinear modeling: S-system approach to understanding complexity.* New York: Van Nostrand Reinhold. This book surveys much of the power-law formalism, including key theoretical developments and a wide selection of applications.

7 Genome Analysis and Global Regulation in *Escherichia coli*

Frederick C. Neidhardt

A central goal in biology is to achieve a cellular paradigm—that is, to solve the structure of some particular cell and describe how it functions. There is little danger of achieving this goal in the near future, so complicated and daunting is the task. Even the primal step of naming the molecular components of a cell has yet to be accomplished, let alone the more advanced steps of discovering the biochemical and physiological role of each component and of developing a detailed mathematical model of how these components function together to produce a unit of life.

A major step toward a molecular inventory of a cell was taken nearly 20 years ago when Patrick O'Farrell (1975; O'Farrell, Goodman, and O'Farrell, 1977) introduced two-dimensional (2-D) polyacrylamide gel electrophoresis to resolve complex mixtures of proteins. O'Farrell demonstrated that resolution of individual proteins from a whole-cell extract by means that utilize size and isoelectric point was capable of presenting a more or less complete picture of the cell's complement of proteins. Subsequent work has confirmed O'Farrell's vision that 2-D gel electrophoresis would be more than just a powerful analytical tool to the cell biologist; it makes possible a new approach to integrative cellular studies.

In this chapter, we first present a synopsis of a research project aimed at creating a total genome expression map of a model cell—that is, a composite 2-D gel map displaying every protein that the bacterium *Escherichia coli* is genetically capable of producing, with each protein matched to its cognate structural gene. The usefulness of such an expression map in a wide range of integrative studies of this organism's physiology will be discussed, including in particular the relevance to such theoretical approaches to integrative cellular function as are presented in chapters 6, 8, and 9 of this volume. The second part of the chapter describes how a protein with important global regulatory properties was discovered and analyzed through the use of 2-D gels.

FUNCTIONAL ANALYSIS OF THE *E. COLI* GENOME

Two-Dimensional Gels Permit Integrative as Well as Analytical Studies

In the field of bacteriology, it was quickly recognized that 2-D gels are not suited solely for reductionist studies. True, they aid important reductionist studies by displaying the protein components of the cell as individual spots, and they offer the means to provide detailed, provocative information about these individual proteins independent of any knowledge of their biochemical identity or genetic origin. However, appropriate radioactive labeling protocols permit one to explore with 2-D gels the integrated operation of the intact cell and its changing patterns of gene expression. Consequently, the most striking outcome of O'Farrell's gel technology has been the revelation of global gene responses in the late 1970s when cell biologists, for the first time, could learn from their cell of interest what proteins *should* be studied rather than having to confine their attention to some arbitrarily chosen protein or set of proteins. When applied to the bacterium *E. coli*, these explorations revealed the workings of complex and overlapping regulatory sets of proteins and led to new challenges in unraveling the intricacies of gene regulation (VanBogelen, Hutton, and Neidhardt, 1990).

Two-Dimensional Gels Facilitate Study of Global Regulation and Gene Networks

The first gel studies on *E. coli* that were of a global nature were performed in our laboratory in collaboration with Steen Pedersen (Pedersen, Bloch, Reeh, and Neidhardt, 1978). We examined the pattern of gene expression during growth in media of different composition, yielding a range of growth rates at 37°C. We were able to recognize five different classes of proteins—those made at constant differential rate irrespective of the media, those made at rates increasing and those at rates decreasing with simple functions proportional to the growth rate of the cells, and those in which the chemical nature of the medium (rather than the cell growth rate) determined the differential rate of synthesis, leading to maxima and minima as a function of growth rate (figure 7.1). This study helped establish what fraction of the cell's proteins are regulated in a manner similar to that of the known proteins of the transcription and translation apparatus of the cell and which are nutritionally induced or repressed. Later, we characterized the effect of steady-state growth in a single medium at different temperatures on the pattern of gene expression and found correlations between the classes described in the two studies (Herendeen, VanBogelen, and Neidhardt, 1978).

These steady-state studies were followed quickly by a large number of studies that examined non-steady-state conditions. The latter led to the discovery and characterization of the heat-shock response, the cold-shock response, the responses to restriction for various macronutrients such as

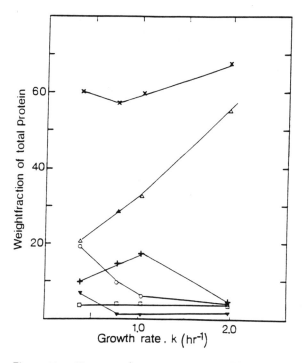

Figure 7.1 Variation of protein amounts in different regulatory groups as a function of growth rate in *E. coli* strain NC3. The ordinate values are the sum of the weight fractions of the 140 individual proteins measured in this study (Pedersen et al., 1978), plus previously measured ribosomal proteins, in each of five regulatory groups in cells growing in steady state in each medium at 37°C at the growth rates shown on the abscissa. The media and the cells' respective growth rates are: acetate minimal (0.38 hr^{-1}), glycerol minimal (0.77 hr^{-1}), glucose (1.03 hr^{-1}), and glucose rich (1.98 hr^{-1}). Individual proteins were assembled into five groups showing similar regulation, even with the realization that within each group there are diverse mechanisms that generate the observed regulatory behavior: Group Ia (33 proteins) decreases in level with growth rate (open circles), group Ib (16 proteins) is constant (open squares), and group Ic (54 proteins plus ribosomal proteins) increases with growth rate (open triangles), whereas group IIa (32 proteins) exhibits a maximum (plus signs), and group IIb (5 proteins) exhibits a minimum (closed triangles). The uppermost line (crosses) indicates the sum for all measured proteins. Group Ia includes several catabolic enzymes of central metabolism, group Ic includes many proteins involved in transcription and translation, and group IIa includes many enzymes involved in amino acid and other biosynthetic pathways. (Modified from Pedersen et al., 1978.)

carbon, nitrogen, sulfur, and phosphorus, and to growth restrictions brought on by various toxic agents or conditions such as high or low pH and osmolarity, oxidants, heavy metals, chaotropic agents, uncouplers of oxidative phosphorylation, and antibiotics. Meanwhile, we came to appreciate the use of gels to analyze mutants in known global gene regulators to discern the pathological consequences to the cell; this approach proved useful for such regulators as *crp* and *cya* (for catabolite repression), *relA* and *spoT* (for the stringent response), *lexA* and *recA* (for the SOS [help!] response), and *rpoH* (for the heat-shock response).

The technology for global studies of this sort has improved continually since 1975. The gel-forming process and the apparatus for electrophoresis have been refined, the measurement of spot radioactivity has been revolutionized by the use of phosphorimaging, the scanning and matching of the images (spot patterns) of gels in an experiment have been reasonably automated, and the management of large volumes of data has been considerably aided by new software. A typical global response experiment involves duplicate control and experimental cultures in which the cellular response to some environmental change (stress) is ascertained by pulse labeling with a radioactive amino acid at two close intervals after the stress is imposed on the experimental culture. In the 1970s, it took 3 months and a materials cost of $2,000 to do one global response experiment gathering data for 200 protein spots. In the 1980s, it took 3 to 5 months and cost $100 per experiment to gather data for 300 to 600 protein spots. In the 1990s, it takes 1 month and costs $100 per experiment to gather data for 1,200 protein spots.

Identification of Protein Spots Aids Studies on Integrative Physiology

From the start integrative purists recognized the value of the protein spot approach to cellular physiology (measuring the integrated system behavior of genome expression during cellular adaptations)—even if spot behavior was unsupplemented by other information. However, most bacterial physiologists, while recognizing the usefulness of studying spot behavior, were much happier when the spots were not simply representations of individual unknown proteins from unknown genes but rather were identified as well-known specific enzymes and structural proteins. Furthermore, an essential step in making sense of the collected spot behavioral data was learning what proteins and what functions were being coregulated.

The value of identifying protein spots on 2-D gels was so obvious that many person-years (part-time) were devoted to making such identifications and providing them to others. This was the birth of what we first called the *E. coli* gene-protein index, which was a referencing of the biochemical or genetic identity of individual protein spots. The ultimate goal we envisioned was a total correspondence between the genome and its products.

A protein database based on 2-D gels of *E. coli* began with publications (Herendeen et al., 1978; Bloch, Phillips, and Neidhardt, 1980) that showed the variation under different growth conditions of 140 individual proteins, introduced the alphanumerical naming system to provide unique designations for 2-D gel spots, identified many other proteins, and resulted, in 1983, in the first edition of the gene-protein database (Neidhardt, Vaughn, Phillips, and Bloch, 1983). Subsequent editions (Phillips, Vaughn, Bloch, and Neidhardt, 1987; VanBogelen et al., 1990; VanBogelen and Neidhardt, 1991; VanBogelen, Sankar, Clark, Bogan, and Neidhardt, 1992) incorporated additional identifications and physiological data and introduced technological improvements (new equipment and protocols) designed to allow other investigators to reproduce the reference protein pattern in their own laboratories and thereby access and contribute to the information in the database.

Edition 5, published in 1992 (VanBogelen et al., 1992), was noteworthy in two respects. First, it contained a set of identifications made using a phage polymerase-promoter expression system to analyze cloned segments from the Kohara library. Second, it announced the availability of an electronic version of the database available as ECO2DBASE from the National Library of Medicine in the repository at the National Center for Biotechnology Information, which made the database more readily available and more easily updated.

The fifth edition contains just fewer than 700 entries, accounting for one-sixth of the protein-encoding genes of this organism. Most of these entries derive from work over the past 15 years in our laboratory using several methods (comigration with purified marker proteins, defective or overproducing mutants, etc.) that allow one-at-a-time identifications of protein spots as known proteins or the products of known genes. Extensive cooperation was received from the community of *E. coli* investigators, in the form of gifts of purified proteins, mutant strains, and cloned genes. Early progress was rapid, but rather soon the inventory of available purified proteins of *E. coli* became exhausted, and the analysis of defective or overproducing mutants proved tedious and sometimes gave ambiguous results. Some assignments of groups of proteins to individual ColE1 plasmids of the Clarke-Carbon library were made by preferential synthesis of these proteins in maxicells and minicells (cf. Neidhardt et al., 1983; Phillips et al., 1987), but cellular controls on gene expression and the unordered nature of this library limited the usefulness of this approach. At the past rate of progress, many decades would be necessary to identify even the products of known genes, with no hope of identifying all gene products by this approach.

New Approaches Permit Rapid Spot Identification

To accelerate the complete identification of *E. coli* protein spots and their cognate genes, methods have been sought that (1) would not depend on purification of individual proteins, (2) would not depend on mutagenesis of individual genes, (3) would produce multiple identifications per experiment, and (4) would not require prior knowledge of inducing or derepressing conditions. The recent congruence of three technical advances has made it possible to satisfy these conditions. *It now is technically possible for a small research group, with the right tools, to elucidate the complete complement of the genes and gene products of* E. coli *in a reasonable length of time.* As a result, tackling this task head-on has become a more economical and efficient strategy for cell biology than waiting for the tally of genes and their products to accumulate.

Three Technical Advances
The ordered chromosomal segment library was established in 1987 by Kohara and his associates (Kohara, Akiyama, and Isono, 1987), who produced an

eight-enzyme restriction site map of the whole chromosome and constructed a library of recombinant lambda (λ) phages carrying overlapping EcoR1-bordered fragments of approximately 15 to 20 kbp in length. A minimal set of 476 of these recombinant phages were selected by them to include the entire *E. coli* genome. This library provided a new possibility for considering a systematic examination of the *E. coli* genome: If one could express totally and selectively the genes on each fragment and could identify their products on gels, one could assign to each fragment the dozen or so gene products it encodes.

The key question then is how to express the genes of the chromosomal fragments. For many reasons, in vitro transcription and translation was considered a poor avenue to explore. In vivo expression of vectorborne genes, however, has always been plagued by high background synthesis of chromosomal products and by the inability to express repressed genes. Fortunately, several workers had been exploring the use of certain phage RNA polymerases in selective gene expression systems. The polymerases of phages such as SP6, T3, or T7 have very useful properties: (1) Their natural promoter sequences are not found in the *E. coli* genome; (2) they largely ignore the transcriptional start and stop signals that govern the polymerase of *E. coli*; and (3) they are naturally resistant to rifampicin, a potent inhibitor of the *E. coli* RNA polymerase. As a result, in a rifampicin-treated *E. coli* cell that possesses, for example, the T7 polymerase, only DNA segments with a T7 promoter are transcribed. Plasmids carrying a T7 promoter are vigorously transcribed largely independent of any requirement for special inducing or derepressing conditions. A number of laboratories have utilized these properties to design *selective and total expression systems* that use plasmid vectors carrying the T7 promoter in a host that also contains the T7 gene*1* coding for the RNA polymerase (Rosenberg et al., 1987; Studier and Moffatt, 1986; Studier, Rosenberg, Dunn, and Dubendorff, 1990; Tabor, 1990; Tabor and Richardson, 1985).

Phage RNA polymerase-promoter systems can be applied to the expression analysis of the Kohara recombinant phages in a number of ways. We have chosen the following strategy: The chromosomal segment is cut from a Kohara recombinant phage and spliced into a plasmid vector containing one or more phage promoters. The resulting recombinant is introduced into a cell containing the appropriate RNA polymerase gene integrated on the chromosome and driven by an inducible *lac* promoter. Expression of the plasmid-carried genes then is achieved in a three-step procedure: (1) induction of the phage polymerase by adding IPTG, (2) addition of rifampicin to inhibit the endogenous *E. coli* RNA polymerase and incubation to permit decay of preformed mRNA, and (3) radioactive labeling of the proteins made in the inhibited cell for subsequent 2-D gel analysis. Even in the absence of further information, this approach allows identification of most *E. coli* proteins and location of their genes to within 10 kb on the genome.

Much more is possible thanks to the third technical advance, *the use of genome sequence information*. The sequencing of the *E. coli* genome was approximately 65 percent completed at this writing, and a conservative estimate is that the job will be done by 1998 (Third International *E. coli* Genome Conference, WoodsHole, MA, 1994). Knowing the nucleotide sequence of the expressed Kohara fragments allows ready assignment of protein spots to open reading frames (ORFs). From the DNA sequence, the amino acid composition of the protein product of an ORF is deduced readily, and from amino acid composition the isoelectric point (pI) and the molecular weight (MW) of the protein can be calculated. The two parameters, MW and pI, allow a prediction of the gel migration of the protein—not with great precision but, for mixtures containing small numbers of proteins that focus on equilibrium gels, sufficiently close to the observed migration to allow matching of products with genes (Neidhardt, Appleby, Sankar, Hutton, and Phillips, 1989). The size of chromosomal segments in the Kohara library indicates that approximately seven to eight proteins, on average, can be expected from each DNA strand. This small number is ideally suited for the use of DNA sequence information to assign protein spots on gels to particular ORFs of the segment being analyzed.

To examine the feasibility of this approach it was necessary to construct an expression plasmid containing a phage promoter and a convenient splicing site, to devise a means to remove the chromosomal segment from each hybrid phage of the Kohara library and splice it into the expression vector, and to design a cell with an inducible phage RNA polymerase to serve as host for the expression plasmid. Details of the procedure had to be designed with special attention to a number of concerns, including minimizing the effects of toxic genes and the opportunity for genetic rearrangements or deletions within the cloned chromosomal segments, optimizing the expression of the cloned genes with minimal expression of cellular chromosomal genes during the labeling period, and achieving simple and easy transfer of the cloned segments to the expression vector. Finally, one had to verify that the DNA sequence of the cloned segments could be used to assign spots to ORFs unambiguously.

The Latest Method

In work initiated by Pushpam Sankar (Sankar, Hutton, VanBogelen, Clark, and Neidhardt, 1993) and continued by Robert Clark and Kelly Abshire (unpublished experiments), these steps are now close to completion. Two independent methods have been devised to transfer the Kohara inserts from lambda phage into expression vectors. One involves in vitro splicing, in which the phage DNA is digested with EcoR1 and the appropriate fragment is gel-purified and then ligated into plasmid pFNT7T3. This plasmid has opposing T7 and T3 promoters, so that both strands of the insert DNA can be transcribed using the same construction (figure 7.2). The second method involves in vivo recombination in which a plasmid, pFNsacB (figure 7.3),

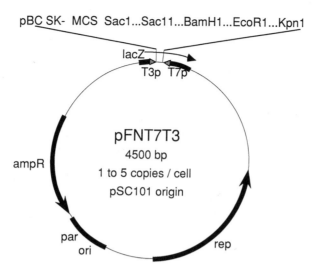

pBC SK- MCS Sac1...Sac11...BamH1...EcoR1...Kpn1

lacZ

T3p T7p

pFNT7T3
4500 bp
1 to 5 copies / cell
pSC101 origin

ampR

par
ori

rep

Figure 7.2 Expression plasmid pFNT7T3. At the top of the circular diagram are shown the two opposing phage promoters (T3p and T7p) flanking the multiple cloning site. The other details of its structure shown contribute to its usefulness as a cloning and expression plasmid.

picks up the Kohara insert by double-crossovers within homologous sequences shared by the plasmid and the Kohara phage. Additionally, to facilitate recovery of the desired recombinants, pFNsacB is a suicide plasmid that can cause the death of host cells unless successful cloning of the insert has occurred, a feature suggested to us by David Friedman of the University of Michigan. This plasmid, like pFNT7T3, carries opposing T7 and T3 promoters that permit transcription of both strands of the insert. Following isolation from an appropriate cloning strain, the hybrid pFNT7T3 or pFNsacB is transferred to each of two expression strains: strain PS1, which carries the phage T7 RNA polymerase gene in its chromosome under control of a *lac* promoter; and strain RC111, which carries the phage T3 RNA polymerase gene in a similar manner. That the expression system can operate in the desired preferential fashion is shown by the data in figure 7.4. The shut-off of host synthesis by the rifampicin treatment is so complete that the assignment of coordinates to the protein spots produced from the inserts in the expression plasmids can be accomplished only by comigrating the extract with labeled whole-cell protein. Preliminary experiments with several sequenced Kohara inserts have provided evidence that the migration properties of the spots can be used to match them with their respective structural genes or ORFs.

The new approach is outlined in figure 7.5. Protein-gene matching begins with removing, one by one, chromosomal inserts from the Kohara miniset phage library of the *E. coli* genome and splicing them into one or another plasmid engineered to permit complete transcription of both strands of the chromosomal segment in a host cell blocked in the transcription of its own chromosomal genes. The labeled proteins expressed from the chromosomal fragment then are resolved on 2-D gels and assigned coordinates on the

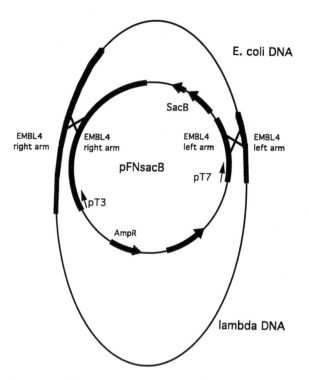

Figure 7.3 Expression plasmid pFNsacB shown in relation to a Kohara recombinant lambda genome. The inner circle depicts the expression plasmid, indicating its two opposing phage promoters (pT3 and pT7), its ampicillin resistance gene (AmpR), the suicide gene (SacB), and the adjacent regions homologous to lambda DNA (EMBL4 right arm and left arm). The outer oval depicts a Kohara recombinant lambda genome, indicating the inserted *E. coli* chromosome segment (*E. coli* DNA) and the regions of lambda DNA duplicated in the expression plasmid (EMBL, right arm and left arm). The two crosses indicate how homologous recombination causes the SacB gene to replaced by the *E. coli* segment in the expression plasmid.

reference gel image for this organism. Each expressed protein spot then is matched to its encoding DNA on the segment by means of the physical properties of the protein and the nucleotide sequence of the segment. In this way, a *genome expression map* will gradually be produced that displays the 2-D gel location of every protein encoded by the genome and matches each protein spot to its gene. Proteins will be sorted into gross metabolic classes as they are identified (i.e., the abundance of those that are produced during growth in normal laboratory medium will be measured in the various media used to generate the range of growth rates shown in figure 7.1). The data will be maintained as ECO2DBASE—a frequently updated, publicly accessible, electronic database—and will be published periodically in hard copy.

The Genome Expression Map Will Be of Great Usefulness

The genome expression map of *E. coli* will display, more reliably than the genome sequence, the complete array of proteins that this organism is capable

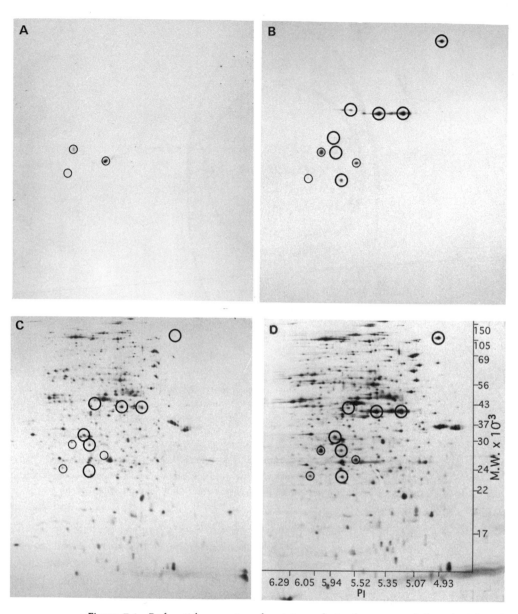

Figure 7.4 Preferential expression of protein products of genes encoded on the leftward strand of Kohara segment 613 labeled with tritiated amino acids following transcription from a T7 promoter on an expression plasmid. (A) Expression from the expression plasmid alone. The three circled proteins are known to be encoded on the vector. (B) Expression from the recombinant expression plasmid containing Kohara segment 613. The spots contributed by the vector are in small circles; those from the 613 insert are in large circles. (C) Expression of total cell protein from strain W3110, wild-type *E. coli* K-12. The locations of the vector and insert spots are indicated as in the preceding panel. (D) Mixture of B and C used to locate the insert spots within the steady-state spot pattern. Approximate scales of molecular weight (MW, in thousands on the *y* axis) and isoelectric point (pI) have been provided for this panel. (Data reprinted from Sankar et al., 1993.)

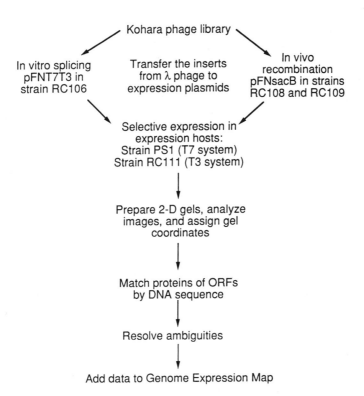

Figure 7.5 Overview of the protocol for finding the protein product of every gene in *E. coli* by sequential, selective expression of overlapping sequenced segments of the genome. ORF, open reading frame.

of making genetically. Given the density of genes in the regions already sequenced, the number of proteins may well be two to three times the number that this cell has been observed to make during growth in any given laboratory medium. For the 1,200 to 1,500 proteins commonly observed in the cells, the map will provide an extremely handy guide to their genes, opening the door to rapid molecular genetic analysis without the need for time- and labor-intensive "reverse genetics." For the very large number of proteins hitherto unknown (because not made in sufficient amounts to be visible on ordinary 2-D gels), the genome expression map will provide an inventory of their genes (complete with sequence) and a sample of the protein for biochemical analysis, making available the materials needed for both mutational analysis and biochemical studies to probe the nature and potential function of each. It should be exciting to discover which are DNA-binding proteins (and therefore possibly transcriptional regulators), which have the characteristics of integral membrane proteins, which are subject to high rates of degradation, and which may be expressed only under conditions not yet tested in the laboratory.

The information provided by the genome expression map will make it possible for investigators to locate quickly the gene of any protein that displays a behavior of interest—that is, an induction or repression in response to some environmental stimulus. A large backlog of proteins observed in response to various nutritional and other stresses is waiting for the kind of analysis made possible by molecular genetics. As interest rises in using 2-D gels to display gene expression in natural environments (including medically relevant situations) and in nongrowth states, this function of the map should be heavily used.

Perhaps most significant in the long run, the genome expression map combined with information obtained from the global regulatory measurements described earlier in this chapter will make it possible to approach cracking the "physiological code" (i.e., looking for patterns in DNA sequences of the regulatory (and coding) regions of, for example, the entire class of proteins with levels that vary directly with growth rate, or that vary inversely with growth rate, or that are growth rate–invariant; those that are prominent during nongrowth or that are never produced in detectable amounts; or those that constitute the most abundant numerically in the cell or those that.... The list of intriguing questions is exceeded only by the list of questions that will not even become obvious to ask until this database is built, patterns begin to emerge, and a new generation of bacterial physiologists is stimulated by its very existence.

Fairly obvious uses of this resource can be recognized for the theoretical approaches to integrative cell biology presented in chapters 6, 8, and 9. The matching of sequenced genes to their protein products on gels will make it possible for the transcriptional regulatory behavior of hundreds of new genes to be observed and then correlated with the structure of their regulatory regions, providing a wealth of information for the linguistic integrative approach of Collado-Vides. The ability of gel technology to generate data on the time course of expression of large numbers of genes operating simultaneously in the same cell offers also the possibility of testing predictions from models of gene circuits generated by the approaches of Savageau or Thomas. Such measurements are complemented also in a valuable way by in situ biochemical measurements. It is worth emphasizing (because it is a point that often is overlooked or misunderstood) that the most powerful use of 2-D gel electrophoresis lies in its ability to resolve total cellular protein *after* some in vivo labeling protocol has been employed. Thus, it enables one to study patterns of gene expression within the undestroyed, living cell.

DISCOVERY AND ANALYSIS OF A UNIVERSAL STRESS PROTEIN

UspA Is Induced During Any Growth Inhibition
Quantitative measurement of radioactive protein spots on 2-D gels can be accomplished in a number of ways (autoradiography, elution and direct determination following dual isotope labeling, phosphorimaging, etc.), but the story of the universal stress protein, or UspA, began simply with gel gazing. Over the course of 10 years of studying stress responses, it was observed

repeatedly that a relatively minor protein spot, designated C013.5 by its position in reference *E. coli* gel images, was induced significantly following the imposition of different environmental stresses. Nevertheless, the protein seemed to belong to no known regulon. For example, although raising the temperature of a growing culture induced this protein along with the normal array of heat-shock proteins, its induction did not depend on a functional sigma-32 (the RNA polymerase transcription initiation factor responsible for induction of heat-shock genes). Because we had no clues to its mode of regulation, protein C013.5 was temporarily set aside while the more central interest of molecular analysis of stress regulons was pursued.

In time, however, the remarkable behavior of this orphan protein demanded attention. No other protein exhibited such a behavior. One environmental condition after another was observed to induce it. In fact, no condition that decreases the growth rate of *E. coli* was found that did not induce protein C013.5, and for this reason it was given the name *UspA*. (One exception that proves the rule is that a shift down to low temperature does not lead to induction, a fact possibly related to other evidence that such a shift induces responses characteristic of growth rate acceleration.) The behavior of UspA clearly generates two significant questions: First, what could be the mode of regulation of a protein that is induced by a variety of disparate condtions? Second, what could be its biological function?

By this time, Thomas Nyström had joined our laboratory from Sweden to initiate a detailed study of this curious situation. The work reported here is summarized from recent publications (Nyström and Neidhardt, 1992, 1993, 1994) of his results and from some of his unpublished observations.

UspA Induction Is Related to Growth Inhibition but not to Steady-State Growth Rate

Table 7.1 lists the disparate conditions known to lead to UspA induction. They have little in common other than being inhibitory to growth. Limitations for the macronutrients carbon (glucose), nitrogen, phosphate, and sulfate all induce UspA equally, producing a six- to ninefold induction peaking within the first 20 minutes of growth rate decrease. Another condition that is instructive is the production of excessive internal levels of guanosine 3'-diphosphate 5'di(tri)phosphate (ppGpp(p)) by overproducing the RelA protein in a strain engineered with *relA* under isopropyl thio-β-D-galactoside (IPTG) control. In this case, the growth inhibition known to occur with excess internal ppGpp(p) resulted in UspA induction. This result made it unlikely that induction was triggered by internal limitation for some single essential metabolite, an inference that is supported by the effectiveness of the nine chemical and physical inhibitors shown also in the table (Nyström and Neidhardt, 1992).

UspA levels are not regulated strictly by growth rate; measurement of UspA levels in cells in steady-state growth in minimal media with different

Table 7.1 Examples of conditions that induce accumulation of the universal stress protein (UspA) in *E. coli*

Nutritional restrictions	*Heavy metals*
Carbon	Cadmium chloride
Nitrogen	*Various chemical and physical agents*
Phosphate	Heat
Sulfate	High osmolarity
Amino acid	Acid
Purine	*Metabolic restrictions*
Pyrimidine	Seryl-tRNA (by serine hydroxamate)
Oxidants and oxidant generators	Unknown (high ppGpp(p) by RelA excess)
Hydrogen peroxide	*Uncouplers of oxidative phosphorylation*
6-Amino-7-chloro-5,8-dioxoquinoline (ACDQ)	Dinitrophenol
Antibiotics and antimicrobials	CCCP
Nalidixic acid	
Streptomycin	
Cycloserine	

carbon sources showed no growth rate effect, although levels of UspA were reduced threefold by supplementation of glucose-minimal medium by a rich supplement of amino acids, nucleotides, and vitamins (Nyström and Neidhardt, 1992). These results are best summarized by stating that UspA induction is brought on by any perturbation of balanced, unrestricted growth at any specific growth rate.

UspA Is Not a Member of Any Known Global Regulatory Network

Many of the conditions listed in table 7.1 activate well-known gene response networks, so it was relevant to learn whether any of their global regulatory loci were involved in induction of UspA. To this end, a series of *E. coli* strains mutant in individual global regulatory genes was examined to test whether UspA induction was impaired (Nyström and Neidhardt, 1992). In each case, the stress relevant to the defective regulatory element was imposed to gauge the involvement in UspA induction. In all cases, there was no impairment of UspA induction. UspA does not appear to be a member of the stringent response system governed by the *relA/spoT* genes, the phosphate limitation regulon governed by *phoB*, the *katF*-encoded starvation-inducible sigma factor system, the leucine response regulatory network governed by *lrp*, the *ompR*-related response to osmolarity shifts, the *rpoH*-governed heat-shock response, or the gene networks affected by *appY* or by *hns*, which include growth phase–dependent genes.

Gene for UspA Was Found by Reverse Genetics, Cloned, and Sequenced

After induction of UspA by causing overproduction of RelA, a cell extract was treated to concentrate soluble proteins, which were resolved on 2-D gels.

The proteins were electroeluted to polyvinylidene membranes, and the UspA spot (identified by autoradiography) was excised and its N-terminal 14-amino acid sequence was determined by automated Edman degradation. A degenerate oligonucleotide probe was made based on part of this sequence and was used to probe a *Sal*I digest of *E. coli* chromosomal DNA. The hybridizing fragment was ligated into a phagemid and introduced into a suitable strain. Subsequent sequencing revealed that only part of the putative *uspA* gene had been cloned, so the cloned fragment was used to probe the Kohara λ phage library. Two clones containing DNA mapping at 77 minutes bound the probe. Following subcloning from one of the Kohara inserts, the entire *uspA* gene was sequenced (Nyström and Neidhardt, 1992).

Gene For UspA Is Monocistronic and Is Subject to Transcriptional Regulation

The nucleotide sequence verified that the cloned gene did encode UspA. An ORF encodes an amino acid sequence that would produce a 15.8-kDa protein with an isoelectric point of 5.2, corresponding neatly with the properties of UspA (C013.5). Following a ribosome binding site, the ORF begins with a methionyl residue followed by the 14 residues found by Edman degradation of the purified UspA protein. The coding sequence ends with a strong rho-independent terminator, suggesting a *monocistronic* operon. The transcription of the gene begins (as determined by primer extension) 131 base pairs upstream of the coding sequence and is mediated by a normal $-35/-10$ promoter. There is little or no indication, from the sequence of the upstream region, of how this gene can be transcriptionally responsive to so many different environmental conditions (Nyström and Neidhardt, 1992).

Using an appropriate fragment of the gene as a probe, measurement of the amount of *uspA* transcript was determined by Northern analysis on a culture undergoing UspA induction by glucose limitation. The *uspA* transcript was found to accumulate rapidly in the cells in response to growth arrest and to have its synthesis suppressed on re-addition of glucose to the cells. Differential stabilization of either UspA or the *uspA* transcript were ruled out as major factors in the induction process (Nyström and Neidhardt, 1992).

From the molecular analysis of *uspA*, then, much of interest was learned but not the answers to the two questions of how *uspA* expression is regulated and for what purpose.

Insertional Inactivation of *uspA* Results In Metabolic and Growth Abnormalities

Turning to genetic analysis, Nyström succeeded in introducing *kan* genes within the *uspA* gene carried on a plasmid and then crossing the inactivated gene into the *E. coli* chromosome following linear transformation (Nyström and Neidhardt, 1993). The mutant strain with the inactivated *uspA* proved extremely useful. Because such a mutant could be isolated (and was verified through 2-D gel analysis to contain no detectable UspA), a functional *uspA* would appear not to be strictly essential for growth. In fact, the steady-state growth of the *uspA::kan* mutant strain (TN1051) in a variety of media with different carbon sources occurred at a rate either indistinguishable from that of its parent or, as in the case of glucose, slightly faster than normal. Nevertheless, the mutant was sick; attainment of exponential growth after inoculation from an overnight culture required 2 to 3 hours longer than for the wild strain. Further, growth on 0.4% glucose or gluconate (but not on fructose, glucuronic acid, glycerol, succinate, acetate, ribose, or L-serine) was diauxic; at a biomass of approximately 60 percent of the final yield of the wild-type culture, the growth of strain TN1051 was arrested for 60 to 120 minutes, after which growth resumed at a slower rate (Nyström and Neidhardt, 1993).

This result suggested that glucose (or gluconate) was being consumed by the mutant faster than commensurate with biosynthesis, leading to excretion of some metabolite that could be used for growth only after the glucose (or gluconate) was exhausted from the medium. This hypothesis makes sense because of the known catabolite-repressing properties of glucose and gluconate. In fact, addition of exogenous cyclic AMP significantly shortened the diauxic lag, and measurement of glucose uptake from the medium confirmed that during the first phase of growth the *uspA::kan* mutant has an excessive rate of dissimilation of glucose. Analysis by high-pressure liquid chromatography of the medium at the time of glucose exhaustion revealed that the major excretion product was acetate. This compound then was shown to be taken up and consumed during the second phase of the diauxic growth of the mutant (Nyström and Neidhardt, 1993).

At least two of the enzymes necessary for *E. coli* to utilize acetate as a sole carbon source (isocitrate lyase and malate synthase A of the glyoxylate shunt) are exquisitely sensitive to catabolite repression, as is one of the enzymes (acetyl-CoA synthetase) involved in forming acetylcoenzyme A (acetyl-CoA) from free acetate (figure 7.6). Therefore, an explanation of the diauxic growth of the mutant on glucose is readily given: Excessive glucose catabolism leads both to acetate excretion and to pronounced catabolite repression, and the latter prevents induction of the enzymes needed for acetate utilization until well after the glucose is exhausted. This explanation points to some role of UspA in controlling the normal rate of dissimilation of glucose so as to match it to the rate of cellular biosynthesis.

Though currently we do not understood how UspA acts to accomplish this, the absence of UspA function is detrimental to the cell. We have mentioned that lags in the transition from stationary phase to exponential growth occur with the mutant in any medium. In addition, compared to the wild-type parent, the mutant is poorly able to survive stresses that cause complete growth arrest, including the addition of such agents as dinitrophenol, CCCP, H_2O_2, NaCl, and $CdCl_2$ (Nyström and Neidhardt, 1994). One might postulate that excessive glucose catabolism could be detrimental under these

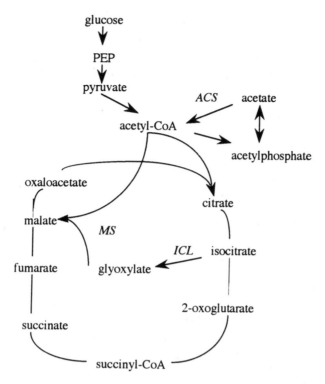

Figure 7.6 Central fueling reactions, including glycolysis, tricarboxylic acid cycle, glyoxylate shunt, and acetate metabolism pathways. The three enzymes of acetate metabolism shown are sensitive to catabolite repression: acetyl-CoA synthetase (*ACS*), isocitrate lyase (*ICL*), and malate synthase (*MS*). PEP, phosphoenolpyruvate.

growth-restricting conditions, but long-term starvation for carbon source leads to a strikingly increased death rate for the mutant (Nyström and Neidhardt, 1994) and, of course, one would suppose there is no excessive carbon utilization in this condition. (Long-term starvation for nitrogen or phosphate with excess glucose leads to no differential effect with the mutant, possibly because the wild strain itself survives so poorly under these conditions.) It must be that prior metabolism in the absence of UspA function when carbon source is available fails somehow to prepare the cells to survive later carbon starvation.

Overproduction of UspA Also Results in Various Metabolic and Growth Abnormalities

Preliminary (unpublished) work by Nyström with a second mutant has added to our information about UspA. A mutant strain has been constructed in which *uspA* on a plasmid has been placed under pTAC control (i.e., UspA production can be induced by IPTG). This strain has a curious phenotype. Addition of IPTG to a culture growing in glucose-minimal medium slows growth; the higher the IPTG concentration, the greater the inhibition. In rich medium, there is much less effect. Cells starved for glucose require a very long time to resume growth on re-addition of glucose, even if an amino acid supplement is included. Anaerobic growth of the mutant in glucose-minimal medium in the presence of IPTG is extremely poor and, once the mutant is growing anaerobically, a shift to aerobic conditions takes a day or so, unless amino acids of the ketoglutarate or succinate family are provided in the medium.

Role and Mechanism of Action of UspA Remain to Be Solved, But Clues Exist

The results just cited support the general picture provided by the *uspA::kan* mutant. Protein UspA appears to be involved in regulating carbon source catabolism, particularly the flow of intermediary metabolites into the TCA cycle, so as to balance catabolism to biosynthetic demand. It appears also to be involved in preparing the cells to survive carbon limitation and stationary phase. No specific site or mechanism of UspA action has been elucidated; this remains for future work. Gel gazing, however, has provided an additional, intriguing clue to UspA's biochemical mechanism: overproduction of UspA affects the pI of several proteins, suggesting a possible role as an agent of posttranslational modification.

CONCLUSION

The UspA story, while incomplete, illustrates the kind of discovery that can best be made by studies in which the global picture of cellular gene expression is monitored. It is unlikely that the existence of a universal stress protein would have been easily detected by any other means. Even retrospectively, it is not easy to construct a specific selection for a UspA-defective mutant, despite the profound effect of such mutations on central carbon metabolism and cell physiology. Likewise, without the discovery and elucidation of cellular factors such as UspA, there is little hope of advancement toward the goal of an integrated picture of cell metabolism and growth.

Two-dimensional gel electrophoresis provides an important tool for integrative studies of the growth of *E. coli*. The plan to construct a complete genome expression map that identifies every protein capable of being made by that organism and that matches each to its cognate structural gene is a resource that is feasible to construct and will aid in the exciting goal of solving this cell.

ACKNOWLEDGMENTS

The work done in the author's laboratory was supported by grants from the National Science Foundation (DMB8903787) and the US Public Health

Service, Institute of General Medical Sciences (GM17892). While in the author's laboratory, Thomas Nyström was supported by a fellowship from the Swedish Natural Science Research Council. Colleagues at the University of Michigan (R. Bender, J. Drummand, M. Koomey, and R. G. Matthews) provided much helpful technical advice, unpublished results, and suggestions during the course of the work on UspA, as has D. Friedman for the genome expression project. The Gene-Protein Database has, for many years, been an ongoing effort of many individuals in the author's laboratory, including K. Z. Abshire, D. A. Appleby, P. Bloch, R. Blumenthal, J. A. Bogan, R. L. Clark, A. Farewell, D. Gage, S. Herendeen, M. E. Hutton, P. G. Jones, P. G. Lemaux, S. Pedersen, A. Pertsemlidis, T. A. Phillips, S. Reeh, P. Sankar, R. A. VanBogelen, and V. Vaughn.

SUGGESTED READING

Neidhardt, F. C., and Savageau, M. (in press). Regulation beyond the operon. In: F. C. Neidhardt, R. Curtiss III, J. Ingraham, E. C. C. Lin, K. B. Low, B. Magasanik, W. Reznikoff, M. Riley, M. Schaechter, and H. E. Umbarger (Eds.). *Cellular and molecular biology:* Escherichia coli *and* Salmonella, 2nd ed. Washington, DC: American Society for Microbiology.

8 Feedback Loops: The Wheels of Regulatory Networks

René Thomas

INTRODUCTION TO REGULATORY NETWORKS

In his letter of invitation to the workshop on integrative approaches to molecular biology, Julio Collado notes: "... [o]ur current period can be characterized as one of increasing accumulation of information and the absence of novel integrative paradigms. One of the central questions we want to adress in this workshop is whether it is time to think of efforts of integration of information in molecular biology. Is there a clear need for people devoted to the task of integrating information? Or is the intuition of experimentalist and his acquaintance with his field enough to provide the required framework?"

Complex regulatory networks certainly constitute one of the fields in which integrative approaches are sorely needed; there is a rapidly increasing amount of data about these systems but apparently little effort toward a global view. In this chapter, the two types of regulation (homeostatic and epigenetic) are described briefly, and the concept of feedback circuits is introduced. These circuits belong to two types—negative and positive—and these two types of circuits are responsible for homeostatic and epigenetic regulation, respectively. Methods that permit the analysis of systems comprising intertwined feedback circuits will be addressed. Finally, we will document the view that complex regulatory networks can be treated without loss of rigor by first focusing on the feedback circuits they contain rather than on each of the individual interactions.

HOMEOSTATIC AND DIFFERENTIATIVE (OR EPIGENETIC) REGULATION

One can distinguish two types of regulation. One, *homeostasis*, is responsible for such processes as the constancy of our blood temperature. It operates like a thermostat, tending to maintain the temperature at an intermediate value, well distinct from both the lower boundary value, which would prevail if the heating device were off, and from the higher boundary value, which would prevail if it were fully on. More generally, homeostasis maintains variables (temperature, concentrations, etc.) at or near a supposedly optimal value, with or without oscillations.

The second type of regulation, in contrast, obliges the system to make a stable choice between the two boundary levels just mentioned—in other words, a choice between alternative steady states. In the case of genes, this essentially amounts to a stable choice between the on and the off positions, which is why we call this type of regulation *differentiative*; and because it is now fairly clear that differentiation is essentially epigenetic, we also denote this second type of regulation as *epigenetic*. Recall that phenotypical differences are called *epigenetic* when, although transmissible from cell generation to cell generation, they are not due to differences in the nucleotide sequences of the genetic material.[1] That epigenetic differences, including differentiation, might be understood in terms of multiple steady states was first suggested by Delbrück (1949) in a short but memorable article. Once available only as a French translation, there is now an English version, with comments, in Thomas and D'Ari (1990).

FEEDBACK CIRCUITS AND THEIR CORRESPONDENCE WITH REGULATION

The interactions between the elements of a regulatory system often form oriented circuits. In these circuits, the level of each element exerts (via the other elements) an influence on its own further evolution—hence the term *retroaction* or *feedback circuit*.[2] For example, if the product of gene A influences the rate of expression of gene B, whose product influences the rate of expression of gene C, whose product in turn influences the rate of expression of A, one says that genes A, B, and C form a feedback circuit. These are essential constituents of regulatory networks and, in particular, they are required for the processes of homeostasis and of multistationarity.

In a simple feedback circuit, each element exerts a direct action on one element only—the next element of the circuit—but an indirect action on all elements, including itself. For example, the *concentration* of a gene product will exert (via the other elements of the circuit) a delayed influence on its own *rate of production* and, consequently, on its concentration at a later time.

There exist two types of circuits depending on whether each element exerts on itself a positive or a negative action; accordingly, one speaks of a *positive* or of a *negative* circuit. Whether a circuit is positive or negative depends simply on the *parity* of the number of *negative* interactions within the circuit. Thus, if a circuit has an even number of negative interactions, it is a positive loop:

In this graph, one can see that x prevents the synthesis of y, which, if present, would prevent the synthesis of x; thus x exerts a positive effect on its own further synthesis; similarly, y exerts a positive effect on its own synthesis.

As will be seen later, negative circuits can generate homeostasis, with or without oscillations, and positive circuits can generate multistationarity. In biological and other complex systems, one usually deals not with isolated circuits but with networks comprising more or less intertwined circuits.

METHODS OF ANALYZING SYSTEMS OF INTERTWINED FEEDBACK CIRCUITS

The most obvious way to approach networks consists of trying to describe them in terms of differential equations in which the time derivative of each element is given as a function of the relevant variables. The problem is that most of the differential equations with which we have to deal are nonlinear. For this reason, these systems usually cannot be treated analytically. The steady states and the trajectories can, of course, be computed numerically, provided one knows (or invents!) all the parameter values and the exact shape of the interactions.

In biological and, as far as we know, other regulated systems, the regulatory interactions usually are sigmoid in shape (figure 8.1). Hence, the effect of a regulator is not perceptible below a threshold range of concentrations, rapidly increases within this region, and levels off for higher values. In other words, interactions are characterized by a threshold value of the regulator, and their effect is bracketed between boundary values. Depending on whether one deals with a positive or negative regulator, the sigmoid is increasing or decreasing.

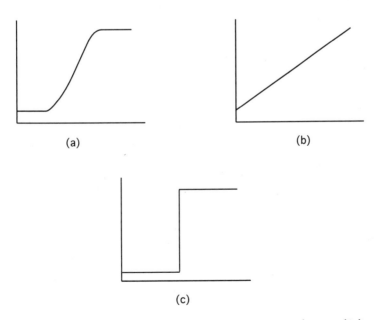

Figure 8.1 A sigmoid curve and its caricatures: (a) sigmoid curve; (b) linear caricature; (c) infinitively nonlinear caricature (step function).

Though we deal with regulatory networks, not all interactions need be regulatory. For example, where substance A transforms into B, which transforms into C, which negatively controls the production of A, only interaction $C \rightarrow A$ is regulatory and, incidentally, the other interactions may be linear.

In view of the difficulties of analysis due to the nonlinearity of the interactions, one may be tempted to use simplifying idealizations. The linear caricature is extremely useful in close proximity to the steady states, because it permits performance of their linear stability analysis;[3] however, far from the steady states, the linear idealization is not adequate, and essential features of the dynamics are lost.

The sigmoid shape of many interactions suggests a diametrically opposite type of idealization, in which the curve is replaced by a step function; this is the so-called logical caricature, which is, in fact, infinitely nonlinear (see figure 8.1). It can be shown (Glass and Kauffman, 1972, 1973; Snoussi, 1989; Thomas and D'Ari, 1990) that this type of simplifying assumption remarkably preserves the essential qualitative features of the dynamics of sigmoid systems. The behavior is not only very insensitive to the exact shape ascribed to the sigmoids but also is not highly sensitive to the slope of the sigmoids.

One way to use this simplifying assumption is to introduce step functions into differential equations, thus generating "piecewise linear" differential equations (Glass and Pasternak, 1978). Another way consists of a logical description (Kaufman, 1969; Thomas, 1973): to a real variable x one associates a Boolean variable x, which (in simple cases) takes the value 0 when x is below the threshold s, and the value 1 when x is greater than s. The state of a system can thus be described by a Boolean state vector xyz... whose components give the logical value of the variables in that order.

The logical method described next is appropriate for the analysis (and particularly for the identification of the steady states, their location, and their nature) in systems whose interactions are sigmoid. If the real interactions correspond to step functions (as in piecewise linear differential equations), the logical description *exactly* fits with the differential one. If the real interactions are sigmoid, the fit is *qualitatively* excellent unless the sigmoids used are *very* flat. In Kaufman and Thomas (1987), starting with a logical description, we have shifted to a differential system and relaxed the sigmoids down to a situation in which the Hill numbers were only two (quite flat sigmoids!). In addition, some of the terms of the differential equations were linear; the essential aspects of the qualitative dynamic picture remained unchanged.

What follows is a "naive-logic" description of two three-element feedback circuits, a negative one and a positive one. Besides demonstrating the basic principles of our logical method, this presentation shows in a concrete way that negative circuits can generate homeostasis, and positive circuits multistationarity.

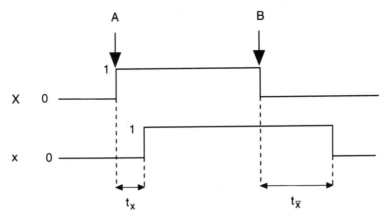

Figure 8.2 The relation between a logical variable x (presence/absence of a gene product) and its image X (gene on or off).

Naive Logical Analysis of Circuits

In simple cases, one can reason as if a gene were on or off, and as if a gene product were present or absent.[4] We write:

x = 0 if the gene product is absent

x = 1 if it is present

X = 0 if the gene is off

X = 1 if it is on

Consider a gene that has been off for a long time; its product is absent as gene products are perishable. Thus, initially $X = 0$ (gene off) and $x = 0$ (product absent) (figure 8.2). If a signal switches on the gene, $X = 1$ (gene on) but, initially at least, $x = 0$; it will take some time (typically, minutes) before the gene product has reached an efficient concentration. Now, we can write $X = 1$ (gene on) and $x = 1$ (product present). Similarly, if a second signal switches off the gene, $X = 0$ (gene off) but, initially, we still have $x = 1$; it will take some time (from minutes to many hours, depending on the stability of the gene product) before the concentration falls below its threshold of efficiency. Now, $X = 0$ and $x = 0$ again.

Note that there are steady periods during which X and x have the same value (see figure 8.2), and these will not change spontaneously. However, if the gene were off and a signal switched it on, there would be a transient period during which the gene would already be on ($X = 1$) but the concentration of its product would still be below the threshold ($x = 0$). In the absence of a counterorder (gene off again), we would have $x = 1$ after a characteristic time delay (t_x). Similarly, when we start from $X = 1$, $x = 1$ (gene on, product present), and switch off the gene, we have $X = 0$, $x = 1$; but after a time delay $t_{\bar{x}}$, x adopts the value of X. Thus, just after a gene has been switched

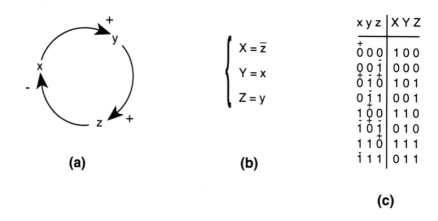

(a) (b) (c)

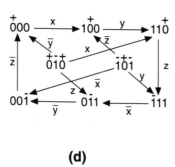

(d)

Figure 8.3 Naive logical treatment of a three-element negative feedback circuit. (A) Graph of interactions. (B) Logical equations. (C) State table. (D) Graph of the sequences of states.

on or off, the values of X and x are momentarily different, and the value of x functions as a *memory* of the preceding value of X.

Now we are ready to treat systems comprising feedback circuits. Let us first consider a simple *three-element negative circuit* (figure 8.3A). The three gene products x, y, z interact as follows: x favors the synthesis of y (i.e., the expression of gene Y), y favors the synthesis of x, and z represses the synthesis of x. Thus we write $X = \bar{z}$ (not z), which means that gene X is on iff (if and only if) product z is absent. Similarly, we write $Y = x$ (gene Y is on iff product x is present) and $Z = y$ (figure 8.3B).

The system comprises three gene products, each of which can be present (1) or absent (0); there are thus eight (2^3) logical states, from 000 (all three products absent) to 111 (all three present). For each of these states, we can ask which genes are on and which are off; this provides us with a state table (figure 8.3C), which gives the value of the "image vector" XYZ as a function of the "state vector" xyz. For example, we see that when xyz is 000 (all three products absent), XYZ is 100 (gene X on, genes Y and Z off), and so on. This can be represented symbolically as 000/100 (xyz/XYZ) or, in a more compact way, $\overset{+}{0}00$, in which the + superscript[5] draws attention to the fact that product x is absent but is being synthesized and that its value thus will presumably switch from 0 to 1.

Starting, for example, from $\overset{+}{0}00$, the system will proceed to $1\overset{+}{0}0$, then $11\overset{+}{0}$, then $11\overset{-}{1}$, then $1\overset{-}{1}0$, and so on. This gives us the temporal sequence of the logical states (see Figure 8.3D).

One sees that in this example all three variables periodically take the logical values 0 and 1; in other words, they oscillate around their threshold values. Depending on the case, one deals with a stable oscillation (the logical counterpart of a limit cycle) or with a damped oscillation (all three variables eventually tend toward their threshold value). In either case, one sees that the negative loop generates homeostasis: The concentrations of the gene products tend neither to their lower boundary value (that of a gene maintained off) nor to their higher boundary value (that of a gene maintained on); rather, they are kept, with or without oscillations, in the vicinity of their threshold value. *In fact, this is a general property of negative loops, whatever their number of elements: Negative loops generate homeostasis.*

Consider now more briefly a *positive three-element loop* (figure 8.4). Here, the situation is completely different. Whereas in the preceding case all the logical states were transient ones (i.e., with at least one + or one − superscript), the state table contains two states, 010 and 101, which have no + or − superscript and thus no order to commute the value of any variable. These are *stable logical states*, and we note them symbolically as [010] and [101]. Another difference with the negative loop is that the transient states each have two superscripts; for example, state $0\overset{+}{1}\overset{-}{1}$ has two commands—continuing to synthesize x and stopping the synthesis of z. Depending on the order in which these commands are realized, the next state will be $1\overset{-}{1}\overset{-}{1}$ or [010]. The exactly synchronous commutation of both variables:

$$\overset{+}{0}\overset{-}{1}\overset{-}{1} \Rightarrow \overset{-}{1}\overset{-}{1}0$$

is conceivable but considered exceptional. We thus have:

$$\overset{+}{0}\overset{-}{1}\overset{-}{1} \quad \begin{array}{c} \nearrow \quad \overset{-}{1}\overset{-}{1}1 \\ \text{or} \\ \searrow \quad [010] \end{array}$$

Which decision is taken depends on the relative durations of time delays.

The major interest of this is to show that a positive loop can generate multistationarity. In the present case, we have two stable steady states— [101] and [010]—and, in addition, an unstable steady state that becomes apparent only when one uses more elaborate versions of the logical approach, or the differential description. Concretely, we see that in the example chosen, a cell in which the three genes considered interact as stated can persist in either of two stable states, one with only gene Y on, and one with genes X and Z on and Y off.

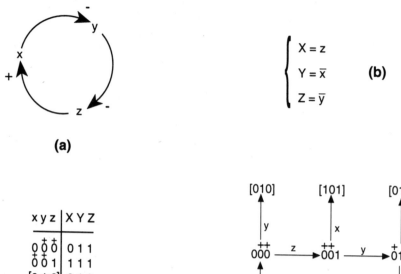

$$\begin{cases} X = z \\ Y = \bar{x} & \textbf{(b)} \\ Z = \bar{y} \end{cases}$$

(a)

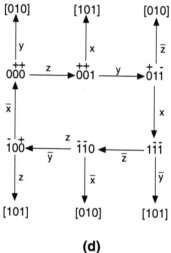

x y z	X Y Z
0 0 0	0 1 1
0 0 1	1 1 1
[0 1 0]	0 1 0
0 1 1	1 1 0
1 0 0	0 0 1
[1 0 1]	1 0 1
1 1 0	0 0 0
1 1 1	1 0 0

(c)

(d)

Figure 8.4 Naive logical treatment of a three-element positive feedback circuit. Circuit A is a positive one, as shown by the fact that it has an even number (2) of negative interactions. (A) Graph of interactions. (B) Logical equations. (C) State table. (D) Graph of the sequences of states.

Generalized Logical Analysis

The preceding section gives a concrete idea of the logical method we introduced more than 20 years ago. This logical method differs from classical ones by its *asynchronous* character; when two or more genes are switched on together, we consider that there is no reason why their products should reach their efficient value at precisely the same time. A decisive advantage of this attitude is that a logical state can have more than one possible follower, thus permitting *choices* (in contrast with synchronous descriptions, in which each logical state has one, and only one, possible follower). Since that time, the method has become increasingly sophisticated. Here we will remind the reader of the main steps of this development.

1. We assume the possibility of self-input; in other words, we describe the image of a variable as a function of the relevant variables *including itself*: $X = f(x, y, z, \ldots)$.

2. Where necessary, we use n-*level logical variables* ($x = 0, 1, 2, \ldots$) rather than limiting ourselves to the boolean description ($x = 0$ or 1). The criterion for using a logical variable with more than two levels is usually simple: If variable x acts in two (more generally *n*) distinct ways, we consider two (more generally *n*) thresholds, and consequently we endow variable x with three (more generally $n + 1$) logical levels. Practically, this means, for example, that if a regulatory gene product (say, a repressor) recognizes two or more related DNA sequences with different affinities, the subtleties of the situation can be readily taken into account in our logical description.

3. The crucial concept of *logical parameter* has been introduced by Snoussi (1989). Consider a gene x with two promoters that are regulated differently; for example, one is positively regulated by y and the other negatively by z. In the naive logical description, we write simply: $X = y + \bar{z}$ (gene X functions if *either or both* of the conditions "y present, z absent" are fulfilled); or we write $X = y \cdot \bar{z}$ (gene X functions if *both* conditions are fulfilled). The introduction of logical parameters allows one to ascribe distinct weights to the effects of the two promoters. This makes the formal description at once more nuanced and more general.

4. Until recently, only part of the steady states that can be identified by the differential description were "seen" in the logical description. This is so because many steady states are located on one or more thresholds (in other words, one or more variables has a threshold value). These states are not seen in classical logical descriptions; when we are below the threshold, we write $x = 0$, and above the threshold, we write $x = 1$, but the situation $x = s$ is not explicitly considered. The remedy simply consists of ascribing logical value to the thresholds. Thus, instead of a scale $0, 1, 2, \ldots$, we have a scale $0, s^{(1)}, 1, s^{(2)}, 2, \ldots$ in which $s^{(1)}, s^{(2)}, \ldots$ are the logical values corresponding to the lowest, second, $\ldots$ highest, thresholds of the variable considered.

Thus, a state can not only occupy a box (*n* dimensions) in the state of the variables, but it can be located on one or more thresholds, in which case it is at the junction between $2, 4, \ldots 2^n$ boxes. A state located on one or more thresholds is called *singular* (versus *regular*).

Once the various sophistications of the logical description are used, all the steady states displayed by the differential description can be identified also on logical grounds. As a matter of fact, for systems using sigmoid interactions, the generalized logical analysis, including the identification (location, nature) of the steady states, is simpler and faster than other methods. In practice, we usually begin with the logical analysis, and subsequently inject the information into differential equations if required.

The beauty of the logical description resides in its simplicity. The tool just briefly described is efficient, but this result has been obtained by introducing an increasing sophistication at the expense of the genuine simplicity. Somewhat unexpectedly, this simplicity has now been regained, thanks to the new concept of *circuit-characteristic state*. A circuit-characteristic state is the logical state located at the thresholds involved in the circuits. For example, in the system represented by the following graph,

$$+2 \left(\; x \overset{-}{\underset{-}{\rightleftarrows}} y \; \right) +2$$

or the following matrix,

$$\begin{pmatrix} 2 & -1 \\ -1 & 2 \end{pmatrix}$$

variable x exerts a negative effect on y above its lower threshold and on itself above its higher threshold; and variable y exerts a negative effect on x above its first threshold and on itself above its second threshold. In this system, one recognizes a positive loop between x and y, both acting above their first threshold; the characteristic state of this loop is $s^{(1)} s^{(1)}$ (at the intersection of the lower thresholds, whatever their real values). One recognizes also two one-element positive loops, one of x on itself and one of y on itself, both variables acting above their second threshold; the characteristic state common to these two one-element positive loops is $s^{(2)} s^{(2)}$ (at the intersect of the higher thresholds, whatever their real value). It was conjectured (Thomas, 1991) and subsequently demonstrated (Snoussi and Thomas, 1993) that *among the singular states of a system, only those that are characteristic of a circuit (or a union of circuits) can be steady and, reciprocally, given a circuit-characteristic state, there exist parameter values for which it is steady.*

These theorems may seem, at first view, esoteric and without practical value. In fact, they were the missing link that now permits us to come back to a major simplification after the long developments that had rendered our logical method more and more sophisticated. Concretely, when we consider systems composed of increasing numbers of variables, each with several logical levels, the number of logical states (and especially of singular logical states) rapidly becomes prohibitive. For example, a four-variable system using three classical levels (0, 1, 2, which becomes 0, $s^{(1)}$, 1, $s^{(2)}$, 2 after one has ascribed logical values to the thresholds) has 625 logical states, 544 of which are singular. Scanning all these states for steadiness would be tedious in the absence of appropriate computer programs. We now have these programs (Thieffry, Colet, and Thomas, 1993) but, even so, it would remain aesthetically impleasant to have to scan all these states. The theorem tells us that among the singular states, we have to consider only those that are circuit-characteristic.

Thus, in practice, the following steps apply:

1. We identify the circuits of the system. This is immediate if there are no more than three variables and, for more variables, it is easily realized by the program.

2. Once a circuit (or a union of circuits) is identified, one knows its characteristic state.

3. For each circuit (or a union of circuits), one computes the constraints on the parameters that render it functional (i.e., that render its characteristic state steady).

4. Can two or more circuits be functional together (i.e., are these circuits compatible)? It suffices to compare the parametric constraints of two or more circuits to determine whether they are compatible and, if so, to find the constraints that render all of them functional.

5. For any situation chosen (say circuits 1, 2, and 3 functional, circuit 4 not functional, circuit 5 indifferent), we can compute the range of logical parameters that will impose this situation. Once these parameters are fixed, one can fill the state table (replacing the Ks by their imposed value), find the classical (regular) steady states directly by the equality between the state and image vectors, and check that the circuit-characteristic states that should be steady are indeed so. One thus has all the steady states imposed by the parameters chosen.

6. As the relation between the logical parameters and the kinetic constants used in the differential description is simple, one can now inject into differential equations real parameters derived from the logical parameters. The resulting differential equations behave as expected from the logical analysis (provided one deals with sigmoids that are not too flat).

The last section of this chapter may seem rather abstract but, in fact, it can be expressed in more "human" terms by a metaphor. Instead of treating individually all the interactions of a system, we identify the feedback circuits and their connections (with one another and with input variables and output functions). This process, which does not result in any loss of rigor, can be compared to an analysis of a clock in which, instead of looking at every individual tooth in the mechanism, one directly focuses on the wheels and their interconnections.

COMMENTS

The concepts proposed in this chapter complement the work of other authors in this volume. In chapter 6, Savageau studies with much higher definition a domain that comprises the regulatory actions but not yet the regulatory retroactions (feedback circuits); no doubt his work will soon extend to this new domain. The approach of Reinitz and Sharp (chapter 13) provides an *automatic* derivation of models from the experimental facts. The models consist of matrices of interactions between genes. The terms of the matrix can be positive or negative, and probably after suppressing those terms whose absolute value seems too low to be significant, the matrix can be used to sort out the relevant feedback loops.

NOTES

1. Those not familiar with the concept of epigenetic differences should read the admirable experimental work of Novick and Weiner (1957) and Cohn and Horibata (1959).

2. Biologists more frequently use the term feedback *loop*. For some time, I have tried to avoid it because, in graph theory, the word *loop* is used only for one-element circuits.

3. Close to a steady state, the higher-order terms of the Taylor development can be neglected. The linear differential system thus produced can be analyzed, and one can tell the nature of the steady state (stable or unstable, etc.).

4. *Present* means that its concentration exceeds the threshold value above which it is efficient. *Absent* means that its concentration is below the threshold.

5. In the tables, the + and − superscripts are redundant but extremely convenient.

SUGGESTED READING

Thomas, R. (1983). Logical description, analysis, and synthesis of biological and other networks comprising feedback loops. *Advances in Chemical Physics, 55,* 247.

Thomas, R. (1991). Regulatory networks seen as asynchronous automata: A logical description. *Journal of Theoretical Biology, 153,* 1–23.

Thomas, R., and D'Ari, R. (1990). *Biological feedback.* Boca Raton, FL: CRC Press.

Thomas, R., Thieffry, D., and Kaufman, M. (1995). Dynamical behaviour of biological regulatory networks: I. Biological role of feedback loops; practical use of the concept of loop-characteristic state. *Bulletin of Mathematical Biology, 57,* 247–276.

9 Integrative Representations of the Regulation of Gene Expression

Julio Collado-Vides

There is no doubt that evolution is the main framework for biology. Evolutionary studies have been enormously enriched by the use of molecular data, stimulating important practical and conceptual modifications to the field. The opposite is not yet evident: Molecular biology still is very experimentally driven, making little use of theoretical integrative paradigms.

Identifying the mechanisms underlying biological functions and describing them at the molecular level involves large amounts of work. In this chapter, discussion revolves around a collection of bacterial regulatory regions. It is not an exaggeration to say that attempts to understand some of these promoters have motivated major research projects in individual laboratories for many years. Nonetheless, it now is commonplace to describe molecular biology as being in the middle of an information explosion, but such a description has no meaning: An excess of information implies that what is in excess is *not* information for we don't know how to understand it (i.e., it cannot inform). Alternatively, if all we have *is* information, it cannot be excessive. This explosion of information is changing the way science is conducted in the field of molecular biology; what a year ago we might have only dreamed of accomplishing is now considered under ambitious. What outcomes will attend *in silico* molecular biology (see chapter 5) not only in practice, but in conceptual possibilities of thinking about biology?

A NEW WAY OF THINKING ABOUT MOLECULAR BIOLOGY

An important conceptual change that might ensure only from the huge flood of *organized* information is as follows: In the near future, computer databases will offer large amounts of relevant biological information, including not only DNA sequences (the dominant feature available currently in databases) but also information on biochemical reactions, their physiology, their regulation, and so forth. This is a central piece of the infrastructure for *in silico* molecular biology.

The organization of such information will represent an important challenge. Biologists tend to think of these practical problems as being the domain of computer scientists, but the organization of databases with such quality information can be an extremely exciting intellectual adventure for the biologist (see chapter 4), assuming he or she views the challenge is as that of organizing biological databases according to, and clearly reflecting, biological principles. Such an adventure of classification is similar to the challenge faced by people such as Linnaeus in classifying plant morphology, but this time at the level of molecules, their structure and functional properties, and their role within an integrated biochemical environment. Boring as it might at first appear, recall that, within science, classification efforts frequently have been the initial window to concept and theory development. Furthermore, given the amount of information and the number of variables available at the molecular level, as well as the array of techniques for comparing, analyzing, and organizing information, the search for biological concepts through database organization may well be (*in silico*) experimental work. Testing alternative conceptual and computational solutions is to be done in a manner similar to the way experimental results are evaluated in the wet laboratory. Constructing a database is like making an experiment, as Peter Karp suggests (Karp, 1992).

Characterizing and evaluating mechanisms in different tissues and species similar to already described mechanisms is becoming at least as common in molecular biology as is the characterization of novel mechanisms. As emphasized by Robert Robbins (chapter 4), a comparative molecular biology is beginning to emerge. The flood of information we are witnessing will result in considerable enrichment of theoretical biology through an emphasis on organizing and understanding complex biological systems at the molecular level. This effort would then be centered around the task of organizing structural and dynamical properties of integrated biological systems and testing biological principles during that organizational work. This work can largely be done, at least at the beginning, without regard for the evolutionary history of such biological systems. This is akin to the study of natural language, or linguistics, which for a period was dominated by studies comparing different languages and the same languages in different periods of time. More recently, the structural school, and particularly that of generative grammar, has shifted the focus, forgetting any historical analyses and concentrating on the analysis of languages as they are now. This shift in focus gives high priority to the understanding of natural languages as complex systems irrespective of their changes through history.

In molecular biology, for instance, the search for a unified description of a collection of regulatory bacterial regions requires no evolutionary assumptions about the origin of such biological structures. It is an empirical approach that searches to extrapolate what is known, taking the collection as it is— that is, as an arbitrary small sample formed, perhaps, by arrangements with quite different evolutionary histories. Some of the regions might be the result of the laws of optimization stated by Darwinian selection; others might merely have survived in the cell by chance, a cataclysm having destroyed the alternative solution. We do not know the individual history of these regions but the dataset is there, and all the promoters can be regulated.

In practice, this new way of thinking about molecular biology starts with the search for adequate representations for describing, in a unified way, large amounts of data. Our presentation (later in this chapter) of the grammatical model of the regulation of bacterial sigma 70 promoters can well be viewed as a long example that shows the importance of neglecting detailed mechanisms and illustrates the plausibility of working based on *selecting and building representations* in molecular biology. Is there a way we can organize this knowledge such that each individual regulatory domain is exemplary of the application of a few principles? Or are we condemned to treat each promoter, each complex protein-protein interaction, and each mechanism of regulation as an individual case, subject to no common rule? These questions will be addressed within the domain of the regulation of gene expression, specifically the initiation of transcription of bacterial promoters. Nonetheless, it is likely that most of the conceptual problems raised here could also be illustrated within other areas of molecular biology.

CURRENT UNDERSTANDING OF REGULATION OF GENE EXPRESSION

Diversity and More Diversity

Regulation of gene expression is at the heart of molecular biology, not only because of the accelerated pace with which new mechanisms are being described, but also because such mechanisms of gene regulation are central to the paradigm that cellular differentiation can be explained in molecular terms. The central regulatory event at the level of initiation of transcription is the binding of regulatory proteins to DNA in a way that enables such proteins to enhance or diminish the frequency with which transcription is initiated at the promoter by the RNA polymerase.

The structural description of the regulatory sites located near (usually upstream from) the regulated promoters has become as rich and important as the dynamics itself. Dynamics and structure often become intermixed. For instance, knowing whether a regulatory protein is dimeric or hexameric in solution is important for understanding the protein's regulatory properties as this characteristic may determine the protein's potential to bind simultaneously to several closely located binding sites, looping the intervening DNA and enhancing repression (Matthews, 1992).

We have collected and analyzed an exhaustive set of regulatory arrangements of bacterial σ^{70} and σ^{54} promoters, for which the binding sites for regulatory proteins relative to the transcription initiation have been experimentally characterized (Collado-Vides, Magasanik, and Gralla, 1991).

Figure 9.1 contains a small fraction of the collection of nearly 120 σ^{70} promoters from *Escherichia coli* and *Salmonella*. Each line describes the regulatory

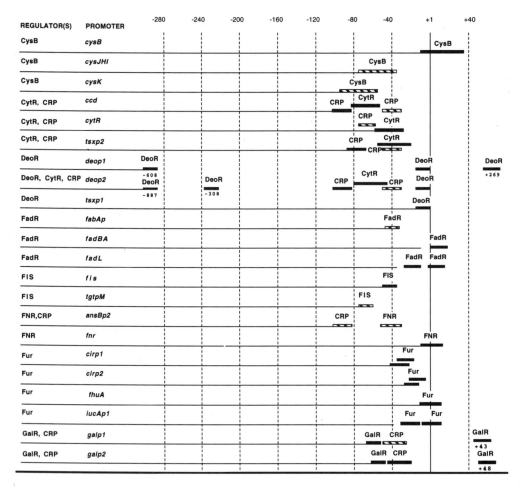

Figure 9.1 A sample of the collection of sigma 70 promoters. Several properties mentioned in the text can be observed, including (1) proteins that activate and repress (cysteine regulator [CysB], fatty acid degradation regulator [FadR], factor for inversion stimulation [FIS], fumarate nitrate regulator [FNR]); (2) multiple promoters (*deop1* and *deop2*, *galp1* and *galp2*); (3) groups of sites that occur together in different promoters, such as the CytR-CRP sites at *ccd*, *cytR*, *tsxp2*, and *deop2* promoters; and (4) groups of sites, or phrases, that are discontinuous, such as the DeoR sites (regulator for castabolism of deoxyribo nucleosides) at *deop2*, with the CytR (cytidine regulator and CRP sites between. Dark bars represent repressor sites; hatched bars represent activator sites.

region upstream from a promoter, with its associated regulatory boxes. All promoters are aligned in relation to the +1 initiation of transcription. Boxes are operator sites where regulatory proteins bind. Black boxes are repressor sites, and white boxes are sites for activators. Although not indicated, a box for the binding site for the RNA polymerase would occupy from approximately −45 to +10 in every single line. It is important to note that this figure oversimplifies the knowledge associated with such a database of regulation; as mentioned later, this knowledge involves protein structure, interactions, and dynamical properties associated with precise mechanisms of action of such activators and repressors over the RNA polymerase. Moreover, on this structural knowledge we can superimpose the physiology of the regulated genes, their precise organization into an operon, and the relative location of regulatory and regulated genes in the whole chromosome.

The following list of issues indicates how a molecular biologist is submerged within this "molecular forest" of excessive diversity, armed currently with very few tools with which to provide a unified understanding of, for example, the collection of bacterial σ^{70} promoters as a whole:

• Catabolite regulatory protein (CRP) is by far the most common activator, present in more than 40 percent of the σ^{70} genes that can be activated. It regulates genes coding for catalytic enzymes that degrade sugar substrates. Given the number of promoters, one may ask why such a collection of genes is not being transcribed by another type of promoter with these genes' specific sigma factor. In fact, if we sum the total number of σ^{70} and σ^{54} promoters that can be activated, the percentage of CRP-regulated promoters (close to 40 percent) exceeds the percentage (no higher than 10 percent), represented by the set of σ^{54} promoters.

• The details of the mechanism of regulation by a single protein varies considerably, and we have no way of predicting when a particular specific mechanism is being used or how many mechanisms a specific protein has at hand. CRP activates by a variety of mechanisms: It can activate by enhancing the binding of the polymerase or by enhancing the kinetic transition of the polymerase from closed to open complex. At another position, in the MalT promoter, it helps the polymerase to escape to elongation. Furthermore, it can activate by promoting positioning that allows interaction with the polymerase of another specific activator. These are well-studied cases, but one may also imagine that, owing to its ability to bend DNA, it might also enhance activation in a manner similar to the so-called architectural proteins (Wolffe, 1994), which, like the integration host factor (IHF), do not interact directly with the polymerase or with a specific activator but participate in modifying the DNA topology to enhance the mutual interaction of other activators.

• The precise positioning of a regulatory protein can have an important effect not only on the details of the mechanism of activation (such as those mentioned earlier) but also on whether the protein will activate or repress the initiation of transcription. One might guess that two similar proteins situated at the same position would be involved in similar mechanisms. There are, however, at least two clear examples that contest such an assumption.

The first example involves CRP and FNR (fumarate nitrate regulator), two structurally very similar proteins. The two of them are σ^{70} activator proteins that bind at approximately -41.5 upstream of the $+1$ initiation of transcription. CRP can also activate when centered at -60 or at -70. These selected positions make sense if one recalls that approximately 10.5 base pairs are required to make a turn of the helix of DNA and, therefore, these three positions are all in the same phase of the DNA—the one that enables CRP and RNA polymerase to interact. FNR, however, can activate only from around -40. Why is this so?

The second example compares the detailed mechanism of repression of LacR and GalR acting at two identical promoters. GalR naturally represses the *gal* promoter by means of two binding sites, one at $+40$ (intragenic operator [OI]) and the other at -60 (extragenic operator [OE]). There is, in addition, a CRP site close to the -60 site. GalR has been shown to bind at OE independently of the close CRP binding but, when the specificity of the GalR operators is modified to enable LacR to bind to the same positions, LacR behaves as if mutually competitive with CRP (Dalma-Weiszhaus and Brenowitz, 1992). This example, in which two similar proteins bound at precisely the same relative positions support different detailed mechanisms of regulation, serves to warn those interested in comparing and generalizing in gene regulation. The question raised here is whether the relevant regulatory distinctions will remain always at the level of idiosyncratic details and whether any attempt at generalization will ever make sense.

• Regulatory proteins can be activators, repressors, or both. When is a particular regulator used? The demand theory of gene expression predicts when a gene is positively or negatively regulated (Savageau, 1983). However, the fact that one gene has to be repressed while another one must be activated does not imply the existence of a protein with these dual functions. Two separate regulators, an activator and a repressor, might as well do the job.

• Taking this one step further, in some cases the same regulator has been found to have different regulatory properties in different organisms. For instance, the regulator of genes involved in nitrogen fixation, NifA, is sensitive to oxygen in *Rhizobium* but insensitive in *Klebsiella*. Within the *Rhizobium* bacteria, there are a fair number of examples that illustrate alternative mechanisms dealing with the physiology of nitrogen fixation.

• Proteins such as IHF protein, can just help regulation without touching anything but DNA; other proteins influence regulation by enhancing dimerization of the DNA-binding protein, and the like. If we move into eukaryotic regulation, the huge diversity of interactions, mechanisms, and potential equivalent solutions is such that the easiest working hypothesis is simply to stipulate that "perhaps the most important principle to emerge out of the study of the regulation of gene expression is that general principles do not exist," or Cove's principle (according to Beckwith, 1987). This seems to be the molecular biologist's version of the same thinking found in linguistics some time ago. When faced with the huge diversity of resources used in different human languages, some linguists reasoned that "languages could differ from each other without limit and in unpredictable ways" (Joos, quoted in Chomsky, 1986).

• As pointed out by Savageau years ago (1977), it is difficult to analyze systems that involve many properties and to design experiments to evaluate the participation of each property under adequate controlled conditions. Take, for instance, the difference between one-operator and two-operator sites. One might assume that two-operator sites occur whenever stronger repression is needed. In one studied case, however, with LexA regulated promoters, it has been shown that two weak operators do not repress as strongly as one single good site (Brent and Ptashne, 1981).

The question we can infer from these examples is: How can we make sense of such a large diversity of alternative mechanisms, alternative protein-protein interactions, and structures? Is there a way to integrate this type of information in some way? So much diversity and so little understanding in molecular biology makes us uneasy.

Unity Within Diversity

Biological systems are not made of disconnected pieces, and a strong argument can be made for the unifying power that molecular biology has exerted on our understanding of the biological world, starting with the fact that DNA, RNA, and proteins are universal molecules. Let us look at the bacterial collection of promoters from another perspective and enumerate some properties that provide a certain unity to the biological diversity found in gene regulation.

First, the chemistry underlying such gene-regulatory mechanisms clearly provides several unifying themes. These are beautifully exposed and illustrated in Ptashne's book (1992), for instance.

Second, there are very few different types of RNA polymerases in *E. coli* compared to the total number of genes and the total number of possible regulatory mechanisms. Thus, there are a number of promoters for which the initiation of transcription is executed by the same RNA polymerase, each one subject to different mechanisms of regulation involving different regulatory activator and or repressor molecules. Within a collection, one expects rules and patterns to be found, as all involve, in one way or another, the participation of a single central molecule, the specific RNA polymerase.

Third, one of the most general rules of gene regulation is the fact that any single regulated σ^{70} promoter must have a proximal site close enough to the promoter site to enable a direct interaction between a regulatory protein and RNA polymerase (Collado-Vides et al., 1991). This contrasts with the σ^{54} collection, wherein regulatory sites clearly are not proximal. The contrast between these two collections provides a clue to understanding this difference. The data suggest that a common step in activating both σ^{70} and σ^{54} promoters is an interaction between the polymerase and an activator protein.

One way to explain activation in σ^{70} promoters is by assuming that the activator protein fulfills two roles. On the one hand, activators enhance transcription initiation either by increasing the binding affinity of the promoter region to the RNA polymerase or by enhancing the kinetics of the transition from closed to open complex—an intermediate step to the elongation phase. On the other hand, given that these effects of the activator on the polymerase occur at positions close to the promoter sites, the activator provides what we could call address or positional information. That these two roles can be separated is illustrated by the activation in σ^{54} promoters.

Activation in σ^{54} promoters can be seen as fulfilling only the first role mentioned, lacking the address or positional role. Activators when bound to DNA can, in principle, activate σ^{54} promoters located around 1 or 2 kb up or downstream. To a first approximation, one could say that the positional role is provided by the polymerase itself, which can stably sit in a promoter waiting to be activated. This is something σ^{70} polymerase cannot do. An address feature and a direction feature are used in a preliminary grammatical model that describes, in a unified way, regulation in the σ^{70} and the σ^{54} promoters (Collado-Vides, 1995).

Fourth, based on evolutionary arguments, the demand theory of gene expression developed by Savageau (1977, see also chapter 6) is able to predict when a gene will be positively or negatively regulated. Furthermore, the dynamical behavior of networks of gene regulation can be predicted based on the logical analysis of Thomas (chapter 8), which classifies feedback loops as positive and negative depending on the number of positive and negative regulatory interactions.

Finally, one could consider our limited resources for understanding—the belief that there must be a logic—as another fact that should provide a unifying picture of gene regulation. Otherwise, if we do not follow such a unifying belief, we are simply excluded from the domains of scientific inquiry. In fact, the opposite belief, that gene regulation varies without limit, implies that molecular biology will remain a discipline that implements only half of the usual strategies of a scientific enterprise; those strategies associated with theory construction would be always missing.

TOWARD A CLEARER UNDERSTANDING OF REGULATION OF GENE EXPRESSION

We are eager to find ways to make sense of the rich structural-dynamical diversity in this field. To begin, we would like to have an idea of which properties of gene-regulatory mechanisms we expect to be predictable and which ones we do not.

In other words, is there a way that we could organize and integrate available collections of well-characterized mechanisms of regulation, such as the set of more than 100 σ^{70} promoters, to generate knowledge? Can we develop an adequate representation that will enable us to extrapolate from what is known—even if these are systems with a large number of variables —and to make sound predictions? Addressing these questions seems a very conservative and logical extension of the entire effort in molecular biology to dissect the large number of promoters and mechanisms of regulation. Paraphrasing Antoine Danchin (chapter 5), several individual stars have been identified; is it time now to start teaching ourselves to make the maps, the charts, to find the connections that may establish a cosmology? What can we learn from looking at these stars all together?

Biological systems are studied at many different levels (i.e., from the cellular one to networks of biochemical pathways, to modulons and regulons, to operons, to individual genes, down to individual chemical groups of amino acids. Relative to the depth of this field, the information available appears anything but excessive. In addition, charting a unified picture of the anatomy of gene regulation is a rather risky enterprise given that knowledge from one case does not necessarily shed much light on closely related cases, as illustrated in some examples mentioned previously. How much information are we able to organize? What scaling or grain of detail (what Mavrovouniotis, in chapter 11, calls "level of analysis") is adequate to build a chart?

In fact, many of the specific structural properties mentioned earlier are too detailed and are not taken into consideration within the formalisms of Savageau (chapter 6) and Thomas (chapter 8). Their formal analyses deal with the distinction of positive and negative systems of gene regulation, irrespective of the detailed organization of such systems such as the number of duplicated sites or the positions of activator and repressor sites in relation to transcription initiation. Their principles hold true regardless of these differences on the anatomy of gene regulation and, as mentioned before, gene regulation has become as much structure as dynamics.

METHODS OF FORMALIZING REGULATION OF GENE EXPRESSION

One central feature of gene regulation is that of the DNA-protein interaction between operator sites and regulatory proteins. Without such recognizer and recognized pair of structures, there is no regulation of gene expression, at least at the level of initiation of transcription. This interaction constitutes Thomas's basic element in the logical analysis of regulatory loops. Thomas began with such elementary interactions and identified general properties in the behavior of multiple interconnected loops (Thomas and D'Ari, 1990). The linguistic approach presented here starts at this same point but addresses the consequences of the structural organization in the DNA of such recognizer and recognizing sites.

The relationship between the recognizer and recognized sites, structurally, can be defined as a dependency relationship between two distantly related strings within the DNA, where two elements, i and j, in a string have a dependency relationship if modifying one of them requires the other one to be modified in order that the string remain acceptable. Certainly, if the fragment of a gene that codifies for a regulatory protein is modified within the domain that interacts with the operator DNA, the recognition event that is fundamental for regulation to work will be destroyed. In some cases, it has been experimentally shown that the protein-DNA interaction can be restored by modifying also the corresponding operator recognized piece of DNA. Therefore, there are pieces of DNA that exhibit a dependency relationship but do not need to be close to one another—in principle, there is no distance restriction along the linear array in the DNA between the regulatory gene and the effective operator site. To take an example from linguistics, this relationship is equivalent to the one between the subject *the man* and the verb *is* in a sentence such as the following:

The man that said that..... is here.

Changing *man* to *men* necessitates a change of *is* to *are* if the new sentence is to remain grammatical (or able to be regulated, in our terms of the regulator-operator dependency). These relationships can occur from far distances for several of such gene regulated-regulator pairs. Consider the following linear array:

$$R_1, R_2, R_3, \ldots, R_n, Op_1, Op_2, Op_3, \ldots, Op_n \tag{1}$$

where R_i represents a gene for the regulatory protein i, which binds to the respective operator site Op_i. This array has dependency relationships similar to those in the sentence:

John, Mary, and Peter are a doctor, a nurse, and an artist.

If we assume that the set of strings, or language of the regulation of gene expression, is infinite, then the fact that strings of the type in expression 1 are a subset of the language of gene regulation enables one to demonstrate that such a language cannot be described by a simple grammar. This proof provides the formal justification to the search for grammatical models of higher levels of generative capacity in gene regulation (Collado-Vides, 1991a).

Without getting into the details, it is important to emphasize the biological facts supporting the search for grammars as a means to sketch a theory of gene regulation. First, the fact that any hypothetical new "sentence" or regulatory transcription unit can be experimentally tested to evaluate whether it is capable of being regulated and therefore belongs to the language of gene regulation. The existence of such a criterion, which is independent of any theoretical consideration, enables one clearly to rule out certain strings as *not* part of the so-called language of gene regulation (see chapter 15). Otherwise, in the absence of such a criterion, there is no way to test hypotheses. The fact that some strings cannot be regulated—either because there is no gene codifying for a regulator or because, in the presence of a large number of repeated operators, it is known that regulation is impaired—was crucial to providing the biological background to this proof (Collado-Vides, 1991a). Whether a criterion exists for determining if a new hypothetical protein or RNA structure is acceptable remains an important open question in the search for any theory of such structures, using any formalization. In this respect, it is important to emphasize that the linguistic formalization presented in the next section does not attempt to become a theory of DNA. Rather, it attempts to become a theory of gene regulation where there is a criterion to evaluate new "sentences." Within this perspective, given the requirement for a membership criterion to accept and reject hypothetical sentences, talking about the language of DNA then makes no sense (Searls, 1992; Mantegna et al., 1994).

These dependency relationships help to emphasize the difference between a grammatical approach and probabilistic approaches. In fact, probabilistic approaches, like those based on information theory (see chapter 5), assume and evaluate the probability of one nucleotide in a sequence assuming either independent occurrence or, when more complex models are used, the estimated probabilities of individual nucleotides taking into consideration some few (one, two, six) preceding nucleotides. However, recall that the relationship between regulator domains and regulated sequences define dependency relationships across very large distances, which may, in principle, be infinite, or even dependency relationships between sequences in different chromosomes.

The formal proof justifying the use of grammatical models in the study of gene regulation has been summarized. In a sense, this represents only half of the story. The other half is to show that a linguistic formalization of the biology can be made in an interesting way and then to show that useful results may come from such a formalization.

The biological intuition that accompanies this method is the assumption that the different regulatory domains are not unrelated objects but are formed by small pieces that evolution might have shuffled, moved, duplicated, and played with in different ways to generate new regulatory arrangements. Irrespective of the evolutionary mechanism behind each regulatory domain, there are large amounts of evidence showing that pieces of DNA can be substituted and the new sequences can direct and regulate transcription initiation. Promoters, as well as specific operator sequences for repressors, can be mutually substituted, and regulation is conserved. This "regulatability" of a DNA sequence can be precisely tested experimentally, with the experimenter observing whether the level of transcription can be turned on and off, under adequate experimental conditions, depending on the presence or absence of the associated regulatory proteins. In other words, a grammar would provide a classification of a large collection of promoters—of their regulatory domains—by means of a reduced set of elements plus a set of combinatorial rules.

GRAMMAR OF THE σ^{70} COLLECTION OF PROMOTERS

Substitutability is at the core of the linguistic methodology. Thus, the first methodological question to address in attempting to build a grammatical model of the collection of more than 100 σ^{70} promoters is that of identifying

Integrative Representations of Gene Expression Regulation

the smallest elements that can be substituted in different contexts and still conserve their function. Elements can substitute mutually if the new combinations can be regulated.

Presenting the Data for Linguistic Analysis

To begin, we should ask whether the individual nucleotides constitute such elements (i.e., small and substitutable). The answer to this question will enable us to define what can be called *the first linguistic representation* of regulatory units. Once we have obtained this representation, we will ask whether such elements can be grouped into clusters because of their properties (position, role in transcription), here again the criterion will be mutual substitution. The iterative process of identifying classes of elements at subsequent higher levels of description that mutually substitute will enable us to organize a system of representations that will constitute the grammatical model.

Individual nucleotides and, in fact, pairs, triplets, or any set of fragments of fixed size will fail the test of substitution just mentioned. This is because, in order for the new combinations to be capable of being regulated, the sites where the regulatory proteins as well as the RNA polymerase bind must be conserved. These operator sites and activator sites have a certain internal structure, either in terms of symmetry (formed by direct or inverted repeats) or in terms of conserved short sequences of DNA. Take, for instance, the consensus sequence for CRP. If a T is exchanged within the operator consensus sequence for a G, and a T is exchanged also for an A, the consensus sequence becomes that for FNR. Regulation thus is altered, and therefore one may consider G and T as two distinctive features: That is, each is a feature that, if changed, conveys a change in regulation (or a change in meaning, if we were talking of natural language). However, when one exchanges a G for a T in the middle domain of a σ^{70} promoter, nothing happens in terms of gene regulation. Sigma 70 promoters contain two conserved domains, the -10, and the -35 boxes, separated by a nonconserved middle domain where the relevant feature is simply the conserved distance of approximately 17 base pairs. One would need to define the classes of substitutable elements in a position-dependent way, which opposes the whole intuition behind a grammatical approach. The result from this analysis is that the complete strings of promoter, operator, and activator sites are the smallest elements that enable a linguistic methodological analysis of these regulatory domains.

An analogy can be made in which these elements or molecular categories are compared with those from linguistics (a dangerous analogy!); in such an analogy, one would think of these elements as the phonemes, the syllables, or the words of gene regulation. It is important to see how irrelevant is the association of these molecular elements with any of these three options. On the one hand, the main idea of a combinatorial algebra is not changed. Furthermore, we do not know which linguistic level (*L-level*) we are discussing. (The use of *L-level* here is purposeful, to distinguish linguistics as a formal method, here applied to biology, from linguistics as a discipline that studies natural languages. Certainly, there is no notion of syntax, phonetics, or semantics in molecular biology. We know only that these are the smallest elements to be used to build a level of linguistic representation—the *first linguistic level*, or *L1*.)

A *grammar* can be understood as a finite set of rewriting rules of the form $x \rightarrow y$, which reads "rewrite x for y," where y can be one or several concatenated symbols. These rules begin with a selected symbol, usually S for *sentence*, and apply subsequently until symbols are reached for which no further rules apply. These are called *terminal symbols* or *preterminal symbols*. In our model, the set of preterminal symbols or molecular categories are four: *Op* for *operator* (a negative site), *I* for *inductor* (or a site for a positive activator), *D* for *dual* (a site with both activator and repressor functions), and *Pr* for *promoter*. The process of symbol substitution, or derivation, continues with another type of rule called *insertion*, which makes use of a dictionary. The dictionary contains all possible "words", each one listed with its associated relevant grammatical properties. Insertional rules substitute at random each preterminal symbol for a specific word in the dictionary. Such a word must have associated with it the properties stipulated in the preterminal symbol. For instance, a symbol *Op* can be substituted by any specific sequence of DNA in the dictionary with which is associated the property *Op*, ("to be an operator"); the same is true for the other preterminal symbols. It should become clear then how a finite set of rules can generate a large number of different "sentences," which in our case constitute regulatory transcriptional domains or extended promoters such as those illustrated in figure 9.1.

Grammatical models or grammars are formed by several components, such as the dictionary and the set of rewriting rules just mentioned. A grammar can have additional components too—for instance, principles that dictate restrictions on the type of rules to be contained in the grammar.

Grammars within the school of generative grammar are conceived as completely explicit systems—that is, no knowledge is required from the reader or user to make them work. In this sense, a grammatical model can be compared to a computer program that generates (or, if used backward, identifies) sentences. Thus, every instruction must be adequately codified. There must be, for instance, a perfect correspondence between the set of properties contained within the preterminal symbols and those listed in the dictionary. In the grammar of σ^{70} promoters, the properties contained in the dictionary include the four categorical properties mentioned earlier, plus the set of specific DNA sequences for each site and for each promoter, as well as additional relevant properties of gene regulation.

The number and diversity of properties of gene regulation is large. Because there is more than one way to represent the same physical information, a selection among alternative representations is required.

Ideally, one would like to identify a set of criteria that applies uniformly in the selection process for every single property. This is not an easy goal because knowledge about gene regulation involves properties of different types. It certainly is an empirical question to determine whether few criteria can be identified that, when applied, provide interesting results, a common procedure within linguistic methodology (Chomsky and Halle, 1968). Alternative properties of the regulation of σ^{70} promoters have been selected based on (and have yielded) the following criteria:

1. Relevance: Measured in terms of their contribution to identifying classes of substitutable elements and, in the case of specific properties, measured in terms of their ability to distinguish any two elements of the collection.

2. Simplicity: Measured in terms of the number of properties.

3. Directness: A description using properties that more directly represent relevant information.

The properties selected on the basis of these criteria are called *distinctive features*. As a result of the exhaustive evaluation of such properties within the σ^{70} collection, a unique *L1* linguistic representation for each regulatory region of transcription units, with the complete minimal set of pertinent properties, has been obtained (Collado-Vides, 1993a,b). The set of properties contained within the regulatory categories Op, D, and I include, among others, properties such as the precise position, the affinity for the binding protein, whether the site works in either orientation, the specificity of the protein-DNA interaction, and the specificity of the sensor metabolite. To obtain a manageable representation, several properties are grouped within the name of the regulatory protein, simplifying the set of properties to three within one "molecular word": the name of the protein, the category of the site (I, Op, D, or Pr), and the position of the site in the DNA sequence.

As a result of the analysis of the collection of regulatory domains, regulatory sites were classified as either proximal or remote. A *proximal* site is one that enables the bound protein to interact directly with the RNA polymerase; in physical terms, this means that a proximal site has to touch the region from approximately -65 to $+20$. *Remote* sites are distant from such an interval. As mentioned previously, σ^{70} promoters must have a proximal site. This distinction is reflected in the coding of the position of regulatory sites in the DNA sequence at the *L1* linguistic representation. Proximal sites, which are obligatory sites, are located by means of their coordinate in relation to the $+1$ initiation of transcription. Remote sites, which usually are duplicated homologous sites, are located by means of their distance in relation to the proximal referential site.

An *L1* representation of a regulatory arrangement uses as many symbols as there are regulatory sites, plus a symbol for the promoter site. For instance, the *lac* promoter (see figure 9.1) is described as follows:

$$\begin{bmatrix} Op \\ d_i = -93 \end{bmatrix} + \begin{bmatrix} I \\ c_j = -60 \\ CRP \end{bmatrix} + \begin{bmatrix} Pr \\ +1 \end{bmatrix} + \begin{bmatrix} Op \\ c_i = +9 \\ LacI \end{bmatrix}$$
$$+ \begin{bmatrix} Op \\ d_i = +402 \end{bmatrix} \tag{2}$$

Every site has information on position and protein specificity. The position of the central nucleotide of the site is indicated either as a c(oordinate) in relation to the initiation of transcription for proximal sites or as a d(istance) in relation to the proximal site. The protein specificity is indicated by the name of the protein in proximal sites and by co-indexing of remote sites; thus all sites with the i index in expression 2 are LacI binding sites. To obtain expression 2, additional principles for the relative location of such symbols, depending on the location of the binding sites in the DNA, were involved. This meant solving the problem of how to represent distances between sites. Certainly, there are various ways to represent these distances. They can be included as additional linguistic symbols, or they can be incorporated as a feature within a complex symbol.

It is the option of incorporating distances as features within symbols of binding sites that offers the most adequate description. Let us compare the description of two related regulatory domains, the one of the *lac* system as illustrated in expression 2, and a regulatory region with the same set of repressor sites at the same relative positions but in which the CRP site is nonfunctional. This modified regulatory region is described as follows:

$$\begin{bmatrix} Op \\ d_i = -93 \end{bmatrix} + \begin{bmatrix} Pr \\ +1 \end{bmatrix} + \begin{bmatrix} Op \\ c_i = +9 \\ LacI \end{bmatrix} + \begin{bmatrix} Op \\ d_i = +402 \end{bmatrix} \tag{3}$$

Observe that the internal properties of repressor sites remain unaltered. In the alternative description, DNA sequences between adjacent sites are represented by means of an extra distance symbol (Dis), with the associated length of the sequence between sites and site sizes (s_i) indicated. Description of the *lac* region would involve four additional symbols for the respective distances between the five sites, as follows:

$$\begin{bmatrix} Op \\ s_i = 20 \end{bmatrix} + \begin{bmatrix} Dis \\ 1 \end{bmatrix} + \begin{bmatrix} I \\ s = 26 \\ CRP \end{bmatrix} + \begin{bmatrix} Dis \\ 3 \end{bmatrix} + \begin{bmatrix} Pr \\ c = -45; \\ +20 \end{bmatrix}$$
$$+ \begin{bmatrix} Dis \\ -20 \end{bmatrix} + \begin{bmatrix} Op \\ s_i = 20 \\ LacI \end{bmatrix} + \begin{bmatrix} Dis \\ 363 \end{bmatrix} + \begin{bmatrix} Op \\ s_i = 20 \end{bmatrix} \tag{4}$$

The first of these symbols would describe the length of the sequence between the upstream operator and the CRP site; the second symbol would describe the distance between the CRP site and the promoter -45 end, and so on. Under this new descriptive system, the representation equivalent to expression 3 would be:

$$
\begin{bmatrix} Op \\ s_i = 20 \end{bmatrix} + \begin{bmatrix} Dis \\ 47 \end{bmatrix} + \begin{bmatrix} Pr \\ c = -45; \\ +20 \end{bmatrix} + \begin{bmatrix} Dis \\ -20 \end{bmatrix} + \begin{bmatrix} Op \\ s_i = 20 \\ LacI \end{bmatrix}
$$
$$
+ \begin{bmatrix} Dis \\ 363 \end{bmatrix} + \begin{bmatrix} Op \\ s_i = 20 \end{bmatrix} \tag{5}
$$

Expressions 4 and 5 differ in the number of symbols as well as in the values of the features within such symbols. Destruction of CRP does not alter the negative system of regulation formed by the three LacI sites. Hence, a system wherein the description of such a negative system remains unaltered is preferred. In other words, the distances between homologous sites are more relevant to the biology of regulation than are the distances between adjacent sites, which participate in different mechanisms of regulation.

In conclusion, a system in which distances between regulatory sites are incorporated as features of the complex symbols representing such sites is more appropriate. In addition, a simple principle is required to define the relative order of precedence from left to right of such terminal symbols. This principle defines the order of symbols in such a way that the relative order of the centered position of sites in the DNA is preserved.

Identifying a single representation of the type in expressions 2 and 3 for each single regulatory arrangement is an intermediate step in the system construction of the grammatical model, as mentioned earlier. It can be said that this representation *presents* the regulatory material for its analysis and organization by linguistic methodology.

Syntax

Once we can present regulatory domains in linguistic terms, we have at hand the linguistic methodology to proceed into the next level of analysis of regulatory domains. Again using the criterion of substitutability, one may ask whether binding sites are naturally grouped into "phrases." They seem, in fact, not to occur as completely unrelated elements. As mentioned previously, for the sake of negative or positive regulation, it does not matter whether the number of say, repressor sites, is one, two, or more. Under this more relaxed evaluation, two or several sites can substitute for one site. One can substitute them in different contexts, and the new productions will make sense in terms of gene regulation, just as substitution of a specific operator sequence for another operator would make a new promoter construction capable of being regulated. Furthermore, one can observe in the dataset that groups of sites occur together in different promoters. Briefly, analysis of the collection lends support to the idea of organizing regulatory sites into systems of regulation. A *system of regulation* is defined as the collection of sites in a regulatory domain whose bound regulatory proteins participate in a single mechanism of regulation. Most of these systems group multiple sites for the binding of the same protein, but there are some that include sites for different proteins (Collado-Vides, 1992).

A negative phrase can be described by a grammatical rule as follows:

$$Op' \rightarrow Op_R + (Op_r) \tag{6}$$

where Op' is a syntactic category identifying an operator phrase, Op_R is a preterminal symbol for a referential (obligatory) operator site, and Op_r is used here to denote a remote operator site. In expression 6, the parentheses around this last site indicate that this is an optional symbol. There are many such cases found in the σ^{70} collection, as well as several positive phrases. Most of the systems of regulation can be adequately described by this type of grammatical rule.

Observe that the set of sites within a phrase as described in expression 6 occur next to one another. However, there also are a good number of more complex systems of regulation with sites interspersed for the binding of different proteins. Certainly, at the higher level of phrases, a set of sites occurring far away from one another may constitute a unit that conserves its function in different contexts. It conserves its function because all the sites of such a unit participate within a mechanism of regulation. The set of *lac* repressor sites is one such example, with three sites for LacI that participate in repression and a CRP site that works separately for activation. In systems of regulation where sites are next to one another, rules of the type used in expression 6 can show that they constitute a unit (Op' in expression 6). However, in a regulatory domain where two or more regulatory phrases occur interspersed, rewriting rules such as the one in expression 6 are not useful. We want the fact that such distantly located sites constitute a unit to be made explicit during the derivation process. The description of this organization of systems of regulation requires a more powerful model.

We do not know what causes multiple homologous sites to be present at certain promoters. A number of promoters occur in the chromosome closely located to another promoter; sometimes the two promoters are regulated by a common protein that binds to a single site located between the two. We called these *multiple promoters*. Multiple promoters offer higher flexibility in gene regulation; if such promoters transcribe the same genes, each promoter may dominate transcription in different physiological conditions, just as transcription of the operon of glutamine synthetase from *glnAp1* and *glnAp2* promoters is regulated in response to nitrogen availability (as described in chapter 14). In addition, binding of a common regulatory protein to a single site necessarily locates this protein at different positions relative to the initiation of transcription for each promoter, a situation most probably associated with a different detailed regulatory effect on each promoter (for instance, the binding of NR_1 simultaneoulsy represses *glnAp1* and activates *glnAp2*). Nonetheless, we did not find a clear-cut preponderant presence of duplicated operator sites within multiple promoters compared to the simple ones (Collado-Vides et al., 1991).

Another important determinant of the number of sites is the nature of the protein involved. Regardless of whether promoters are simply or multiply organized, promoters regulated by a specific protein were found frequently to have the same number of homologous sites. This is one of the justifications for locating the number of duplicated sites as a property associated to each protein in the dictionary. Another is that the precise adequate positions for regulation are protein-specific, as illustrated earlier. Furthermore, given the sparsity of information, we preferred to assume that the relative distances adequate for regulation are also protein-specific, and to avoid in this way the risk of building a grammar that may overgenerate many false positive regulatory arrangements.

The grammar is built assuming that the number, and the relative position, of duplicated homologous sites is determined by the protein. This assumption implies a precise order in the process of a derivation. The protein must be selected in the derivation before the set of possible duplicated sites are generated.

A *derivation* is the process of specifying a particular regulatory arrangement (methodologically related to the way sentences are generated in the study of natural language) by rules of successive substitution of symbols, starting from an initial symbol. A derivation in the model begins with a set of rewriting rules. The initial symbol for the whole σ^{70} grammar is Pr''' — technically called a *projection* of the Pr(omoter) category; similarly, Op' is a projection of Op. The initial symbol is a projection of the promoter, as the promoter is the only strictly obligatory category in a regulatory domain because negative and positive phrases do not always occur, as in constitutive promoters. In the example given in figure 9.2, the derivation continues with specification of positive or negative phrases and their associated obligatory sites. Once the obligatory sites are generated in a particular derivation, the set of proteins involved must be selected. If the proteins selected by the insertional rules do not specify properties associated with duplicated sites, the derivation ends here (see figure 9.2). Insertion at this level, the so-called protein level or *P-level*, occurs only at proximal referential sites. A regulatory phrase has a proximal referential site as an obligatory site in the derivation. Representations at the *P-level* can be considered stable or acceptable because duplicated and remote sites are optional.

Insertional rules might as well select from the dictionary proteins that come with additional information indicating the presence of duplicated sites and their associated distances in relation to the proximal referential site, as well as information on the protein that binds to these remote sites (in most of cases, this is the same as the one binding the referential site). When those words are being selected, a new type of grammatical rule, the *transformational*

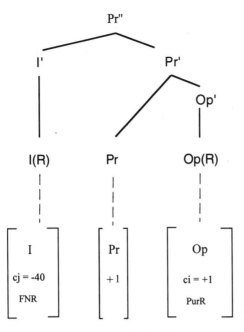

Figure 9.2 Derivation (*P-level*) of a putative array with an activator FNR site and a PurR repressor site.

movement rule, must be activated. This is a powerful type of rewriting rule that allows a symbol to be moved from one place to another within a sentence. For instance, figure 9.3 depicts the *P-level* of a derivation involving the symbol for the LacI repressor. The LacI molecular word includes information on the relative position of two remote sites (see figure 9.3A). These properties, the distance co-indexed features, are moved to the categories for duplicated sites that accept such values by means of transformational rules, generating the structure shown in figure 9.3B. The result is the derived level, or *D-level*, where obligatory and optional sites are described in their relative positions in a regulatory domain. This derivation makes explicit the fact that, at the *P-level*, these distantly located sites constitute a system of regulation.

In a sense, transformational rules can be seen as a syntactical consequence of the existence of certain properties associated with words found in the dictionary. These properties indicate the ability of certain proteins to bind to multiple sites located at relatively remote positions. Transformational rules use the *P-level* to generate a new representation with all the duplicated and remote sites, the *D-level*.

A reduced set of rewriting rules (table 9.1) that generate positive, negative, and dual regulatory phrases, with their associated referential and duplicated sites, along with the transformational movement rule completes the syntactic component of the σ^{70} grammatical model. These rules are sufficient to generate all the collection of known regulatory domains, as well as many more domains that are consistent with the biological principles encoded in the model.

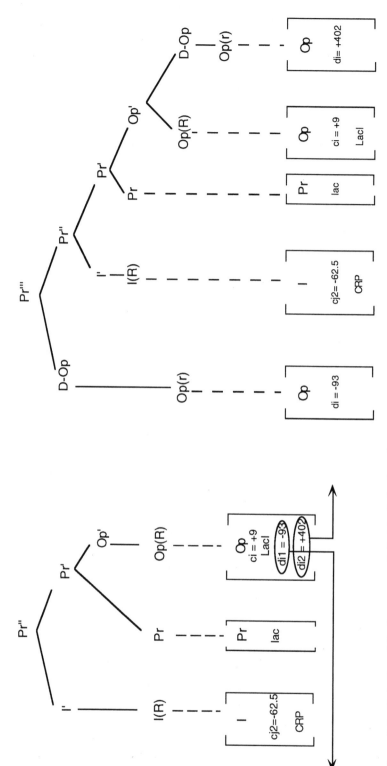

Figure 9.3 Derivation of the *lac* promoter. (Left) *P-level*, after lexical insertion. Note that the "word" for LacI contains information for additional sites (the distance co-indexed features). Transformational rules are represented by arrows. (Right) *D-level* after application of transformational rules that moved the d-features generating additional duplicated sites. Note that this is one of three possible descriptions at the *D-level*, the other two having only one duplicated site. Dashed lines represent insertional rules.

Table 9.1 Set of rewriting rules of the grammatical model of σ^{70} promoters

Pr‴	→	Pr″ + (D-Op)
Pr″	→	(I′) + Pr′
Pr′	→	Pr + (Op′)
I′	→	D-I + I$_R$
Op′	→	Op$_r$ + D-Op
D-I	→	I*
D-Op	→	Op*

Note: Rewriting rules are represented as $x \rightarrow y$ which reads, "rewrite x as y." Pr‴ is the initial symbol of the grammar. Optional categories are indicated in parentheses. D-Op and D-I refer to categories for duplicated operator and activator sites, respectively. All other symbols are defined in the text.

This grammar has an additional component not previously mentioned here: It is a set of principles that characterize the set of rewriting rules used, as well as the set of nonterminal symbols and their hierarchical organization in any derivation. These principles, though not strongly backed empirically, bear some similarity to some of the principles found in the study of natural language (Collado-Vides, 1992). The notion of c-command, central to the theory of natural language (see chapter 15), is useful in a unified description of biological properties of gene regulation.

PREDICTIONS AND EXTENSIONS

In biology, we can study and analyze available structures (sequences, metabolisms, regulatory circuits, organisms) but are limited mostly to making predictions about novel or theoretically equally plausible structures. To extrapolate from a dataset, we need, on the one hand, to be able to articulate the underlying biological principles of the dataset and, on the other hand, to have a formal tool flexible enough to integrate these principles adequately into an explicit and predictive system.

The grammatical model we describe was built taking into account approximately 100 promoters and, most importantly, the biological principles that start to emerge from their integration. The ability of the grammatical formalism to deal with complex systems supports the main result of this formalization: a model made on the basis of knowledge of slightly more than 100 promoters that generates a much larger collection of novel arrangements which are predicted to be consistent with the main principles of the circuitry of regulation of the bacterial σ^{70} promoters.

If adequately implemented, the grammatical model should stipulate a computer program capable of automatically generating all the promoters of the σ^{70} collection, as well as the detailed set of predicted arrangements. In collaboration with David Rosenblueth at Instituto de Investigaciones en Matemáticas Aplicadas y en Sistemas, Universidad Nacional Autónoma de México,

[[I,d,j,-19.5],[I,c,j,PhoB,-31.5],[Pr,lac],[Op,c,i,MetJ,-63.5]]

[[I,d,j,-44],[I,d,j,-21.5],[I,c,j,PhoB,-30.5],[Pr,lac],[Op,c,i,MetJ,-63.5]]

[[I,d,j,-38],[I,c,j,RhaR,-38.5],[Pr,lac],[Op,c,i,MetJ,-63.5]]

[[I,c,j,Ada,-42],[Pr,lac],[Op,c,i,MetJ,-8.5]]

[[I,c,j,Ada,-36],[Pr,lac],[Op,c,i,MetJ,-8.5]]

[[I,c,j,CRP,-41.5],[Pr,lac],[Op,c,i,MetJ,-8.5]]

[[I,c,j,CRP,-60],[Pr,lac],[Op,c,i,MetJ,-8.5]]

[[Op,d,i,-879],[Op,d,i,-300],[I,d,j,-189],[I,d,j,-160],[I,d,k,-95],[I,d,k,-61],[I,d,k,CRP,-32],
[I,c,j,MalT,-44],[Pr,lac],[Op,c,i,DeoR,-8]]

[[Op,d,i,-879],[Op,d,i,-300],[I,d,j,-122.5],[I,d,k,-90.5],[I,d,k,-61.5],[I,d,j,-30.5],[I,d,k,CRP,-27.5],
[I,c,j,MalT,-37.5],[Pr,lac],[Op,c,i,DeoR,-8]]

[[Op,d,i,-879],[Op,d,i,-300],[I,d,l,NarL,-149],[I,d,k,IHF,-84],[I,c,j,FNR,-41],[Pr,lac],[Op,c,i,DeoR,-8]]

[[Op,d,i,-879],[Op,d,i,-300],[I,d,k,-152.5],[I,d,k,NarL,-55.5],[I,c,j,FNR,-41.5],[Pr,lac],[Op,c,i,DeoR,-8]]

Figure 9.4 A sample of predicted regulatory arrays. All arrays assume a *lac* promoter (Pr,
lac). The first set is a small sample of combinations of activator phrases with a MetJ repressor
site at -63.5. The second set is a small sample of combinations of activator phrases with a
DeoR negative phrase that has the two upstream remote sites.

we are implementing the model in Prolog. An estimate of no fewer than
7,000 "sentences" or predicted regulatory arrays are generated by this implementation. A few samples of these appear in figure 9.4.

For each of these arrangements, one can substitute a specific sequence for
the binding of one site. The number of specific known DNA sequences varies
from approximately 2 to 10 to nearly 30 alternative sequences for CRP. Thus,
the number of DNA sequences predicted to be capable of being regulated is
much larger than 6,000. This large collection of predicted regulatory DNA
sequences can be used to analyze the structure of the space of regulatory
regions, as opposed to that of coding and junk DNA.

One way to test whether these predictions are true would be to synthesize
the DNA regulatory sequences and test for transcription in vitro in the
presence of the associated regulatory proteins and additional components
necessary for transcription to occur. The test for regulation would consist of
evaluating whether, in the presence of such proteins, transcription can be
turned on and off.

There is another, more practical use of the grammar. If each terminal
symbol of the grammar is associated with an algorithm for identifying the
specific possible DNA sequences bound by each regulatory protein, the

grammatical model can be used as a syntactical recognizer for regulatory domains in unannotated DNA sequences upstream of coding regions.

Two important extensions of this initial grammatical modeling of gene regulation are the inclusion of additional sets of regulatory mechanisms, such as the set of σ^{54} promoters and eukaryotic promoters, and the incorporation of regulatory properties and plausible correlations with physiological properties of the regulated genes. Sigma 54 promoters are bacterial promoters that share some properties with regulation in higher organisms, such as their capacity to be activated from sites located at remote distances from the promoter—and in the absence of any proximal site, contrary to σ^{70} promoters. Implementing a grammatical model able to describe this and other properties of the σ^{54} collection is clearly an intermediary step to modeling further gene regulation in higher organisms, where regulation is considerably more elaborate (as illustrated in chapter 10). As mentioned previously, one possible way to start modeling such systems is to classify regulatory proteins and the different types of RNA polymerases depending on two properties or features, a directional and an address feature. Proteins that bear a particular feature may participate in different detailed mechanisms of gene regulation. A unifying model of the σ^{70} and the σ^{54} classes of regulatory mechanisms can be envisioned, here again provided that some differences in the details of the mechanism are neglected (Collado-Vides, in press). The dramatic case of the LacR-GalR comparison, as well as other examples mentioned earlier, point to the unavoidable requirement of neglecting some details if one is interested in obtaining models capable of describing large datasets.

We have thus far contemplated only one level of biological organization, the one of mechanisms of gene regulation. An additional level dealing with physiological properties of the regulated genes might well enrich the grammatical model with an additional layer of biological properties. The first step in this direction is to collect the data, organize it, and search for possible biological correlations. Once such correlations are found, then the work of how to formalize and select linguistic alternatives can be initiated. We are currently working on constructing a database with this type of information gathered from the literature.

DISCUSSION

Grammars are complex objects, as are biological systems. Bringing pieces of these two disciplines together in an understandable way is therefore particularly difficult. No doubt the grammatical model appears rather complicated but, considering the complexity of the data, it is a simple model—simple in the precise sense that each single step in the grammar was evaluated and selected based on the criteria outlined. A strict construction of a grammatical model of another collection of promoters would require one to justify each of the representations involved, starting from the *L1* level of nucleotides, as summarized here for the collection of σ^{70} promoters.

Analogies between linguistics and molecular biology are abundant. Methodological applications are much more sparse. For this reason, it is convenient to emphasize some results obtained here that are not those usually taken for granted within such analogies. For instance, the notion of a word is far from restricted to strings of DNA. As can be seen clearly in the set of properties in expression 3, DNA sequence is one among a set of equally important properties within a word. If we wanted to make a simple association, it would be more precise to associate words to regulatory proteins rather than to strings of DNA. Incidentally, note that the predictive capacity of the grammar depends strongly on the information contained in the dictionary—that is to say, the set of binding sites for the known proteins and their associated regulatory function. The grammar—as is true natural languages—is unable to predict which sequences of DNA are part of the dictionary.

There is no doubt that the linguistic formalization of the regulation of gene expression is still in its infancy, and the amount of work ahead is immense. Linguistics is a method of analyzing large collections of data or, more precisely, knowledge. At this moment, because of the large amount of time that it requires, a bottleneck for this modeling is the obtainment and organization of sufficiently large databases on the regulation of gene expression.

Development of the grammar of the regulation of σ^{70} promoters relied on a number of rich ideas or techniques from linguistics. Sometimes grammatical descriptions (symbols, rules, or principles) might appear as ad hoc complex solutions, even if most of them were well inspired by the data. Assuming some biology is known about both the principles underlying such a dataset and the differences with respect to the first collection, developing this second grammatical model will be subject to additional restrictions compared to the first model. Elaborating a grammatical model of an additional collection of regulatory mechanisms subject to different biological restrictions should enrich considerably the grammatical approach. The differences between the two grammars must reflect, in some reasonable way, their known relevant biological differences. This is precisely the case of the preliminary attempts toward a unifying model of the σ^{70} and the σ^{54} classes of regulatory mechanisms. As more similar, biologically homogeneous knowledge bases become collected and subject to a linguistic formalization, the differences of each collection with respect to the previous collections multiply, providing a richer soil on which grammars can be built.

Perhaps one of the most attractive attributes of the grammatical approach is that, complex as the σ^{70} grammar might be, we have used a relatively small fraction of the methods, ideas, representations, and solutions available within linguistic theory. Many linguistic resources remain and provide ideas to test in such formalizations. Likewise, large amounts of data await formalization.

ACKNOWLEDGMENTS

This work was supported by grants from Consejo Nacional de Ciencia y Tecnología and Dirección General de Asuntos del Personal Académico, Universidad Nacional Autónoma de México.

SUGGESTED READING

Chomsky, N. (1957). *Syntactic structures*. The Hague: Mouton and Co. The classic in generative grammar.

Haegeman, L. (1991). *Introduction to government and binding theory*. Cambridge, MA: Basil Blackwell Ltd. A detailed textbook on the theory of natural language.

Jackendoff, R. (1977). *X-bar syntax: A study of phrase structure*. Cambridge, MA: MIT Press. On the domain of the linguistic theory dealing with syntactical categories as projections of lexical categories.

Ligthfoot, D. (1982). *The language lottery: Toward a biology of grammars*. Cambridge, MA: MIT Press. An exciting book for a biologist, arguing that natural language is part of biology.

McKnight, S. L., and Yamamoto, H. R. (Eds.). (1992). *Transcriptional regulation* (Cold Spring Harbor Monograph Series 22). New York: Cold Spring Harbor Laboratory.

Tjian, R. (1995). Molecular machines that control genes. *Scientific American, 272* (Feb.), 38–45.

10 Eukaryotic Transcription

Thomas Oehler and Leonard Guarente

Transcription in eukaryotic and prokaryotic cells is carried out by RNA polymerases. In prokaryotic cells, RNAs are synthesized by one polymerase. In contrast, in eukaryotic cells, transcription falls into three broad classes, depending on the type of RNA polymerase that is used. Ribosomal RNA genes, usually as a tandem repeat of many copies, are transcribed by RNA polymerase I. Genes encoding small RNAs, such as tRNAs or 5S RNA, are transcribed by RNA polymerase III. The remainder of genes, representing the majority of genes in a cell, are transcribed by RNA polymerase (pol) II. This last class of genes encodes proteins, and their RNA, termed *messenger RNA* (mRNA), contains a polyadenylic acid tail (polyA) at the 3′ end. Messenger RNA is much less stable than ribosomal or small RNAs.

The hallmark of genes transcribed by RNA pol II is that their transcription can be regulated by different mechanisms. This regulation is responsible for rapid physiological responses in a single-cell system and for cell type identity in multicellular organisms. In bacterial systems, a high degree of regulation can be achieved by the action of activators and repressors that bind in the vicinity of promoters and modulate their activity. An additional level of regulation is imposed by using different sigma factors to divide genes into separate transcriptional classes.

COMPONENTS OF ACTIVATED TRANSCRIPTION

In eukaryotes, some of these same regulatory mechanisms apply, but there are also differences. In this chapter, we focus on the mechanisms that regulate transcription by pol II and compare these to the situation in bacteria. Our current knowledge is that transcription in eukaryotes has adopted many of the mechanisms present in bacteria, but has added novel mechanisms that greatly increase the complexity of this process. We begin by describing components of the machinery that give rise to activated transcription in eukaryotes. These components belong to three different classes: basal factors, transcriptional activators, and coactivators.

Basal Factors

Eukaryotic promoters contain two essential elements. The first is the TATA box (Benoist and Chambon, 1981). Like the bacterial promoter (Pribnow, 1979), this element determines the position of transcription initiation. Unlike bacterial promoters, the TATA box is not bound by RNA polymerase but by a component of the basal machinery, TFIID (Matsui, Segall, Weil, and Roeder, 1980; Buratowski, Hahn, Guarente, and Sharp, 1989). TFIID consists of the TATA-binding protein (TBP) and a set of associated proteins (Dynlacht, Hoey, and Tjian, 1991; Tanese, Pugh, and Tjian, 1991; Zhou, Lieberman, Boyer, and Berk, 1992). In fact, eukaryotic polymerases do not bind to specific DNA sequences. While the ability to recognize the promoter is conferred by polymerase-bound sigma factors in bacteria (Helmann and Chamberlain, 1988), recognition in eukaryotes always requires proteins that are separate from the polymerase and bind to the promoters of pol I, II, and III genes. In general, this function is carried out by TBP, but there are a few exceptions. In promoters that do not contain a TATA box, the binding of the transcription complex might be conferred by another DNA element, the so-called initiator (Inr), which is located downstream of the TATA box (Smale and Baltimore, 1989). The Inr is bound by proteins different from TBP. In the case of pol II, a multitude of other basal factors adds to TFIID bound at the TATA box (Zawel and Reinberg, 1992). In yeast, there is evidence that most of these other basal factors exist in a holoenzyme complex associated with pol II in cells (Koleske and Young, 1994). The exception to this is TBP, which is not found associated with pol II apart from the pre-initiation complex at the promoter.

Some basal factors are themselves multiprotein complexes, such as TFIID (Dynlacht et al., 1991; Tanese et al., 1991; Zhou et al., 1992), TFIIE (Ohkuma, Sumimoto, Horikoshi, and Roeder, 1990; Inostroza, Flores, and Reinberg, 1991), and TFIIH (Feaver, Gileadi, Li, and Kornberg, 1991; Schaeffer et al., 1994). In the case of TFIIH, some of the subunits are involved also in DNA repair (Schaeffer et al., 1993). In the case of TFIID, the TBP-associated factors, or TAFs, are not relevant at all for basal transcription but are required for activated transcription (Dynlacht et al., 1991; Tanese et al., 1991).

Interestingly, while pol II contains subunits homologous to the β, β', and α subunits of the *Escherichia coli* enzyme (Sweetser, Nonet, and Young, 1987), the yeast enzyme contains an additional seven subunits (Sentenac, 1985). Moreover, the other basal factors, such as TBP, are not found in *E. coli*. Thus, the machinery used to assemble the preinitiation complex at an eukaryotic promoter is much more complicated than on the bacterial counterpart.

Activators

Eukaryotic transcriptional activators are a varied array of factors that bind to the second important element in the promoter, termed the *enhancer* or upstream activating sequences (UAS) (Banerji, Rusconi, and Schaffner, 1981; Moreau et al., 1981; Guarente, Yocum, and Gifford, 1982). Activators typically contain two separable domains, one for DNA binding and a second for transcriptional activation (Brent and Ptashne, 1985). Activation domains will function when tethered to heterologous DNA-binding domains, such as the DNA-binding domain of GAL4, a yeast transcriptional activator, or lexA, a bacterial repressor (Hope and Struhl, 1986). The ability of eukaryotic activators to function when bound hundreds or thousands of bases from the TATA box is different from the case of most bacterial promoters. The one case in bacteria that is similar to eukaryotes is the promoter that is activated by the nitrogen-responsive activator, NR_I. This activator can function over large distances (Reitzer and Magasanik, 1986). Importantly, all genes activated by glnA contain promoters that use the nitrogen-specific sigma factor, sigma 54 (Hirschman, Wong, Sei, Keener, and Kustu, 1985; Reitzer et al., 1987).

The flexibility in spacing between UASs and the TATA box means that there is no mechanistic constraint to the complexity of eukaryotic promoters. Hence, it is not surprising that many promoters contain many UASs that bind different combinations of transcriptional activators. Thus these promoters are regions where many different signals can be integrated to give a single transcriptional response. The large size of eukaryotic promoters probably is a necessary consequence of the higher complexity of these organisms.

Coactivators

A third class of factors aids in transcriptional activation without binding to the promoter at all. These coactivators must derive their specificity from interaction with activators or basal factors. Obviously, a higher level of specificity is gained by binding to activators because basal factors are essential for every pol II transcription. Among such factors that have been described is the yeast alteration-deficiency in activation (ADA) complex (Marcus, Silverman, Berger, Horiuchi, and Guarente, 1994; Horiuchi et al., 1995), which binds directly to acidic activation domains (Silverman, Agapite, and Guarente, 1994). In mammalian cells, additional coactivators have been described that bind to specific activation domains. The yeast suppressor of RNA polymerase B (SRB) complex, which binds to the C-terminal repeats of the large subunit of pol II (Thompson, Koleske, Chao, and Young, 1993), and the TAFs (Dynlacht et al., 1991; Tanese et al., 1991; Zhou et al., 1992) mentioned earlier are examples of coactivators that bind to the basal machinery.

In principle, at least two general mechanisms can be imagined for the function of coactivators. The first is the direct bridging of interactions between activators and basal factors. In fact, it has been proposed that the factors PC4 (Ge and Roeder, 1994; Kretzschmar, Kaiser, Lottspeich, and Meisterernst, 1994) and the ADA complex (Marcus et al., 1994; Horiuchi,

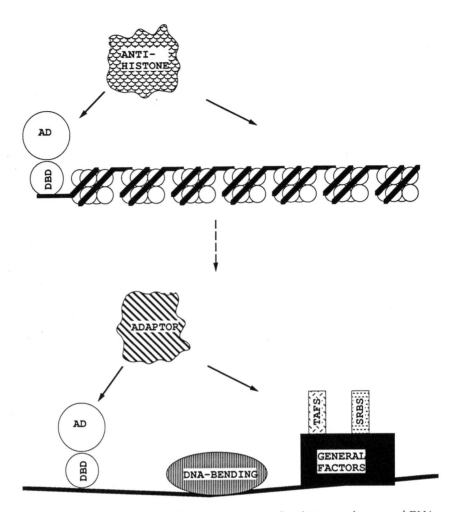

Figure 10.1 Possible mechanism for the participation of antihistones, adaptors, and DNA-bending proteins in the activation of transcription. (Top) DNA is packed into chromatin, making genes inaccessible for transcription. Antihistones could destroy the chromatin structure, allowing other factors to access the DNA. (Bottom) After disassembly of the chromatin structure, adaptors bind to transcriptional activators and basal factors, bridging these components. DNA-bending proteins support transcriptional activation by bending the DNA and thereby favoring a correct positioning between transcriptional activators and basal factors. DBD = DNA binding domain of a transcriptional activator; AD = activation domain of a transcriptional activator; TAFs = TATA-associated factors; SRBs = suppressors of RNA polymerase B.

Silverman, Marcus, and Guarente, 1995) function in this manner. No bacterial proteins have been found that function as transcriptional adaptors. A second possible mechanism for coactivators is to counteract repression by histones, because in eukaryotes the DNA exists in a chromatin complex. In fact, the switch–sucrose nonfermentable (SWI/SNF) coactivator complex identified in yeast has been shown to function in this manner (Hirschhorn, Brown, Clark, and Winston, 1992; Cote, Quinn, Workman, and Peterson, 1994). Evidently this coactivator complex is targeted to promoters by interacting with transcriptional activators in a fairly nonspecific way. There are additional effects on transcription that can be mediated by proteins that bind to DNA and cause DNA bending (Wolffe, 1994; Bazett-Jones, Leblanc, Herfort, and Moss, 1994). Among these proteins are high-mobility group–like proteins, such as lymphoid enhancer binding factor 1 (LEF-1). Figure 10.1 sketches a plausible multistep process by which two coactivators function serially in activation.

CONCLUSION

In bacteria, there is no chromatin. However, the structure of the DNA is an important factor in determining transcriptional activity at a promoter (Goodman and Nash, 1989). In this context, the DNA-bending protein complex integration host factor (IHF) can influence transcription levels by binding to DNA between an NR_I site and the promoter (Claverie-Martin and Magasanik, 1992).

In eukaryotes, mechanisms to regulate transcriptional activation have been expanded. The number of components involved in basal transcription also exceeds the number of factors in bacteria. Many transcriptional activators orchestrate complex processes such as tissue specificity. Coactivators add another layer of complexity to eukaryotic gene expression. The mechanistic basis for the activities of coactivators will continue to provide fertile ground for future studies.

SUGGESTED READING

Maldonado, E., and Reinberg, D. (1995). News on initiation and elongation of transcription by RNA polymerase II. *Current Opinion in Cell Biology, 7,* 352–361.

Paranjape, S. M., Kamakaka, R. T., and Kadonaga, J. T. (1994). Role of chromatin structure in the regulation of transcription by RNA polymerase II . *Annual Review of Biochemistry, 63,* 265–297.

Roeder, R. G. (1991). The complexities of eukaryotic transcription initiation: Regulation of preinitiation complex assembly. *Trends in the Biochemical Sciences, 16,* 402–408.

Zawel, L., and Reinberg, D. (1995). Common thesis in assembly of eukaryotic transcription complexes. *Annual Review of Biochemistry, 64,* 533–561.

11 Analysis of Complex Metabolic Pathways

Michael L. Mavrovouniotis

In chapter 9 of this book, Julio Collado-Vides discusses the illusion of the information volume generated by experimental investigation: Although the information we have accumulated is extensive in absolute amount, it remains sparse in comparison to the depth and breadth of biological systems and phenomena. An algebraic or differential model of a complex metabolic pathway is likely to contain a large number of parameters describing kinetics and regulation. Usually, only a small fraction of these parameters is known. Is it possible to study the behavior of a complex metabolic pathway without a priori knowledge of all the quantitative parameters appearing in a detailed model?

The theory of the kinetics of individual enzymatic reactions has been thoroughly investigated, and techniques are available for the derivation of a model from the reaction mechanism and the study of the dynamics of individual reactions. Much also has been accomplished in the study of the behavior of small pathways, for which detailed and reliable experimental measurements are available.

From the practical viewpoint, however, the problem of analyzing complex metabolic pathways or networks remains unresolved. Methods tailored to individual reactions or very small pathways do not suffice. We need methods that are scalable to pathways or metabolic networks of arbitrary size. Consequently, we need methods that can be applied even when many model parameters are unknown.

Naturally, in the absence of a complete quantitative description, we may not be able to produce a quantitative simulation. Nonetheless, can we at least determine some rough qualitative features of the metabolic pathway? In the derivation of qualitative properties of a large metabolic network, we are particularly interested in understanding how an entire metabolic pathway's behavior is influenced by properties of its building blocks (particular bioreactions, metabolites, or pathway segments); our hope in the study of any complex system is that only a small subset of the building blocks exercise a very strong influence on any particular property. This chapter will consider some ideas that are potentially useful for coping with the incompleteness of quantitative information and for deriving qualitative features of pathways.

Let us place some limits on the scope of our discussion. Our study of pathway analysis in this chapter will be limited to the behavior of enzymatic reactions and will exclude direct modeling of enzyme synthesis through gene expression. Cohen and Rice (see chapter 12) might argue further that our very notion of pathways (and the manner in which we will assume spatial homogeneity, concentrations, mass-action kinetics, and quasi–steady state) cannot possibly describe the behavior of an individual cell. The description presented here is indeed unsuitable for the very likely oscillatory and hetero-geneous behavior of single cells. However, we consider it an acceptable first approximation of the aggregate behavior of a population of cells in a long time scale, an approximation that can undoubtedly be refined if more detailed data are available and finer time scales and spatial features are of interest. We also note that the validity of the assumptions made about con-centrations or kinetic laws is less relevant in our view of the problem, because we will seek only *bounds* of the behavior, *not* a simulation or prediction of the behavior itself. We further argue that complex natural phenomena usually have *limits* that take a form much simpler than the original phenomena themselves. In this chapter, we try to exploit the potential of such simplified limits. We consider the traditional view of pathways as one that continues to be useful and *commensurate with current experimental capabilities*. A detailed spatial view of the type advocated in chapter 12 is not incorrect but may very well be premature, much like the alchemists' efforts to convert one chemical element to another. We do not doubt that a transition will take place toward detailed spatial descriptions, taking into account individual cell pecu-liarities; this will be fruitful and practical *at some point*. As it is, the time-averaged, position-averaged, and population-averaged view is, for most systems, already stretching the limits of current quantitative information and experimental measurements.

Simulation and Analysis

What is the expected or desired outcome of the analysis of a metabolic pathway? Quantitative predictions certainly are welcome. These usually entail the formulation of an algebraic or differential model and simulation of the model to derive the predicted behavior of the system (figure 11.1). The formulation of the model may present considerable difficulty if it describes biological processes whose mechanism of operation is not fully understood. For the activity of enzymes, the formulation of the model is not difficult; it involves rate expressions with well-established forms. However, the model will entail a number of parameters ($\underline{w}$ in figure 11.1A), in addition to the usual model inputs ($\underline{x}_0$ in figure 11.1A) or initial conditions [$\underline{u}(0)$ in figure 11.1A]. The values of all these parameters are necessary for deriving the quantitative behavior. However, unlike an isolated enzymatic reaction, a complex meta-bolic pathway interacts with an intricate biological environment, which will give rise to an external parameter vector $\underline{w}$ with very high dimensionality.

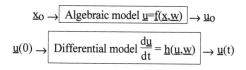

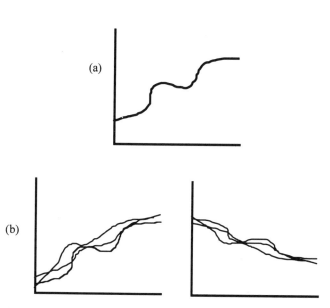

Figure 11.1 (a) Quantitative predictions through simulation. (b) Analysis of a system might encapsulate a number of quantitative behaviors, providing only qualitative distinctions, such as between the family of rising profiles and the family of decreasing profiles.

The information that is available (from experimental data) on the interaction between the pathway and its environment is invariably incomplete and qualitative. Many of the necessary parameters usually are not known at all. Even input parameters or initial conditions referring to the pathway itself might be undetermined. The consequence of the incompleteness and qualitative character of data is that realistic simulation with a quantitative model is difficult.

Analysis, however, does not presuppose quantitative simulation. In fact, in order to gain insight into the behavior of a system it might be preferable to have an aggregate, qualitative description that encapsulates a range of behaviors [a range of possible $\underline{w}$, $\underline{x}_0$, $\underline{u}(0)$] (figure 11.1B). Qualitative distinctions can shed light on the role of the structural components of a metabolic pathway, showing, for example how individual reactions, metabolites, or pathway segments influence the feasibility or the flux of the pathway. Such qualitative evaluation of the role of pathway building blocks can be interpreted in terms of *hypothetical variations* in the pathway: If a particular step in the pathway is limiting the flux, the hypothesis is that an anaplerotic branch bypassing that step would result in higher flux through the pathway.

Analysis of Complex Metabolic Pathways

Julio Collado-Vides, in chapter 9, asks whether we are condemned to treating each biological building block as an individual case, without common rules governing large classes of biological systems. He provides a grammar for promoters, to counteract this notion. The present chapter delineates perhaps a skeletal framework for metabolic resource-management processes and their limits, in the hope that eventually we can capture the metabolism's common rules within this framework. The derivation of common rules governing classes of biological systems requires that we uncover qualitative features of biological behavior and not just quantitative simulation of a specific instance of a system.

Qualitative analysis is thus both a necessity (because data usually are insufficient for quantitative predictions) and a virtue (as it can provide insights that are not apparent in a single quantitative scenario). The success of a qualitative study depends on using a suitable level of abstraction. An extremely qualitative view would have the advantage of requiring little external information, but it would lead to results that have little or no usefulness. This would be the case, for example, with a simple graph indicating the presence or absence of interactions and influences within a metabolic pathway, without additional information on the direction and magnitude of influences.

One way to avoid difficulties in interpreting qualitative results is to focus on bounds and limits of quantitative or quantifiable behavior (figure 11.2). The utility of the results will depend on a natural choice of form for the bounds. In figure 11.2A, the first form tries to follow an exact behavior too closely and might require more information than is available in practice; the second form should be easier to attain, and the third form might be the most natural. For a given form, bounds can be made tighter if more information is available (figure 11.2B); the definition of distinct regions might be more desirable in other cases (figure 11.2C), and it might particularly suit Thomas's method (see chapter 8). One should strive generally to minimize the data requirements while ensuring high utility for the results. Note that it may be necessary to use aggressive bounds that are not formally correct; this is represented in figure 11.2B as incomplete inclusion of the system's behavior in the bounds.

PROPOSED APPROACH

The qualitative analysis suggested in this chapter relies on bioreaction stoichiometries, a study of the thermodynamic feasibility of pathways, and idealized kinetic analysis based on diffusion limits of enzymatic reaction rates. The estimation of the Gibbs energy from molecular structures is the foundation of this approach. The Gibbs energy provides a first crude bound on bioreaction behavior; it determines the direction in which a bioreaction may take place. However, the actual Gibbs energy depends on exact concentrations for each bioreaction's substrates and products. To remove this

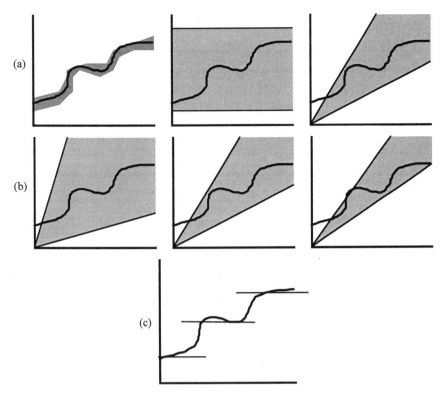

Figure 11.2 (a) Forms of bounds on the behavior of a system. (b) Bounds of a fixed form but narrowing gap; the bounds are aggressive, as they miss a portion of the system's behavior. (c) Another form of approximating behavior, seeking regions of different qualitative behavior.

impractical data requirement, we suggest an approach that relies on concentration bounds rather than exact concentration. To provide tighter bounds than just the direction of bioreaction rates, we discuss an idealized kinetic interpretation of thermodynamic constraints.

The methods sketched here target the scale of a whole pathway; they aim to produce thermodynamic and kinetic bounds for the behavior of the pathway and to identify specifically the bottlenecks (individual bioreactions or pathway segments) that limit the pathway. Ultimately, this approach extends to the global allocation of metabolites and enzymes, which are limited resources. Additional techniques determine what transformations are possible in a metabolic network and what sequences of steps can accomplish them.

We should emphasize that these ideas do not comprise a complete and polished framework. None of the methods is, in itself, a finished product, and there are many deficiencies and gaps that remain; it would be fair to state that this chapter raises more questions than it answers. One should view the specific methods merely as examples. The chapter's aim is to describe general *kinds* of methods and an *overall philosophy* that, in the author's view, hold considerable promise in the study of the metabolism.

THERMODYNAMIC FEASIBILITY

Consider a general biochemical reaction pathway, involving A metabolites, designated as $a_1, a_2, \ldots, a_A$, and S pathway steps (bioreactions), designated as $r_1, r_2 \ldots, r_S$. Let α_{ij} represent the stoichiometric coefficient of metabolite a_j in step r_i, with the usual convention that $\alpha_{ij} = 0$ if a_j is a product of r_i, $\alpha_{ij} < 0$ if a_j is a reactant (substrate) of r_i, and $\alpha_{ij} = 0$ if a_j does not participate in r_i. The stoichiometry of the biotransformation accomplished by step r_i can then be written as:

$$r_i = \sum_{j=1}^{A} \alpha_{ij} a_j \tag{1}$$

To evaluate the thermodynamic feasibility of the reaction r_i, one needs the reaction's Gibbs energy of reaction, ΔG_i. Generally, ΔG_i depends on the conditions of the reaction (i.e., reactant and product concentrations, temperature, pressure, pH, and concentrations of other solutes present in the aqueous solution). When all actual conditions, except the metabolite concentrations, are the same as the standard conditions, we can model the effect of the concentrations through the following equation:

$$\Delta G_i = \Delta G_i^{\circ\prime} + \sum_{j=1}^{A} [\alpha_{ij} RT \ln(\phi_j C_j)] \tag{2}$$

where $\Delta G_i^{\circ\prime}$ is the standard Gibbs energy of reaction, R is the ideal-gas constant, T is the temperature, ln is the natural logarithm, C_j is the concentration of a_j (expressed as moles per liter), and ϕ_j is the activity coefficient (which is equal to 1 for ideal solutions).

Thermodynamic Evaluation with Known Concentrations

To evaluate the thermodynamic feasibility of a metabolic pathway, in a system with known concentrations C_j and activity coefficients ϕ_j, we simply calculate ΔG_i for each bioreaction. The step r_i is thermodynamically feasible in the forward direction if $\Delta G_i < 0$. The pathway is therefore feasible if $\Delta G_i < 0$ for all i.

In infeasible pathways, it is likely that most reactions are feasible and only some particular isolated reactions are infeasible, posing a thermodynamic bottleneck. Note that in this analysis each bioreaction is considered separately and classified as feasible or infeasible. Thus, bottlenecks are always localized to single bioreactions.

Obstacles

In the practical application of this type of thermodynamic analysis, the first obvious problem is that the standard Gibbs energy $\Delta G_i^{\circ\prime}$ might not be known. This is very likely to be the case for most pathways beyond the

Mavrovouniotis

thoroughly studied core of the energy metabolism. As we show in the next section, this obstacle can be circumvented if the structures of the reactants and products are known. In this case, the Gibbs energies can be estimated through contributions of chemical groups occurring in the metabolite structures.

The second difficulty is that exact concentrations C_j and activity coefficients ϕ_j usually are not known. In the spirit of analyzing a family of quantitative behaviors, as stated in the introduction, we expect many results to be insensitive to exact concentrations. Many steps will remain feasible or infeasible over a broad range of concentrations. Thus, we can use bounds (ranges) for the product $C_j\phi_j$ to describe inaccuracies or to analyze qualitatively a family of behaviors of the metabolic pathway. This approach will be discussed in a subsequent section. The use of ranges gives rise to nonlocal thermodynamic features in a metabolic network: In addition to thermodynamic bottlenecks localized to a single step, we now have distributed thermodynamic bottlenecks that involve a sequence of steps.

A final difficulty is that the relationship of the thermodynamic analysis to observable bioreaction rates or pathway fluxes is unsatisfactory. The thermodynamic analysis derives the direction of the rate but gives no information about its magnitude. As a result, there is a glaring discontinuity at $\Delta G_i = 0$. A slight change of the ΔG_i in either direction gives a change in the direction of the rate; intuitively, we expect a smooth transition from the forward rate ($\Delta G_i < 0$) to zero rate ($\Delta G_i = 0$) and then to the reverse (or negative) rate ($\Delta G_i > 0$). This outcome would be observed if we had exact kinetic expressions available for each bioreaction (assuming that the expressions are indeed consistent with the thermodynamics). Detailed kinetic expressions are hard to obtain for complex pathways. We will instead translate thermodynamic restrictions to rate restrictions by deriving diffusion-based rate limits, which can be computed solely from a qualitative description of the enzymatic reaction mechanism.

The resolution of these difficulties will be covered in the following sections. We then will present a more integrated view that casts all of these considerations as tradeoffs and constraints in the allocation of metabolic resources, proposing the global application of rate limits for complex metabolic networks. The mathematical and computational treatment of this view as a whole is currently under way.

GROUP CONTRIBUTION METHOD

Let the standard Gibbs energies of formation of each metabolite a_j be represented as $\Delta G_f^{\circ\prime}(\alpha_j)$. The standard Gibbs energy of a reaction is directly related to these:

$$\Delta G_i^{\circ\prime} = \sum_j \alpha_{ij} \Delta G_f^{\circ\prime}(a_j) \tag{3}$$

We have developed an approximate group contribution approach for the estimation of these standard Gibbs energies of formation (Mavrovouniotis, 1990, 1991), enabling the thermodynamic treatment of biochemical reactions and pathways when direct equilibrium data are not available. In group contribution methods (Reid, Prausnitz, and Poling, 1987), one views a compound as composed of functional chemical groups and sums amounts contributed by each group to the property of interest—in this case, the standard Gibbs energy of formation of the compound. Corrections for special structural features of the compound can also be added to the Gibbs energy. The contribution of each group or feature must be multiplied by the number of its occurrences in the compound structure. Thus, the standard Gibbs energy $\Delta G_f^{o\prime}$ of compound a_j is given by an expression of the following form:

$$\Delta G_f^{o\prime}(a_j) = \sum_k m_{jk} g_k \tag{4}$$

where m_{jk} is the number of occurrences of group k in a_j, and g_k is the contribution of group k. For example, figure 11.3 and table 11.1 show the estimation of the Gibbs energy of formation of cystathionine, using the

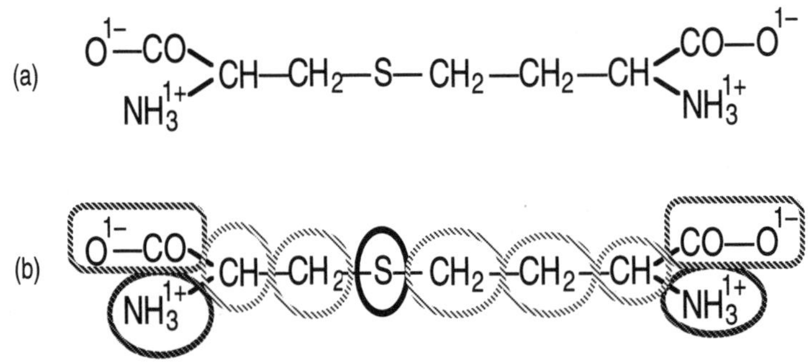

Figure 11.3 The structure of cystathionine (a), and its decomposition into groups (b), for the estimation of its standard Gibbs energy of formation (see table 11.1).

Table 11.1 Calculation of the Gibbs energy of cystathionine from group contributions

Group or Correction	Number of Occurrences m_{ik}	Contribution g_k (kcal/mol)	Total Contribution $m_{jk} g_k$ (kcal/mol)
Origin*	1	− 23.6	− 23.6
–S–	1	9.5	9.5
–NH$_3^{1+}$	2	4.3	8.6
–COO^{1-}	2	− 72.0	− 144.0
> CH$_2$	3	1.7	5.1
> CH–	2	− 4.8	− 9.6
Total Gibbs energy			− 153.9

*The *origin* term is a fixed contribution that serves as the starting point for all compounds.
Note: See also figure 11.3.

current version of the group contribution method (Mavrovouniotis, 1990, 1991). Even a structure as complicated as adenosine triphosphate (ATP) can be decomposed into the appropriate groups (figure 11.4). The contributions g_k for this method were estimated from a database of known Gibbs energies and equilibrium constants (from sources such as Thauer, Jungermann, and Decker, 1977) through multiple linear regression.

Though the current preliminary method has established the feasibility of a group contribution approach for biological Gibbs energies, many avenues for potential improvements remain. The accuracy of the method will be significantly improved if the regression (for estimating the contributions) is carried out over a larger database. A more comprehensive database would also facilitate experimentation with different sets of groups, to derive a set with improved accuracy and easier application. Nevertheless, there is an interesting contrast between the thermodynamics of small molecules, which is considered here, and the far more intractable complexity of macromolecules, which are considered in other chapters of this book.

THERMODYNAMIC EVALUATION WITH BOUNDED CONCENTRATIONS

Because the metabolite concentrations C_j and activity coefficients ϕ_j usually are not prespecified, one must account for the permissible ranges of these

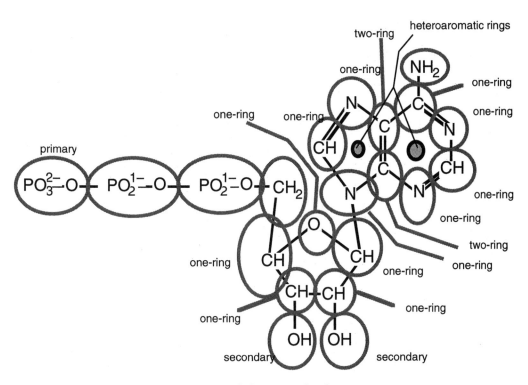

Figure 11.4 Decomposition of adenosine triphosphate into groups.

parameters. Bioreactions share metabolites, and so each metabolite's concentration affects a number of bioreactions simultaneously, coupling the thermodynamic characterization of the bioreactions. If different combinations of metabolite concentrations can be selected to make each bioreaction feasible separately, but no one combination can make them be feasible simultaneously, we have a distributed bottleneck (Mavrovouniotis, 1993).

For each metabolite, we assume that we have an upper and lower bound either for the product $\phi_j C_j$ or for the quantities C_j and ϕ_j separately. We indicate these bounds with the superscripts ^{max} and ^{min}. Thus:

$$\phi_j^{min} C_j^{min} \leq \phi_j C_j \leq \phi_j^{max} C_j^{max} \tag{5}$$

Scaling

The analysis can be simplified by rescaling the variables. For each metabolite a_j, we replace the concentration by a new scaled activity parameter f_j, defined as:

$$f_j = \left(\ln \frac{\phi_j C_j}{\phi_j^{min} C_j^{min}} \right) \Big/ \left(\ln \frac{\phi_j^{max} C_j^{max}}{\phi_j^{min} C_j^{min}} \right) \tag{6}$$

For each bioreaction r_i, we replace the standard Gibbs energy by the new parameter g_i, defined as:

$$g_i = \frac{\Delta G_i^{\circ\prime}}{RT} + \sum_{j=1}^{A} [\alpha_{ij} \ln(\phi_j^{min} C_j^{min})] \tag{7}$$

We transform the stoichiometric coefficients a_{ij} into w_{ij} as follows:

$$w_{ij} = a_{ij} \ln \frac{\phi_j^{max} C_j^{max}}{\phi_j^{min} C_j^{min}} \tag{8}$$

Finally, in place of the Gibbs energy ΔG_i, we define the function H for a reaction r_i:

$$H(r_i, f) = g_i + \sum_{j=1}^{A} w_{ij} f_j \tag{9}$$

where f denotes all the f_j collectively, as a vector. Given the standard Gibbs energies, the stoichiometries of the steps, and the concentration bounds of all metabolites, we can compute the parameters g_i and w_{ij} uniquely. The scaled activity parameters f_j cannot be computed, but they are constrained by the permissible concentration values:

$$0 \leq f_j \leq 1 \tag{10}$$

The thermodynamic condition $\Delta G_i < 0$ for a pathway step to be feasible in the forward direction becomes:

$$H(r_i, f) < 0 \tag{11}$$

Localized Bottlenecks

The thermodynamic evaluation of a single reaction entails using the selected concentration bounds to compute the upper and lower bound of the Gibbs energy. The signs of these bounds reveal quantitatively the feasibility and reversibility of the reaction. For the scaled parameters, let $H_{max}(r_i)$ and $H_{min}(r_i)$ be defined, for each reaction r_i, as the constrained optima:

$$H_{max}(r_i) = \max_f H(r_i, f), \qquad \text{subject to } 0 \le f_j \le 1 \tag{12}$$

$$H_{min}(r_i) = \min_f H(r_i, f), \qquad \text{subject to } 0 \le f_j \le 1 \tag{13}$$

If we are viewing one reaction in isolation, each variable f_j can vary between 0 and 1, completely independently from the others. Therefore, the maximum value of H is obtained by selecting $f_j = 1$ whenever $w_{ij} > 0$ and $f_j = 0$ whenever $w_{ij} < 0$. This corresponds to the selection of the minimum permissible concentrations for the reactants and the maximum permissible concentrations for the products—the least favorable conditions. Similarly, the minimum value of H can be obtained by selecting $f_j = 1$ whenever $w_{ij} < 0$, and $f_j = 0$ whenever $w_{ij} > 0$. This corresponds to the selection of the maximum concentrations for the reactants and the minimum concentrations for the products (i.e., the most favorable conditions). Thus:

$$H_{max}(r_i) = g_i + \sum_{\substack{j \\ (w_{ij} > 0)}} w_{ij}$$

and

$$H_{min}(r_i) = g_i + \sum_{\substack{j \\ (w_{ij} < 0)}} w_{ij} \tag{14}$$

A simple criterion for the thermodynamic feasibility of an isolated metabolic step r_i is:

$$H_{min}(r_i) < 0 \tag{15}$$

The quantity $H_{min}(r_i)$ represents the scaled Gibbs energy when all concentrations take their most favorable values (for that step). Thus, $H_{min}(r_i)$ is the most favorable distance from equilibrium that the reaction can attain. If the condition is expression 15 does not hold, then r_i is a localized bottleneck. Regardless of the thermodynamics of other reactions in the pathway, step r_i cannot take place in its forward direction within the specified concentration bounds.

The quantity $H_{max}(r_i)$ is the least favorable distance from equilibrium for the step r_i. If it is also negative, the reaction is thermodynamically feasible (and irreversible) for all concentrations in the specified bounds.

Note, however, that for a whole set of bioreactions to be feasible simultaneously (i.e., under the same metabolite concentrations), the inequalities $H(r_i, f) < 0$ must be satisfied simultaneously using the same vector f. Therefore, testing the $H_{min}(r_i)$ of each reaction is a necessary condition, but it is not sufficient. If a step has $H_{min}(r_i) < 0 < H_{max}(r_i)$, then the feasibility may be influenced by surrounding steps.

Distributed Thermodynamic Bottlenecks

To define bottlenecks, we require that the bounds on f always be satisfied, and we consider subsets of the set of inequalities $H(r_i, f) < 0$. If the pathway as a whole is not feasible, then some of these subsets will be feasible (i.e., will be satisfiable by some f) and some will not. A set of steps B forms an infeasible subpathway if, for each f in the range $0 \le f_j \le 1$, there is at least one step in B that is infeasible.

If a subpathway contains (as a subset) an infeasible subpathway, then it is also infeasible. A bottleneck is a minimal infeasible subpathway (i.e., a set B that is infeasible and does not have any infeasible proper subsets). Thus, bottlenecks are defined by minimality (with respect to inclusion) in the class of infeasible subpathways. A bottleneck B is localized if it is a singleton set and is distributed if it involves two or more steps. The reason that distributed bottlenecks exist is that a number of inequalities of the form in expression 11 may be unsatisfiable as a set, even though each inequality is separately satisfiable: The sets of vectors f that satisfy the individual inequalities are disjoint.

An infeasible pathway can have several bottlenecks, with different numbers of reactions, because a bottleneck is minimal only with respect to inclusion, not cardinality. It can be shown (Mavrovouniotis, 1993) that a bottleneck exists if and only if there is a nonnegative linear combination of the steps in the pathway $\sum_i e_i r_i$, with $e_i \ge 0$, such that the feasibility condition (expression 15) is violated by the reaction combination as a whole. In other words, for a bottleneck we have:

$$H_{min}\left(\sum_i e_i r_i\right) > 0 \qquad \text{for some } e_i \ge 0 \qquad (16)$$

This property plays an important role in the algorithm for the systematic identification of distributed bottlenecks, whose details appear elsewhere (Mavrovouniotis, 1993).

As an example, we analyze the glycolysis pathway. Table 11.2 lists the bioreactions and also supplies the necessary standard Gibbs energies (Lehninger, 1986). We note that the commonly used way of assessing thermodynamic feasibility, through the signs of the standard Gibbs energies in table 11.2, would be misleading here: If all the positive $\Delta G^{\circ\prime}$ were interpreted as infeasibilities, then this central pathway would be ruled infeasible.

To carry out the thermodynamic analysis, we must postulate upper and lower bounds for the concentrations and activity coefficients of each metabolite. We assume here, for simplicity, that the temperature is T = 298.15K

Table 11.2 Standard Gibbs energies of the reactions in glycolysis and computations individual reaction parameters for identifying bottlenecks

Index	$\Delta G_i^{\circ\prime}$ (kcal/mol)	(1) g_i	(1) $H_{min}(r_i)$	(1) $H_{max}(r_i)$	(2) g_i	(2) $H_{min}(r_i)$	(2) $H_{max}(r_i)$
1	−4.000	−8.781	−11.083	−6.478	−8.781	−15.912	−1.650
2	0.400	0.675	−1.627	2.978	0.675	−6.456	7.806
3	−3.400	−7.768	−10.070	−5.465	−7.768	−14.899	−0.637
4	5.730	0.464	−1.839	5.069	−2.755	−9.886	11.507
5	1.830	3.090	0.787	5.392	3.090	−4.041	10.220
6	1.500	3.867	1.564	7.017	3.867	−3.264	11.845
7	−4.500	−5.570	−7.872	−3.267	−5.570	−12.701	1.561
8	1.060	1.790	−0.513	4.092	1.790	−5.341	8.921
9	0.440	0.743	−1.560	3.045	0.743	−6.388	7.874
10	−7.500	−10.635	−12.937	−8.332	−10.635	−17.766	−3.504
11	−6.000	−6.623	−9.773	−4.321	−6.623	−14.601	0.508

Note: The first interval has bounds $C_{min} = 0.1 \times 10^{-3}$ M, $C_{max} = 1 \times 10^{-3}$ M. The second interval has bounds $C_{min} = 0.004 \times 10^{-3}$ M and $C_{max} = 5 \times 10^{-3}$ M.

(25°C) and that all activity coefficients f_j are equal to 1. We take pH = 7, and we use fixed concentrations for the currency metabolites adenosine triphosphate (ATP), adenosine diphosphate (ADP), nicotinamide adenine dinucleotide, oxidized (NAD), and NAD reduced (NADH) (as described by Mavrovouniotis, 1993, based on data from Lehninger, 1982, and Ingraham, Maaløe, and Neidhardt, 1983). Thus, we focus on the concentration bounds of the remaining metabolites; we assume that these share the same concentration range, C_{min} to C_{max}. Let us examine three different combinations of values for the bounds C_{min} and C_{max}.

First we take $C_{min} = 0.1 \times 10^{-3}$ M and $C_{max} = 1 \times 10^{-3}$ M; the initial computations are shown in table 11.2. Note that steps 1, 3, 7, 10, and 11 have $H_{max}(r_i) < 0$; thus, the thermodynamic prediction is that they are always feasible in the forward direction. On the other hand, bioreactions 5 and 6 are localized bottlenecks, because they have $H_{min}(r_i) > 0$ (i.e., they cannot be made feasible in the forward direction for any combination of concentrations in the specified range). The remaining steps (2, 4, 8, and 9), for which $H_{min}(r_i) < 0 < H_{max}(r_i)$, are candidate locations for distributed bottlenecks. The algorithm identifies the combination $r_8 : r_9$ as a distributed bottleneck. Viewed in isolation from each other, r_8 and r_9 are feasible: That is, for each reaction, there is a portion of the permissible concentration space resulting in a negative ΔG and making the forward direction of that reaction thermodynamically feasible. However, their combination is not feasible: There is no overlap between the portions of the concentration space that make them feasible. Algorithmically, this is detected as $H_{min}(r_8 + r_9) > 0$ for the linear combination of r_8 and r_9. Thus, the bottlenecks (figure 11.5) are r_5, r_6, and

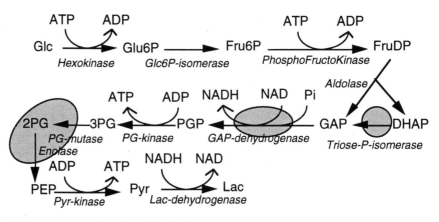

Figure 11.5 Bottlenecks in the glycolysis pathway, for the concentration interval $C_{min} = 0.1$ mM, $C_{max} = 1$ mM.

the combination (distributed bottleneck) $r_8:r_9$. Note that r_4, which had the largest (most unfavorable) $\Delta G^{\circ\prime}$, actually does not occur in a bottleneck at all. We note that a statement such as "r_5 is a bottleneck" is rather curious: The only experimentally observable conclusion is, strictly speaking, that the pathway is infeasible for the specified concentration interval. The attribution of the infeasibility to specific bottlenecks should be interpreted as counterfactual: If the standard Gibbs energies of the bottlenecks were somehow lower (algebraically), which clearly is impossible, the pathway would turn feasible. Such counterfactual attributions are always present, if not as apparent, in our understanding of complex situations.

The second interval we consider is $C_{min} = 0.004 \times 10^{-3}$ M and $C_{max} = 5 \times 10^{-3}$ M. With this relaxation of the concentration bounds (table 11.2), there are no localized bottlenecks (no $H_{min}(r_i) > 0$), but there is a larger set of candidates for formation of distributed bottlenecks, through steps 2, 4, 5, 6, 7, 8, 9, and 11. The algorithm identifies the combination $r_4:r_5:r_6$ as the only distributed bottleneck (figure 11.6).

If the interval is relaxed further to $C_{min} = 0.0025 \times 10^{-3}$ M and $C_{max} = 5 \times 10^{-3}$ M, then all bottlenecks are eliminated. This suggests that the composite transformation $r_4 + r_5 + 2r_6$, which yields FruDP + 2 NAD + 2 Pi → 2 PGP + 2 NADH, is thermodynamically difficult as a whole. To take place, it requires FruDP to have the maximum and PGP the minimum possible concentrations. Furthermore, it points to the crucial role of the catabolic reduction charge, which is the ratio NADH/(NAD + NADH). If this ratio is too high, it may shut down the pathway (thermodynamically) at this bottleneck.

The entire analysis based on Gibbs energies is relevant for native cell conditions only to the extent that we can sufficiently define intracellular concentrations and activity coefficients. Cohen and Rice (chapter 12) might question whether we can even meaningfully refer to concentrations and

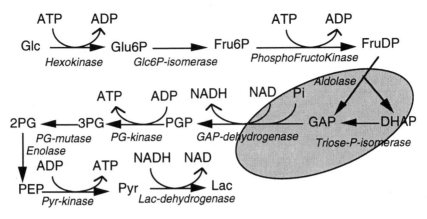

Figure 11.6 Bottlenecks in the glycolysis pathway, for the concentration interval $C_{min} = 0.004$ mM, $C_{max} = 5$ mM.

activity coefficients, given the small numbers of molecules per cell and the complex interaction of the three-dimensional liquid solution with two-dimensional surfaces and macromolecular or mesomolecular assemblies. This chapter takes the view that some measure of the amount or activity of a metabolite will always be applicable, even if only as a first approximation. The fact that the thermodynamic framework just outlined depends on *intervals* rather than exact concentrations and activity coefficients is, in this regard, a significant advantage.

RATE-LIMIT INTERPRETATION

In this section, our objective is to relate the thermodynamic analysis to limits in the kinetic behavior of biochemical pathways. We will start by seeking a kinetics-based justification for the concentration bounds that are essential to the identification of thermodynamic bottlenecks. We then will seek a broader combination of bioreaction kinetics and thermodynamics. This leads to a method for the estimation of maximum rates (or minimum enzyme requirements) for bioreactions and pathways.

Justification of Concentration Bounds

One question that may arise in using thermodynamic techniques is the origin of the concentration bounds used in the identification of distributed bottlenecks. An upper bound on a concentration can be attributed to the limited total soluble pool of the cell. An exceptionally high concentration for one metabolite might exceed the total soluble pool available or might occupy too large a fraction of that pool, leaving insufficient room for the rest of the metabolites. How can we justify the lower bounds on the concentrations? Note that this question arises even in judging the reversibility of a single

Analysis of Complex Metabolic Pathways

reaction, as every reaction is thermodynamically feasible if we allow its product concentrations to be infinitesimal.

The only fundamental consequence of an extremely low concentration is that, by virtue of mass-action kinetics, the rate of any reaction involving that metabolite as a substrate will become extremely small. The lower bound on metabolite concentration can be understood as the lowest amount of substrate required for reasonable reaction rates; this assumes that an estimate or bound for rate expressions can be derived.

Diffusion or Encounter Limits

Enzymatic reaction mechanisms involve attachment of the reactants (substrates) to the enzyme, transformation of the substrates to products and, finally, detachment of the products. Because a mechanism must involve the attachment of the substrates on the enzyme, the rate of a single-substrate irreversible enzymatic reaction has an upper bound, equal to the diffusion-controlled rate of encounter between the enzyme and the substrate (Hammes and Schimmel, 1970; Fersht, 1977; Hiromi, 1979):

$$r_{max} = k e_{total} [A] \tag{17}$$

where e_{total} is the concentration of the enzyme (excluding any amounts that are deactivated or inhibited), [A] is the concentration of the substrate, and k is the bimolecular encounter (diffusion) parameter. Maximum values for k are reported as roughly $10^8 - 10^9$ $M^{-1}s^{-1}$. Naturally, most enzymatic reactions cannot attain this rate because they are limited by additional substrates, dissociation of the products from the enzyme, slow intramolecular rearrangements within one or more substrate-enzyme complexes, and various regulation phenomena. This equation nonetheless yields a valid (conservative) upper bound for all enzymatic reaction rates, and it could be achieved by the most efficient enzymes under optimal conditions (Albery and Knowles, 1976).

Given an expected (or minimum) magnitude for the rate, equation 17 can be inverted to yield a minimum substrate concentration [A]. One can thus address the question of the origin of lower bounds for concentrations.

The equation ignores many complications, however. An enzymatic reaction with many substrates and products should theoretically have a slower rate, because the enzyme must encounter each of the substrates (often in a predefined sequence) before the reaction can be completed. This effect of multiple substrates and products on the reaction rate is not reflected in the equation.

Thermodynamic Consistency

Continuity in the Interpretation of Thermodynamic Considerations
Another shortcoming of equation 17 is that it does not take into account the displacement of the reaction from equilibrium; it predicts for a reaction that

is very close to equilibrium the same high rate it would predict for an irreversible reaction. This is directly related to an aspect of the purely thermodynamic analysis that appears counterintuitive—namely, the discontinuity that occurs at $\Delta G_i = 0$.

Although classical thermodynamics has nothing to say about kinetics, there really is only a single set of underlying microscopical phenomena responsible for both kinetics and thermodynamics. This should lead to some consistency between the thermodynamic and kinetic components of a bioreaction's behavior.

We can restate the problem, equivalently, in terms of the mass-action ratio Q and the equilibrium constant K', as $\Delta G_i = \ln(Q/K')$. For a value of Q only slightly smaller than K' and a value of Q slightly larger than K', we reach sharply different conclusions, because we have an inversion of the direction of the rate. However, the thermodynamic driving force of the transformation stems from the difference between Q and K'. Based on our common sense about how reactions operate (or, more generally, about thermodynamic driving forces), starting from Q < K' we would expect the rate of the transformation to be gradually reduced as Q approaches K', become zero when Q = K', and then smoothly become negative for Q > K', meaning that the reverse transformation now becomes thermodynamically feasible.

In other words, although the feasible *direction* of the biotransformation does change at Q = K', the magnitude of the rate should undergo a smooth change. Purely thermodynamic arguments apply to the direction of the rate of a biotransformation but do not, by themselves, give information on the magnitude of the rate. Ideally, thermodynamic concepts should be combined with other physical and chemical considerations to yield information on permissible bioreaction rates; conforming to the idea that the thermodynamic driving force is determined by Q and K', a quantitative prediction of the range of permissible rates should have the form depicted in figure 11.7.

Thus, we encounter an incongruity between thermodynamic arguments and rates, which we propose to bridge later. It should be emphasized again that this analysis requires some nonthermodynamic considerations; we will

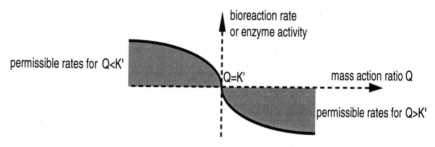

Figure 11.7 The range of permissible bioreaction rates or enzyme activities (vertical axis) should, ideally, vary smoothly with the mass-action ratio Q (horizontal axis) and the equilibrium constant K'.

show in the next section how the analysis can be accomplished with diffusion-based rate limits.

Maximal Rate Analysis Combining Diffusion and Thermodynamic Considerations

Simple thermodynamic analysis alone (specifically, a comparison between the mass-action ratio and the equilibrium constant) can predict whether or not a reaction is feasible. Unfortunately, it can say nothing about the maximum permissible rate at which the enzymatic reaction could take place. As was mentioned earlier, we would like to have a maximum permissible rate that smoothly approaches zero as the reaction approaches equilibrium (figure 11.7). How could we obtain rate bounds without experimental information on the actual enzyme kinetics?

The diffusion-based estimation of the maximal rate can be extended to multisubstrate, multiproduct reactions, reversible or irreversible, in a thermodynamically consistent way (Mavrovouniotis, Stephanopoulos, and Stephanopoulos, 1990b). The key for the integration of diffusion-based maximal rates and thermodynamic arguments lies in examining a complete enzymatic reaction mechanism. For example, in the mechanism of figure 11.8, the unknown kinetic parameters k_i must be consistent with the equilibrium constant (which can be estimated through group contributions to the Gibbs energy). Hence:

$$K' = \frac{k_1 k_2 k_3 k_4}{k_{-1} k_{-2} k_{-3} k_{-4}} \tag{18}$$

At the same time, each step that involves the encounter of two species is subject to the bimolecular diffusional limitation. This applies to either forward or reverse steps. Hence, k_1, k_2, k_{-3}, and k_{-4} have upper bounds determined by the diffusion-controlled rate of encounter.

We have shown (Mavrovouniotis et al., 1990b) that the simultaneous utilization of the diffusion-based upper bounds of bimolecular steps in the

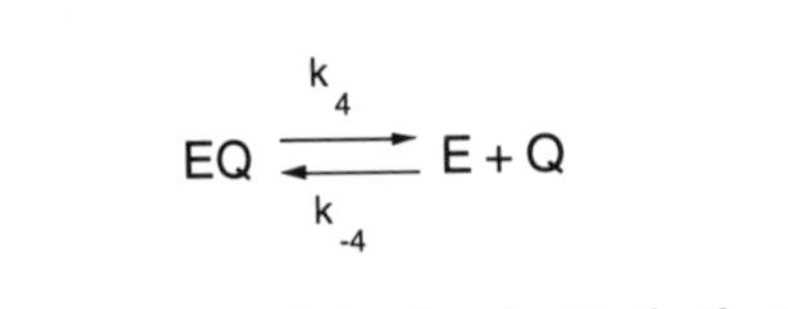

$$E + A \underset{k_{-1}}{\overset{k_1}{\rightleftharpoons}} EA$$

$$EA + B \underset{k_{-2}}{\overset{k_2}{\rightleftharpoons}} EAB \underset{k_{-3}}{\overset{k_3}{\rightleftharpoons}} EQ + P$$

$$EQ \underset{k_{-4}}{\overset{k_4}{\rightleftharpoons}} E + Q$$

Figure 11.8 Ordered mechanism for the two-substrate two-product reaction A + B⇌P + Q.

mechanism *and* the thermodynamic consistency relating the equilibrium constant to rate constants leads to an upper bound for the overall bioreaction rate (Mavrovouniotis et al., 1990b). This upper bound follows the thermodynamic requirement depicted in figure 11.7 (i.e., it smoothly approaches zero as the mass-action ratio approaches the equilibrium constant). Thus, we obtain the desired continuity at $\Delta G_i = 0$ for reversible reactions.

The diffusional arguments generally depend on the type of mechanism through which the enzymatic reaction takes place. In some instances, the mechanism of the enzymatic reaction under consideration is known to be ordered, and the specific order in which substrates are bound (and products are released) also is known. The approach given above is straightforward for reversible or irreversible ordered mechanisms. The maximal-rate treatment for random-order mechanisms can be carried out similarly, although the mathematical expressions it produces are likely to be cumbersome.

We complete our discussion of rate limits for isolated bioreactions with two observations that are important in the extension of the analysis to entire pathways. First, for a bioreaction with multiple substrates and products, the total enzyme will distribute itself among the various enzyme-containing complexes to achieve the optimal rate. Thus, the computation of this rate involves the allocation of the enzyme, as a limited resource, among the enzyme-complex pools.

Second, we can show that the resulting relationship for the maximum rate (for the two-substrate two-product case) can be written in the form:

$$r_{max} = Ef(A, B, P, Q) \tag{19}$$

where A, B, P, and Q are concentrations (or activities) of the metabolites that serve as reactants and products of the bioreaction. Because the enzyme appears proportionally in the right-hand side, the form can also be written as a relation for the maximum specific activity of the enzyme, as a function of concentrations:

$$(r/E)_{max} = f(A, B, P, Q) \tag{20}$$

Finally, we can focus on the enzyme concentration and write the relation to show the minimum enzyme requirement, for a given flux or per unit flux:

$$E_{min} = r/f(A, B, P, Q)$$

or

$$(E/r)_{min} = 1/f(A, B, P, Q) \tag{21}$$

The thermodynamic consistency is guaranteed for all these relations because $f \to 0$ as the mass-action ratio approaches the equilibrium constant (i.e., as $\Delta G_i \to 0$).

Global Maximal Rates

The diffusion-limit computations just given for a reaction assume that the concentrations of reactants and products are fixed. Thus, this is the equivalent of examining one reaction at a time as a candidate localized bottleneck (except that we are focusing on the rate manifestations of thermodynamic obstacles). This does not address fully the maximal-rate analysis of a whole biochemical pathway. Estimation of the separate maximal rate of each bioreaction in the pathway is insufficient, one reason being that, if we assume that the pathway is operating at quasi—steady state and is producing a single final product, then the pathway *as a whole* generally is characterized by a rate. This rate may depend on characteristics of all the individual bioreactions, as concentrations of intermediates shared by reactions are interdependent. More importantly, if we examine reactions simultaneously, we must have consistent concentrations for the intermediate metabolites. This is precisely what gives rise to distributed bottlenecks in the purely thermodynamic analysis; the same kinds of tradeoffs with intermediate concentrations will give rise to rate manifestations of distributed bottlenecks.

A pathway might be constrained by just one limiting bioreaction or by a small number of bioreactions; this will depend on the maximal-rate behavior of all the bioreactions in the pathway, over a range of concentrations for the metabolic intermediates. The relevant mathematical problem and solution methods must draw on all the previous arguments.

A preliminary example is shown in figure 11.9, calculating enzyme requirements for a pathway for a fixed rate. There is a close relationship between the enzyme concentration and the rate. Instead of fixing enzyme concentrations and maximizing the rate, we can fix the rate of the whole pathway and calculate the minimum enzyme required to achieve this rate. The mathematical problem is essentially the same; we are merely shifting our biological interpretation. The figure shows individual enzyme requirements for the steps. The large variation in the values is typical; usually, only some of the steps are constraining (thermodynamically). The figure also shows a nonlocal calculation of the minimum enzyme requirement over a small section of the pathway; the enzyme requirement is much larger than the sum of the individual requirements of the reactions because of the constraint of consistency for the intermediate metabolite concentrations. This pathway segment coincides with the region of distributed bottlenecks derived previously (in figures 11.5 and 11.6) without any rate arguments. We can thus think of the global minimum enzyme requirement (or maximum-rate) computations as a refinement of the distributed thermodynamic bottlenecks discussed earlier.

Although there can be no doubt that calculations of the type outlined previously can be carried out, practical utility of any bound computed from the maximum-rate methodology is likely to vary a great deal. For many systems it may turn out to be so far from observed values as to be useless. The antidote needed for such systems is the incorporation of additional constraints (from partially observed parameters) into the calculation, to bring the bound closer to realistic values.

Savageau (chapter 6), in his critique of the Michaelis-Menten formalism, points out that interactions among enzymatic reaction mechanisms (e.g.,

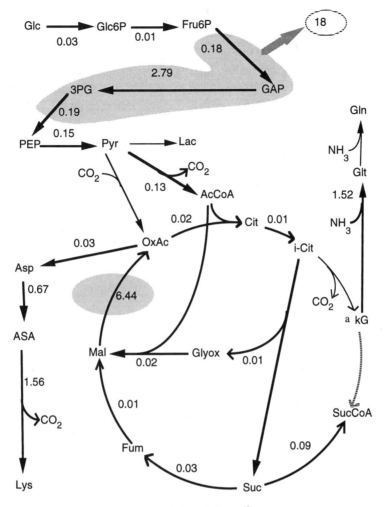

Figure 11.9 Enzyme requirements for a lysine pathway.

the sharing of some intermediate complexes) alter the rate expression. The nonlocal effects discussed earlier can track this effect, but on rate *bounds* rather than the actual rate. Although we do make the quasi-steady-state assumption in studying maximal rates (because we are not interested in short-time-scale behavior), we make it only for the entire system, not one enzyme at a time. This avoids some of the pitfalls pointed out in chapter 6.

GLOBAL RESOURCE ALLOCATION

In this section, we speculate on the global features of thermodynamic and diffusion-limit kinetic analysis, within a finite-resource allocation view. Much of the discussion in this section aims to make more explicit and concrete various arguments that already have been presented in the previous sections.

The nonlocal character of the analysis proposed here is emerging from the previous section, where we showed that the enzyme requirement of a whole pathway must involve a consistent choice for the intermediate concentrations in the entire pathway, in a manner similar to distributed thermodynamic bottlenecks. Metabolites thus are limited resources. Whereas we focused on individual bounds in our previous analysis, $\phi_j C_j \leq \phi_j^{max} C_j^{max}$, we mentioned that the origin of these bounds is the total available soluble pool. A global view would make explicit this constraint as an overall bound for all the metabolites collectively:

$$\sum_j \phi_j C_j \leq (\phi C)^{total} \tag{22}$$

An additional, and quite subtle, resource-allocation interaction is created among enzymatic reactions when the same metabolite participates in many reactions. The concentration of the metabolite that is available for the kinetics of any one reaction is only the free metabolite; but each reaction requires, for optimal operation, an amount of metabolite bound in an enzyme complex. The allocation of the metabolite A among the shaded complexes in figure 11.10 is a nonlocal constraint; satisfaction of this requirement couples the kinetics of each reaction to the overall distribution of the metabolite.

In our presentation of diffusion-limited rates, the enzyme was similarly a limited resource to be allocated among the complexes. We can have a global constraint of the form in expression 22 here as well, induced by the total enzyme (or protein) pool.

These types of global constraints are the silver lining of pathway complexity: They make a more complex pathway likely to yield tighter bounds. In a very long pathway, a limited amount of ADP, for example, must be

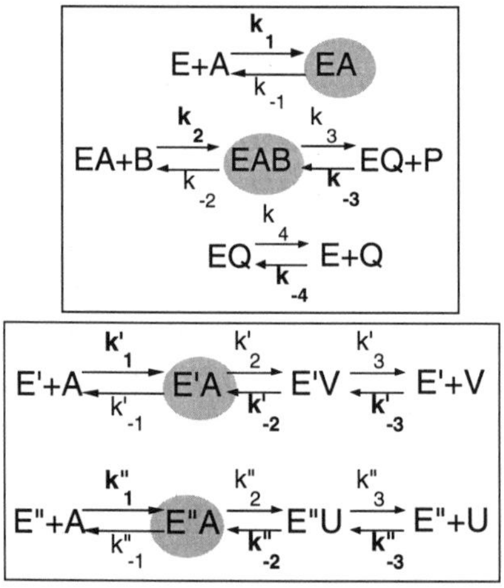

Figure 11.10   Additional resource allocation interactions among enzymatic reactions are introduced when the same metabolite participates in many reactions: The allocation of the metabolite A among the shaded complexes is a nonlocal constraint.

allocated among many reactions that involve ATP and ADP. In a shorter pathway, the same amount is distributed among only one or two steps that involve ADP, and the resulting allocation per step is higher. As a result (and aside from other constraints), rates will be slower in the longer pathway, which is more constrained than the shorter pathway.

Unified Framework

We believe that the judicious use of the arguments and methods presented thus for permits the study of global allocation of metabolic resources. The optimization objective of the allocation is a maximum metabolic flux (for a given enzyme pool) or minimum enzyme requirement (for a given flux). To reach this objective, we can manipulate the distribution of finite resources: A finite soluble pool (within which individual metabolite concentrations may vary) and a finite enzyme pool (within which individual enzyme activities or required amounts may vary). Within these, there might exist further restricted subpools.

Constraints and limitations on the operation of the metabolic network are induced by the finite character of the resources, through physicochemical phenomena interactions. The following physicochemical factors, in particular, are inherent to the operation of the metabolism: thermodynamic feasibility of steps and pathways and thermodynamic consistency of kinetic parameters; diffusion limitations on bimolecular steps of reactions, propagating to rate limitations on reactions and pathways; and metabolites and enzymes bound into complexes and unavailable for other mechanism steps. Additional constraints can be imposed based on other data (such as known fluxes, concentrations); physicochemical phenomena not studied here (such as diffusion along gradients within the cell, membranes, and compartments); and limitations stemming from gene expression and regulation. Some of these considerations might be modeled as higher-level alternatives: Analysis of alternative metabolic networks arising from specific subsets of the genome, and subsequent comparison of their bottlenecks and behavior bounds, shed light on regulatory phenomena.

The procedure for posing this resource optimization problem might begin with the formulation of the complete metabolic network to be analyzed. The final products of the network usually are secreted or accumulated. If we are studying growth processes but do not wish to model polymerization directly, then building blocks for biological macromolecules would represent final products of the metabolic network; the same is true for processes in which macromolecular products are secreted or accumulated. In the case of macromolecules, the proportion of utilization of different monomers determines the proportions of pathway fluxes in the metabolic network.

The next phase would establish the values of as many parameters as possible. This would include expected or observed growth rates or industrial bioprocess production rates; fixed concentrations for currency metabolites;

and other regulated concentrations (including enzyme amounts). The Gibbs energies can be determined from the structures of the metabolites in the network, through the group contribution approach.

Upper bounds for concentrations can be enforced by the total available soluble pool, whereas lower bounds are induced by diffusion limits of rates. Additional bounds on individual metabolites (or sets of metabolites) might be provided; for example, one bound may require a base level of amino acid concentrations for the metabolites' polymerization into proteins.

With these constraints, one would then carry out a local analysis to establish the feasibility of individual steps thermodynamically and through maximal rates (or enzyme requirements). This will identify localized bottlenecks and might restrict the need for global analysis to only a portion of the original network. It also will provide useful initial guesses for any parameters for which iterative numerical solution is needed in the global scale. Finally, global solution of the optimization problem would yield distributed thermodynamic bottlenecks, rate bottlenecks, and global enzyme requirements. The results define the most efficient possible operation of the cell, subject to the imposed constraints; they represent a bound on the behavior of the metabolic network. The results are most sensitive on that subset of the data that gave rise to the most constraining bottlenecks; this warrants careful reexamination of key data that played a role in bottlenecks.

Biological Interpretation

Certain assumptions underlie the proposed thermodynamic analysis. First, the use of concentrations is permissible only if the transformation actually involves dissolved (intracellularly or extracellularly) substrates and products. For example, if an enzyme is part of a multienzyme complex, and it receives its substrate not from the bulk of the aqueous phase but directly from an active site of another enzyme in the multienzyme complex, then the use of the substrate concentration is misleading. Second, the thermodynamic analysis assumes that the stoichiometry is fully known and that there are no additional energy-currency metabolites (ATP, etc.) participating in the transformation.

These considerations do not prohibit the thermodynamic analysis, but they alter the way in which it is applied: If, under some conditions, a biotransformation is known to occur but the thermodynamics appear unfavorable, one may postulate that the transformation is assisted by additional ATP; one can even calculate, from Gibbs energies or from rates, the minimum necessary number of moles of ATP.

With respect to multienzyme complexes, note that the thermodynamic analysis is difficult for individual enzymes within the complex but is perfectly legitimate in its usual form for the enzyme complex as a whole. Thus, if one suspects that in a series of bioreactions (i.e., a pathway) multienzyme complexes may exist, one can apply the thermodynamic analysis for various hypothetical multienzyme complexes and determine which ones make sense thermodynamically. Such hypotheses can help to focus and guide subsequent experimental investigation.

Thus, the methodology we propose enables more extensive study of the consequences of these phenomena, by assisting in the formulation and examination of alternative hypotheses. Other biological processes, which might at first appear to impede the application of this framework, may also be amenable to study through hypothesis generation and testing.

CONSTRUCTION OF PATHWAYS TO BE ANALYZED

An assumption in the analysis of a metabolic network, as presented earlier, is that the network is an assembly of definite pathways that convert substrates to products in known proportions, with fluxes through bioreactions in known proportions. It may, however, be necessary, to *construct* the pathways that accomplish a given (known or desired) transformation, from a database of bioreactions. The construction of pathways allows the determination of what transformations are possible, purely at a stoichiometric level. The constructed pathways can then be analyzed as discussed earlier.

The pathways sought might not be restricted to those commonly considered as the pathways a cell uses for its own needs. Nonobvious alternatives might be of interest as possibilities for new genetically engineered strains or as alternatives a microorganism would use under extreme conditions. Consider, for example, the problem of identifying a mutant strain lacking a particular enzyme. One may wish to identify those sets of substrates on which the mutant cell should be able grow and those sets of substrates on which the cell should not be able to grow. These characteristics depend primarily on the presence of suitable pathways (despite the lack of a particular enzyme) to consume the substrates in question; especially for mutant strains, such pathways may differ significantly from the standard well-known routes. Conversely, if the target is not elimination of an enzyme but absence of growth on specific sets of substrates, one must pinpoint the enzymes that should be eliminated to block all the pathways for the catabolism of the substrates. A similar argument can be applied to anabolic routes for the presence or absence of pathways for specific products.

A method for the construction of pathways can be a useful starting point for generating pathways for subsequent analysis and experimentation. The method presented by Mavrovouniotis, Stephanopoulos, and Stephanopoulos (1990a) allows the construction of the complete set of pathways for any transformation, although it carries high computational requirements. Figure 11.11 shows two pathways constructed by this algorithm.

CONCLUSIONS

The crucial problem in qualitative analysis of biological systems or, indeed, of natural systems of any type is the definition of an appropriate level of

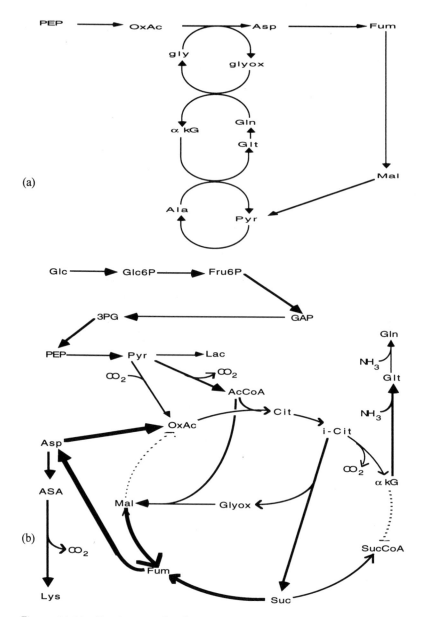

Figure 11.11 Construction of pathways. (a) A complex pathway that accomplishes the simple transformation of phosphoenol pyruvate to pyruvate. (b) An alternate pathway for the synthesis of aspartate or lysine without using malate dehydrogenase (the shaded reaction that derives oxaloacetate from malate).

qualitative analysis: The appropriate level avoids details about which information is sparse or unreliable, but it must not become so abstract that the conclusions it produces are biologically irrelevant. These criteria are met, for example, by the description of regulation put forth by Thomas in chapter 8. Had he omitted the sign of the regulatory interaction from his description, his conclusions would have been too weak. Had he included the magnitude or details of functional forms, the unavailability of data would have limited the application of his method.

Julio Collado-Vides (chapter 9) points out that experiments remain the primary driving force in biology, with little use of theoretical paradigms. This may be attributable in part to cultural differences between experimentalists and theorists; but the fact that theoretical models have so often relied on parameters and data that are very difficult (often nearly impossible) to obtain experimentally has undoubtedly contributed to this deficiency. The difficulty of data unavailability occurs to varying degrees; often the issue is data *accuracy*: That is, the theory implies, or relies on, far more accurate values than the experiment can provide or test. The gap will be bridged as experimental capabilities advance, but it is also a challenge for theorists to select problems and develop techniques that are compatible with practical experimental measurements. Clearly, complications such as compartmentalization, enzyme complexes, vectorial reactions, and so forth will challenge the useful application of the ideas discussed in this chapter.

The qualitative analysis advanced here relies on bioreaction stoichiometries, thermodynamic feasibility study of pathways, and idealized kinetic analysis based on diffusion-limits of enzymatic reaction rates. The framework includes the estimation of the Gibbs energy from molecular structures, as this parameter is essential for the entire subsequent analysis. We note that the molecular structures of metabolites usually are known, even for new or less studied pathways, enabling the application of the group contribution method. The framework complements the traditional way of determining thermodynamic feasibility with an approach that relies on concentration bounds rather than exact concentration and an idealized kinetic interpretation of thermodynamic constraints. The idealized diffusion-limit kinetics provide a gradual transition from the forward reaction to zero net-rate and then to the reverse reaction. Analysis on the scale of a whole pathway produces thermodynamic and kinetic bounds for the behavior of the pathway and specifically identifies the bottlenecks (individual bioreactions or pathway segments) that limit the pathway. Ultimately, this involves a global allocation of metabolites and enzymes, which are limited resources, either as entire pools or in specific subpools. There are many nonlocal interactions, such as the allocation of a substrate among substrate-enzyme complexes from all the bioreactions in which it participates. One can also construct pathways from a complex metabolic network to gain insights into what transformations are possible in the network and what sequences of steps can accomplish them.

This proposed approach has the potential to scale well toward pathways of arbitrary size and complexity. The data required for its application either are commonly available or can be estimated. The approach, fundamentally, places no limit on the size of pathway, and it does not require that the pathway take any particular form. Linear segments, branching, cyclical subpathways, or any other arbitrary topologies of the metabolic network are acceptable. Thus, the entire metabolism of a cell, if it is known, could be analyzed as a whole. The qualitative features obtained by this method have broad validity: Pathway feasibility and diffusion-based rate limits are unaffected by many of the details of enzyme structure and behavior that are difficult to obtain and model. The framework is extensible; it allows many avenues for incorporating additional data in the form of constraints on rates, metabolite concentrations, or enzymes. As additional constraints are introduced, the results of the analysis are likely to become tighter. The results are always bounds rather than precise predictions of behavior. These bounds can nevertheless be used for the formulation and testing of hypotheses. For example, if a hypothesis on pathway regulation affects the amount of an intermediate metabolite, then it must be consistent with the thermodynamic requirements placed on that metabolite by other pathways in which it participates.

ACKNOWLEDGMENTS

This work was supported in part by the National Library of Medicine grant R29 LM 05278.

SUGGESTED READING

Cornish-Bowden, A. (1989). Metabolic control theory and biochemical systems theory: Different objectives, different assumptions, different results. *Journal of Theoretical Biology, 136,* 365–377.

Davies, D. (Ed.) (1973). *Control of biological processes.* Cambridge, UK: Cambridge University Press.

Fell, D. A. (1992). Metabolic control analysis: A survey of its theoretical and experimental development. *Biochemical Journal, 286,* 313–330.

Kohn, M. C., and Lemieux, D. R. (1991). Identification of regulatory properties of metabolic networks by graph theoretical modeling. *Journal of Theoretical Biology, 150,* 3–25.

Reder, C. (1988). Metabolic control theory: A structural approach. *Journal of Theoretical Biology, 135,* 175–201.

Savageau, M. A. (1976). *Biochemical systems analysis.* Reading, MA: Addison–Wesley.

Sen, A. K. (1990). Topological analysis of metabolic control. *Mathematical Biosciences, 102,* 191–223.

Voit, E. O. (1992). Optimization of integrated biochemical systems. *Biotechnology and Bioengineering, 40,* 572–582.

12   Where Do Biochemical Pathways Lead?

Jack Cohen and Sean H. Rice

One way that we judge the success of a science is by its ability to generate questions and find answers to them. In this regard, molecular biology has been second to none. The fusion of the mature sciences of genetics and biochemistry has produced a kind of scientific revolution, driving new discoveries at an accelerating rate.

There is another criterion, though, one that is less important to a budding science but essential if it is to mature. A science that seeks to explain a particular part of the world should, ultimately, do so in such a way that its results can be combined with those of other fields and integrated into a larger picture. If two different fields of research are to share and elaborate on one another's results, they must share a common underlying model. Here lies one important area of discord between molecular and organismal biology. We shall argue that it is this second criterion, integration, that molecular biology must now take on as its challenge.

There are, in principle, at least three levels of integration. The first is to integrate the information structures within one subdiscipline—say, DNA informatics—so that it provides one clear message. There can also be integration of subdisciplines into a discipline, by tying them to an agreed on basic structure. The third and most ambitious integration is to link several disciplines, each providing context for the others, so that explanation is found for emergent properties as well as for historical contingencies. In this volume, Robbins (chapter 4) addresses the first level; Savageau (chapter 6) and Kutter (chapter 2) address the second. We will attempt to address the second and third.

Molecular biology has set itself the task of looking for the fundamental pieces with which the biological jigsaw is to be put together. Not surprisingly (but with surprising efficacy), it has found many of them, and there are certainly more to come. Once found, these pieces can be arranged on a page next to one another in a reasonable sequence, and...Behold! An organism! Well, not quite. Nonetheless, it is tempting, given the elegance of the biochemical pathways chart that results from this exercise, simply to sketch in a few pulleys and gears, stick a handle in one end, and start cranking (figure 12.1). This model assumes that the cell is a simple machine and that molecular biology and cell biochemistry are its mechanisms.

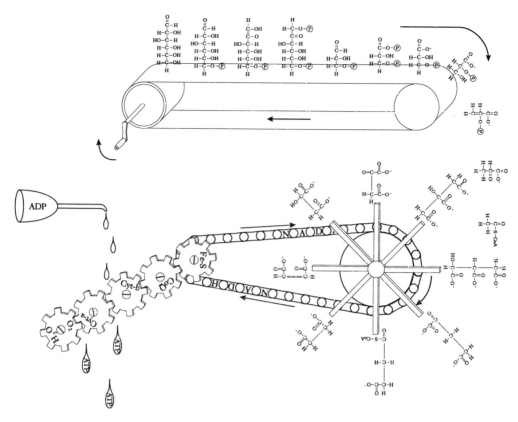

Figure 12.1 The wrong kind of integration.

However, this image is inadequate—indeed absurd—for a number of reasons. First, the various parts of the process do not all occur in the same cellular compartments but rather in different contexts. The cell has a real geography, and the different processes take place in different "habitats." Even within these habitats, the reality is neither as clockwork-mechanical as shown in figure 12.1 nor as smooth as if it were occurring in a test tube, where the mechanics of molecular motion can be approximated by measures such as concentration and pH. Biochemistry takes place in a test tube; molecular biology takes place in a cell. Interestingly, as research has proceeded on different fronts, it has become apparent that these are very different worlds.

Most of the attention in molecular biology is focused either on the biochemistry of metabolism or the consequences of our new-found sequencing skills. Most of the chapters in this volume have this same focus. This is consistent with the fact that these are the most high-profile areas of investigation and the ones that attract the most popular interest.

While this work has been proceeding, molecular biologists, along with specialists in some other fields such as electron microscopy, have been quickly advancing on another front as well—understanding the actual physical structure of the cell and the mechanics of the ways macromolecules do their jobs.

This work has begun to give us a view of the physical environment in which metabolism and DNA processing take place, and it is because of these discoveries that we know that natural cellular biochemistry cannot be the chemistry of solutions. Wachterhauser (1988), among others, has taken this lesson to heart in his modeling of origin-of-life scenarios. Rather than any "primordial soup," he models more of a "primordial pizza" of recursive chemistry on catalytic surfaces.

CHOICE OF MODEL

So successful have concepts such as concentration, pH, and equilibrium constant been in chemistry that often we forget they are approximations. Although useful in a chemistry of equilibrium and large volumes, these terms do not always have much meaning *within* a cell. For example, some important cellular processes, often involving free oxygen or other "toxic" substances, take place within peroxisomes. A typical peroxisome is roughly spherical and approximately 1 μm in diameter (though they come smaller or larger). At pH 7, such a structure would have approximately 30 H+ ions in it. A peroxisome half that diameter, still well within the natural size range, would have 4 such ions (and only 40 at pH 6). In a similar volume within a mitochondrion, there are some 30,000 enzyme molecules, reputedly sensitive to pH. Structure at this scale lies at the edge of our confidence in our models, between microcosm and macrocosm. Though our approximations may work on one side, on the other side pH is no more relevant than is the gross national product to the owner of a corner store or the average population density of the United States to someone living in a Manhattan apartment.

We have two very different models of the amounts of substances involved in living chemistry. The first uses the biochemical assumptions of large numbers and enormous surfaces and talks in terms of concentration and rate; the other speaks of single molecules engaging with single enzymes, or individual antigen molecules engaging with antibodies. Neither works as an integrated approach to molecular biology. Within a cell, the important calcium ions are generally 10^{-9} to 10^{-11} M. There are calcium ions in the cell, but it is their distribution, not their concentration, that matters. Even the gastrointestinal tract, apparently a volume, is not a test tube, as digestive enzyme molecules do most of their work when adsorbed onto the brush border, with its complicated microenvironment, where macroscopical concentrations are inappropriate. Most biochemical processes do not occur in solution when performing molecular biological processes, rather they occur on a surface or a filament.

There is another important criticism of the use of concentrations in models of biochemical pathways. The theory of mass action, and the seemingly more sophisticated a Michaelis-Menten formulations, assume instantaneity. As early as 1963, Goodwin pointed out that there are delays in the real cell.

He showed us, for example, that the simple DNA-RNA-protein model, the protein being an enzyme whose product fed back to control transcription, took approximately 3 minutes to implement. Therefore, such systems are nonlinear oscillators. The various transcription oscillators would interact with one another in the cell, at least by competing for tRNAs, and longer-phase oscillations would be generated. The music of cellular biochemistry would not be smooth and continuous but at least "vibrato" and probably "staccato": Goodwin showed that each reaction would be off for at least half the time. In any individual cell, all concentrations would fluctuate wildly as each reaction interacted with many others. One turn of the handle in figure 12.1 and all the gears would jam up. Of course, averages from a population of such oscillating cells might approximate to a smooth, mass-action-like model. That is why such models have been at least effective—and publishable—if not useful, in explaining the biochemistry of large populations of cells. Nonetheless, such a measure of concentration, or reaction rate, stands for a global result of a process, not for any meaningful part of any cellular mechanism. The molecular biology of a cell is not usefully represented by the averaged biochemistry of many cells. Such apparent integration overlooks the most important elements of cellular control.

Our usual alternative to thinking of cell concentrations is to think of single molecules being passed from enzyme to enzyme. Larger molecules, such as Peter von Hippel's polymerase, riding like an automobile along the RNA, can pick up successive pieces (like traffic cones along the highway) and rearrange them into multiple stacks. Such images are common in models of hormones and their receptor molecules, antigen-presenting arrays on immune-cell surfaces, actions of antibiotics and lytic enzymes. Even as early as the 1960s, the blood-clotting cascade was presented in these single-molecule terms. Such single-molecule models do Boolean algebra, as in Thomas's nice visualizations of cellular cascades (chapter 8). In Boolean systems, feedback networks have to be specific built-in circuits (allosteric enzymes might have the right properties in many cases). Each element of the circuit must be a switch, a threshold, rather than a quantitative response. At best, such circuits have some of the indeterminate, flexible properties of neural net models; much more usually, they have the rigid determinacy of robot factories manufacturing complex products.

There is also the question of the connection of such single-molecule Boolean chemistry to other systems, perhaps outside the cell or in other cell compartments, that operate by analog rules (such as mass action). When driving down a country road, there are discrete cars behind and in front, passing is rare, and the sequence of cars can be thought of as a Boolean system, at least so far as stoplights and crossings are concerned. The kind of statistical traffic reports heard on the car radio are irrelevant to a given driver. Our driver, however, should pay more attention to the gross generalities of the traffic report as he or she approaches a city. In the same way, molecules on cell membranes (the back roads of biological chemistry) may be Boolean, but the aggregate behaviors of small-molecule pools form sinks and sources at the ends of these reaction chains. Even the apparently large volumes within peroxisomes and lysosomes are not, in any real sense, liquid solutions. Very nearly all the water molecules are constrained by protein-ionic interactions. They are like traffic jams and, as in a real traffic jam, interactions are again Boolean.

It is very difficult to understand how the models of biological chemistry *can* apply to molecular biology. Certainly, if we wish to integrate molecular biology with other biological areas such as physiology, endocrinology, growth, or behavior, the biochemical models should be replaced by specifically molecular biological ones.

UNIVERSALITY OF THE MODEL

There is also a problem with the diversity of life. A general belief exists, as exemplified in this book by Robbins (chapter 4) and Danchin (chapter 5), that DNA-programmed molecular mechanisms are essentially universal or, at worst, variations on a few themes. Our present understanding of the functional molecular biology of life is derived almost exclusively from extensive studies of a few model organisms. Some defend this by arguing that the molecular processes being studied are indeed the universals of life on Earth: Understand a fly and, except for a few details to be filled in later, you understand everything else. This argument sounds perfectly reasonable; it just happens to be untrue.

Even among the few organisms on which we have focused, there is variation in molecular function at the most fundamental levels. Because sequences are related and comparable, we can construct phylogenies; however, study of the molecular machinery relating these sequences to functions reveals vast differences across phylogenies. For example, the mechanics of meiosis would seem as general as anything among eukaryotes; but there sits *Caenorhabditis* without any centromeres, which are a crucial part of the "universal" story as it is told in most textbooks. The DNA-RNA-protein story is so general as to merit being called *dogma*; indeed it is the plot link between sequence and function. Reverse transcriptase and introns add a little suspense but can be accommodated without changing the basic script. Even time, and Goodwin's oscillations, can be accommodated. This is not so for RNA editing, though. When specific and predictable base substitutions are made to mRNA in the gastrointestinal tract and liver of mammals on one branch and in trypanosomes on a distant branch, we lose the link. This process is a kind of information processing, and potentially a kind of inheritance, that is not even hinted at by our studies of other organisms.

Once we realize that there are many different ways to organize living systems, we must view functional molecular biology as much more diverse than molecular sequences might suggest. Conveniently, for the last 3.5 billion years, organisms have been exploring and mapping the space of functional molecular systems. We have glanced at a very few of these maps and have been encouraged by congruence rather than dismayed by diversity. Assembling the phylogenetic atlas is the next step. It is not an atlas only of genes; it must also have pages of context, as well as the basic maps, if an integrated picture is to emerge.

Until now, studies of molecular function have been a natural history of cell biochemical processes. These biochemical processes are listed in school biology textbooks like the wild flowers in a Victorian flora. Now that molecular function can be correlated with the phylogeny of gene families, we can start to develop and test general theories. This requires that we put these molecular processes into the real microgeography of cells, taking them out of solutions and putting them onto membranes and filaments, fleshing out the cytoskeleton.

DNA MAPPING TO PHENOTYPE

The most obvious and dramatic successes of molecular biology are with DNA itself, the central molecule of inheritance. The DNA sequences are almost always modeled as unique, with the mRNA copies seen as more susceptible to population studies, and tRNAs, ribosomal RNAs, and even the ribosomes themselves seen in terms of concentrations. Functionally, events on the DNA are on or off. Multiplication of copies enables the cell to sum these Boolean events and to control diffusion, concentrations, and competition further out in the cell. This image of the DNA as the organism plan, sitting in the central administration and issuing orders to workers at the periphery, the workers being "products" of the DNA sequence, has made us see organisms in a new way. This picture of the cell appears tailor-made to be our test case for molecular biological integration and its further integration with other areas of biology. Molecular biology is seen as having provided a simple model of the relation of the unique DNA molecule to the rest of the cell machinery: In this simple view, the DNA sequence is the preformed information according to which an organism is built and that produced and controls the machinery. The media say that DNA is a blueprint, so that to know the DNA is to know the organism. In this view, phenotype is seen as simply the molecular biological mapping of the genotype. The biochemical pathways chart provides the gears that link the information in the DNA to the final organismical pattern, just as the machinery of the modern knitting machine links the knitting pattern to the final patterned cloth or the cardigan, or as the lines of programming code produce the computer screen-saver.

We have shown that our biochemical models cannot support the pathways chart. How then can we use such models as our imaginary connection between DNA and organisms? We must try to achieve our second kind of integration: the integration of molecular genetics with the other processes involved in the development of an organism.

DNA as Program Versus Part of a Dynamic System

A popular current line is that DNA functions much as a computer program, providing the instructions for how to go about building an organism. These instructions are read by the cellular machinery and direct and control development and metabolism. It is this view that leads some authors to conclude that the complete DNA sequence provides a "chemical definition of a living organism" (chapter 5). Were this true, integrating molecular biology would involve nothing more than sequencing. In fact, though, the computer analogy leaves out most of what is interesting about development.

Try running a computer program with the wrong operating system and you will not get even a partially formed output; you will get an error message. By contrast, change the environment in which an organism develops (holding the genome constant) and you will usually get a different organism. This is because the environment, inside the cell as well as outside, does not read the instructions contained in the DNA but rather interacts as a full partner.

The view of DNA as program derives from the idea that it is what is passed from one generation to the next. In light of this, it is worth remembering that organisms go to great lengths to pass on more than just their DNA. Everything from maternal inheritance to internal gestation, or just being picky about where to lay those eggs, represents a passing on of, or control over, the other players in development.

It also is worth keeping in mind that such analogies change with the times and may be influenced more by the available technology than by anything else. Soon after the rise of photography, some psychologists concerned of the brain as working like a camera; when digital computers came on the scene, they became the preferred model for the brain. DNA Informatics is the fashionable, but not the true, model for development.

The program model sometimes is actually a good approximation. It depends on which organism is being considered. The bacteriophage T4 is easy to think about even if, as Kutter (chapter 2) has shown, we still have 90 percent of its physiology to explain in these terms. *Escherichia coli* is more difficult but probably is going to be possible. The enterprise falls to pieces with eukaryotes. The phenotype of T4 is very close to the transcripts and translations of all its DNA sequences (including overlapping and out-of-frame elements and so forth), and is responsive to the *E. coli* physiology that it invades; turning off, then subverting, the bacterial systems requires the sequential expression of most (perhaps all) of the T4 (see chapter 2). The genotype and phenotype of T4 are not essentially different, and the DNA really does (with a few conventional permissive or recursive steps) determine what happens with the multiple copies of downstream molecules.

Building Cells

For a cellular organism, such as *E. coli*, there is more conceptual distance between the DNA sequence and the final phenotype. There are many possibilities for each of its gene systems even at the basic level of the operon. Consider the situation from the point of view of one of the daughter DNA rings, which has inherited half of the mother bacterium. The bacterial "cytoplasm" will elicit different gene effects, depending on the history of the bacterium, *even if the present environment is the same*. Let us imagine that lactose is at an intermediate concentration in the environment. Then one cannot tell whether or not the new *lac* operon is repressed: If there has been much less lactose during the previous cell cycle, the new DNA ring will be repressed, whereas if lactose levels have been adequate, the ring will be expressed. The DNA is the same, but there is a historical constraint because of the self-locking nature of many operons (they show hysteresis). History determines which genes will be expressed, so the phenotype is conceptually much further from the genotype; contingencies affect phenotype.

Eukaryotes such as *Paramecium* show a further distancing of phenotype from genotype: One cannot simply interpolate the machinery of the biochemical pathways chart to deduce phenotype from genotype, as one (nearly) can for T4, because there is much hereditary continuity outside the genome. The beautifully organized rows of cilia and other organized surface structures are constructed with molecules that are produced at two or three removes from the DNA instructions. However, the architecture of this surface, the pellicle, results from two other antecedents: the packing rules used by the animal for assembly of the supermolecular structures, and the accidents of previous lives (shaking of a *Paramecium* culture produces a few animals with reversed patches of pellicle, which are replicated in progeny). The packing rules are "free." In other words, it is not necessary to specify, in the DNA, that NaCl crystals should be made cubical or that fats should be hydrophobic (though it may be specified that water should freeze at a lower temperature, by producing an antifreeze protein).

Building Multicellular Phenotypes

Eukaryotes that develop are more complex still. They nearly always control the first steps of development via mothers' phenotypical construction of the egg (incorporating the mothers' history) but also access far more of these external rules, those of viscosity or lattice packing. The rules apply to the diffusion of gene products that set up the initial gradients and thresholds as the zootype (e.g., of the *bicoid* product in *Drosophila*), to the movements of gastrulation leading to the phylotypic stage, and to the whole geometrical system by which the adult morphology is achieved.

Once again, development is not the execution of a computer program. Rather, it is a dynamical system in which, if all goes well, the information in the DNA will be expressed at the right times and in the right contexts to nudge the system toward an organism that resembles its parent. When all does not go well, the same genome will not give the same phenotype. (A number of seemingly odd molecular genetic phenomena, such as extensive redundancy, may well be adaptations to deal with this.) The DNA, the cytoplasm, the internal and external environments of the embryo interact with one another in the dynamical system of development as the landforms, jet stream, solar radiation, and evaporation interact in the formation of a complex weather system. Just as certain kinds of weather systems consistently produce tornadoes, so do certain kinds of embryos—namely, birds—consistently produce feathers. The DNA is no more the source of the feather's complexity than any other of the parts of the dynamical system.

To the extent that other parts of the system are held constant, variation in feather morphology and color does correlate with variation in DNA sequences. This does not, however, mean that the DNA makes the feather. Likewise, there is no general mapping that takes DNA sequence to feather structure in all environments, any more than all the genes whose mutations affect wing structure in *Drosophila* actually *make* the wing. By extension, there is no general mapping from genotype to phenotype.

INTEGRATION AND CONTEXT

We have argued that the use of biochemical models is less and less sufficient if we wish to produce a molecular biological explanation of morphogenesis as we proceed from T4 to tree ferns. Any causal link between part of the DNA and the adult morphology is so interactive and recursive that the geometrical context may play more part in the final structure of, say, a feather than any single gene. It is useful to remember here that there is a reproductive system (much used for illustration in biological teaching in the 1920s) that has a very familiar, well-defined morphology, yet that is completely context-defined (i.e., there is no heredity at all): This is a flame. Flames, and comparable structures such as whirlpools and tornadoes, show us that reproducible morphology can arise entirely from context, without any contribution from internal instruction. Perhaps the feather, equally, owes its structure as much to contextual rules as to a certain DNA sequence.

Consequently, the context in which a molecular biological process takes place deserves as much study as the biochemical content of that process. This is a tall order. Because anything that one might study exists in an infinite number of successively larger contexts, we cannot possibly consider all of them. We thus need some criteria for deciding which context and, more important, when to change context. When is it time to start focusing less on finding new pieces and more on understanding how the pieces that we have found fit together? When is it time to integrate?

One hint that it is time to start focusing on context is when we find that two different structures—say, two different kinds of molecules—are interchangeable in the sense that it does not matter to the explanation which one is present in a biological process; the outcome is the same. When two structures are interchangeable in this sense—that is, we can substitute one for the other in a particular contextual explanation—they are said to be *fungible*. When we explain how a bridge lets us cross a river, steel or reinforced concrete are fungible substructures; which material the bridge is made of does not affect our crossing. Indeed, a tunnel may be fungible with a bridge if getting to Manhattan is the problem.

Fungibility implies two things, constituting both good news and bad news for our understanding of a molecular biological system at two levels. The bad news is that a strict reductionist approach is not sufficient; if one subsystem (e.g. an enzyme molecule, a process, concrete or steel) can substitute for another, then we need to know why this is so, in order to understand how the subsystems contribute to the system. The good news is that a reductionist approach also is not necessary. The system is telling us that, to the extent that two components are fungible, substructure is irrelevant to function (see Cohen and Stewart, 1994, for an extended discussion).

That is to say, if changing the composition of a structure within a system does not change the output of that system, then simply breaking that structure down into its component pieces will not tell us how or why it works. In such a case, either the system simply ignores certain properties of the structure (the ones that we changed) or else other parts of the system change their behavior to compensate. (Note that this is a major general criticism of the use of "defect experiments" in biology: many gene knockout experimenters have not understood this lesson.) In either case, we must go up a level and study the system as a whole, in order to figure out what is going on. The flip side of this is that we need not always know everything about each piece of a system in order to make progress in studying the system as a whole. This is not to say that reductionism provides no useful information; rather, it emphasizes only that we should start studying context well before all the details of content have been worked out. In fact, contextual study informs our future study of content: Context distinguishes crucial content from that which is irrelevant. Contextual thinking makes reductionism more effective.

One of us (Cohen) has studied several problems whose solutions have turned out to be contextual. Many nonalbino animals, such as Light Sussex chickens and hooded rats, have some white plumage or pelage; what is the melanocyte system doing in those areas? It used to be thought that these areas lacked pigment because they had no pigment cells. Surprisingly, the pigment cell system is present and competent, but local tissue cues turn it off. They do this also, and more interestingly, in hairs and feathers that are white as a consequence of x-irradiation; this had previously been explained as an intrinsic sensitivity of the pigment cells themselves to radiation (see review by Cohen, 1967).

It had been known from 1935 to the 1960s that transplanted feather papillae produced feathers of the donor tract kind if the papilla carried its epidermal coat with it: Breast feather papillae transplanted to the saddle feather tract produced breast feathers on the bird's back. However, if only the dermal papilla was transplanted, a feather of host tract kind appeared (i.e., a saddle feather). It was argued, therefore, that the epidermis carried the specificity. However, in 1965, there was embryological reason to believe that the surrounding dermis could be controlling the system. Indeed, in a three-way set of transplants—dermal papilla from one feather tract and epidermis from a second into dermis of a third—the dermis turned out to control the epidermal differentiation; what we had dismissed as just context was, in fact, the controller (Cohen, 1969).

Nearly all spermatozoa in the mammalian female genital tract fail to get to the site of fertilization, and this has classically been attributed to differences among the sperms. They have been seen as competitors (the Woody Allen view) or as representatives of competing males and to have different swimming or egg-finding abilities (Cohen and Adeghe, 1987). However, it soon became clear that antibodies in the female tract discriminated among spermatozoa and that the tract sequentially filtered out the great majority, it was turned on to do so by the presence of sperms at copulation (Cohen, 1992). Spermatozoa are not simply salmon swimming upstream; the stream itself— context—is the major determinant of which get to the top.

Another example of fungibility in biomedical science is seen in some "knockout" experiments. Here, a gene that is known to be transcribed is rendered inactive, and the resulting phenotype is observed. As Wolpert (1993) has documented, such perturbations often bring about no change in the phenotype of the adult organism. It may seem paradoxical that a gene product could be an active part of a biological system but that knocking out that gene would not change the output of that system. Such a result does not fit in the wall-chart view of life but should not be surprising given the demands on a developing embryo, which call for a "balanced genome" and canalized development. The strength of selection for developmental stability, in variable environments and polymorphic populations, should put a high value on redundancy and correcting mechanisms in development. Note also that these mechanisms become most relevant when something goes wrong.

DEFECTS AND DISEASE

A problem that afflicts all sciences is the fact that once you define the kinds of answers that you expect to get, it is very difficult to know what you are missing. Though we tend to define fields of science by the broad kinds of subjects that they investigate, most field specialists are very good at defining intractable problems as being outside of their field's domain. Physicists, for example, banished the study of turbulent flow when the mathematics to study such flow was lacking. Only when new mathematical techniques and, of particular importance, computers made the subject amenable to study in a way that physicists appreciated was it retrieved from meteorology and mechanics.

There is another reason that we often overlook context. *Within* an experiment, it is important that context be controlled. This often builds up in a good scientist a natural assumption of equivalent contexts for the control and the experimental situations: The idea of all things being equal specifically (and appropriately) excludes context from consideration. This works very well for a simple experiment; it is, however, not the right way for a research program to proceed. Within an experiment, context *should* be controlled, but within a progressive research program, it needs to be actively explored. Only in that way can new questions be asked, to be answered by new controlled experiments.

It is difficult to assess what we are missing by ignoring some processes, simply because we do not spend much time looking at them. There is one kind of evidence, though, that hints that molecular biology is missing a lot that we would like it to include. This particular kind of evidence, which brings itself to our attention whether we look for it or not, is the prevalence of different kinds of human diseases.

A disease is, in a sense, a kind of perturbation experiment. Something has gone wrong in the workings of the system, and we have to figure out what. Because of the humanitarian and social need to deal with diseases, we cannot get away with ignoring those that do not conform to our model. Disease forces us to look at context, so our success at understanding diseases is a measure of the ability of our model to explain the organism as a whole, not just to answer the questions that we have prepared for it. Unfortunately, the reductionist approach has not scored well on this test. As Strohman (1994) has pointed out, only 2 percent of our disease load is accounted for by single genetic or chromosomal aberrations. Further research aimed at identifying particular genes will increase this value a bit, but probably not by much. Thus, 98 percent of the things that can go wrong with this system (us) call for an explanation based on complex interactions of many components in the context provided by a developing human. Savageau (chapter 6) makes the same point from a different direction.

Throughout this discussion, we have argued that it is time for molecular biology to move beyond the stage of simply locating the pieces out of which organisms are made. Though this sort of molecular natural history will always have its place, the time has come to devote as much work to understanding what these pieces actually do and why they do it.

Developing integrative approaches to molecular biology means more than just filling in the gaps between the things that we know. It means more than just finding a particular intermediate in a chemical reaction or the second gene turned on in a sequence. Molecular biology has provided exquisite detail for many other biologists: The evolutionist has told whole new confident stories about the ancient and very recent history of life on Earth; the ecologist looks at populations with new eyes for their genetic diversity, and the reproductive biologist discovers new answers about questionable paternity.

Despite this, our models of chemical kinetics are still those derived from well-mixed solutions. In light of what now is known about the cellular environment in which chemical species compete, cooperate, and prey on one another, it is time to go back and study the basics of biochemical processes as they would happen on surfaces and in very small volumes. It is time, indeed, for the molecular biologists to find their contexts in all this new work that they have made possible, from biochemistry to ecology.

This call for third-level integration is a call to study the dynamics of biochemical processes in complex environments, cellular, organismal, and ecological. Such integration would essentially take place within molecular biology itself: Molecular biology must show that it works in the real world of the organism. This means more than producing scenarios that work on a wall chart or even in a test tube. It means studying the complex systems that result when such processes are in the microgeography of real cells that they are continually rebuilding. It also means asking why a particular simplification works rather than just being glad that it does.

The problem with the biochemical pathways view is not that it ignores the importance of context but rather that it assumes that understanding context is so much easier than understanding content. We do understand that when one knows a great deal about a ribosome, the many proteins and RNAs constitute a world of complexity that takes up all one's thoughts; it is very difficult, then, to see ribosomes as minor actors in someone else's play. However, we must try. We believe that molecular biology is now at a stage where it is ripe to undertake this sort of integration and that, in so doing, we will raise not only the level of understanding of this field but that of biology as a whole.

SUGGESTED READING

Cohen, J., and Stewart, I. N. (1994). *The collapse of chaos*. New York: Viking.

Rollo, C. D. (1995). *Phenotypes*. London: Chapman and Hall. A very good evolutionary or developmental text that takes the integrated molecular biology approach to development and makes genetic and ecological sense of it.

Strohman, R. (1994). Epigenesis: The missing beat in biotechnology? *Bio/Technology, 12,* 156–164.

13 Gene Circuits and Their Uses

John Reinitz and David H. Sharp

The development and application, over the past decade, of experimental techniques such as single and double labeling using fluorescence-tagged antibodies and in situ hybridization has produced abundant data on gene expression in model systems such as *Drosophila, Xenopus*, and mice. These data are of major importance, as they indicate when and where genes act to control development. The interpretation of these data is leading to enormous progress in understanding the regulatory interactions among genes that control important developmental processes in these organisms. In *Drosophila*, the emerging picture of how the circuitry of segmentation genes establishes localized expression domains that become progressively refined over time, leading to the determination of this insect's pattern of body segments, is a good example of such progress (Lawrence, 1992). The conservation of genes that control development across species is opening the way to cross-species studies of gene regulation and, hence, to new avenues for the study of evolutionary development. A recent study (Patel, Condron, and Zinn, 1994) comparing segmentation mechanisms in beetles to those in *Drosophila* is an exciting example of such work.

The interpretation of gene expression data is presenting major new challenges as well as opportunities. In *Drosophila* as well as other organisms, the complexity of gene expression patterns often prevents interpretation of such patterns in terms of the underlying regulatory circuitry. To obtain the biological information implicit in gene expression patterns will require new methods for the analysis of these data, based on new ideas. The recently proposed method of gene circuits is a promising step in this direction. The purpose of the gene circuit method is to provide a way to infer from gene expression data how concentrations of products of a given gene change with time and how these changes are influenced by the activating or repressing effects of other genes. The utility of the gene circuit method has been demonstrated by its application in solving a number of problems concerning segment determination in *Drosophila* (Reinitz, Mjolsness, and Sharp, 1995; Reinitz and Sharp, 1995). Our aim in this chapter is to describe briefly how the gene circuit method works and to illustrate how it has been used to gain new understanding of the regulatory actions of gap and pair-rule genes in *Drosophila*. The issues raised are not unique to this organism, however. They will arise in any analysis of gene regulation in multicellular organisms. For further discussion, see chapters 6 and 8.

THE BIOLOGICAL SYSTEM

The body of the fruit fly *Drosophila melanogaster*, like that of all other insects, is composed of repeated units called *segments*. We have selected the process of segment determination in *Drosophila* as a subject of investigation by the gene circuit method because of its biological importance and also because computational investigations of gene regulation can be done in an exceptionally clean way in this system.

After fertilization, a *Drosophila* embryo undergoes 13 virtually synchronous nuclear divisions, without the formation of cells. Each of these constitutes a cleavage cycle (Foe and Alberts, 1983). For example, cleavage cycle 7 constitutes the period from the completion of the sixth nuclear division to the completion of the seventh. Zygotic genes do not become activated until cleavage cycle 11, by which time the embryonic nuclei have formed an ellipsoidal shell known as the *syncytial blastoderm*. Although all the synchronous mitoses take place rapidly, the interphase periods between them gradually lengthen. Cleavage cycle 14 is substantially longer than the previous cleavage cycles. During this cycle, cell membranes invaginate around the nuclei, sealing them off into cells. At the instant when cellularization is complete, gastrulation begins, terminating the blastoderm stage. A larva hatches approximately 22 hours later.

Cleavage cycle 14 is extremely important. At the beginning of this cleavage cycle, the blastoderm nuclei are totipotent in the developmental biology sense: They are not determined to follow any particular pathway of differentiation. By the end of cleavage cycle 14, the nuclei are stably determined at a spatial resolution of a single nucleus, and the segmental pattern of the animal has been laid down by a chemical pattern of differentially expressed transcription factors. The genes that code for these factors are known as *segmentation genes* and can be divided into four classes. The maternal coordinate genes are expressed from the maternal genome in the form of gradients, only two of which affect the generation of segments: *bicoid* (*bcd*) and maternally expressed *hb* (hb^{mat}). The gap and pair-rule genes are expressed beginning in cleavage cycle 11: They are the main players in the process of segmental determination. Gap genes are expressed in broad domains 10 to 20 nuclei wide that are localized from the time they are first detected, although they undergo some refinement as the blastoderm stage proceeds. Pair-rule genes are expressed in a spatially uniform pattern for most of the blastoderm stage but, midway through cleavage cycle 14, resolve into patterns of seven stripes, each approximately three nuclei wide. The pair-rule genes acting together activate the segment polarity genes at the onset of gastrulation in expression domains only one cell wide that stably specify segments (Akam, 1987; Ingham, 1988).

Understanding how gap and pair-rule genes form patterns during the blastoderm stage is thus essential for understanding segmentation. This is primarily a problem of gene regulation. The disruption of the pattern of expression of one segmentation gene in mutants for another shows that the patterns form as a result of regulatory actions among a network of genes. At the same time, the segmentation gene network does not couple to the morphology of the embryo until after gastrulation. Because the state of expression of the segmentation genes can be monitored by antibody methods, all the state variables in the segmentation system are observable. This is a very unusual situation in biology; it allows us to use the gene circuit method as a precise analytical tool.

THE NEED FOR GENE CIRCUITS

The Problem of Complexity

The need for gene circuits can be illustrated in several ways. We first note that the regulatory actions of genes generally are not observed directly. Rather, they must be inferred from in vivo experimental data, which often cannot be accomplished solely by visual inspection of expression patterns. This can be quite confusing in situations where gene products coarsely distributed in space act in concert to create more spatially refined expression of other genes. For example, *even-skipped* (*eve*) is a pair-rule gene expressed in seven stripes by the end of cleavage cycle 14. It is easy to see that *eve* stripes 4 through 6 are under the control of *knirps* (*kni*), a gap gene whose expression approximately overlaps with these stripes, which are disrupted in *kni* mutants. However, that does not tell us how, for example, stripe 6 in particular is generated. It also does not explain why stripe 6 is disrupted in mutants for *Kruppel* (*Kr*), another gap gene; *Kr* is not expressed between stripe 4 and a region posterior to stripe 7! In fact, gap genes not only regulate pair-rule genes such as *eve*; they regulate one another. A mutation in a single gap gene can alter the expression of several others.

More generally, assertions of the form "gene *a* activates gene *b*" usually are based on the observation that expression for gene *b* is reduced in mutants for gene *a*. Of course, gene *a* usually is acting as well on genes *c*, *d*, *e*, and so on, which in turn act back on gene *b*. Thus, there is the critical question of separating direct from indirect actions of gene *a*. Answering this question requires a method of analysis that keeps explicit track of each individual gene's contribution to the expression of a given gene, which is the purpose of the gene circuit method. Note that even if a direct mechanism of action, such as a segmentation gene product binding to a specific site, is established by biochemical methods in vitro, the role of this regulatory mechanism in embryogenesis must still be demonstrated in vivo using transgenic embryos.

Hence, the use of biochemical methods does not remove the necessity of considering how the intact regulatory circuitry acts in an embryo, and the need for gene circuits remains.

Quantitative Problems

Many questions about gene regulation are essentially quantitative and so cannot be answered by qualitative inferences from data. An example of such a problem concerns the shifts in cell fate markers that occur when the *bcd* gradient is rescaled by changing the number of copies of the gene. Analysis of the quantitative behavior of these shifts using the gene circuit method led to the insight that a second gradient also was involved (see under "Simultaneous Specification of Positional Information by Two Gradients"). Another example of such a question is how the same set of gap gene domains can establish *eve* and *hairy* stripes that are displaced from one another by as little as one nucleus.

Timing Problems

Understanding the temporal coordination of developmental events and its link to the precise timing of regulatory actions of genes defines a new class of problems. As we discussed earlier, segmental determination in *Drosophila* is a result of a coordinated cascade of zygotic and maternal gene products. During the blastoderm stage, mitoses (nuclear divisions) and nuclear movements are under the control of a maternal clock (Foe and Alberts, 1983; Wieschaus and Sweeton, 1988; Merrill, Sweeton, and Wieschaus, 1988). When gastrulation begins at the end of the blastoderm stage, the embryo goes through the midblastula transition. At this time, a large number of zygotic genes become active, and control of cell division is transferred to the zygotic genome. After gastrulation, the zygotically controlled cell divisions occur in spatially discrete domains that are under the control of the *string* (*stg*) gene (Edgar and O'Farrell, 1990). The *stg* gene is under the control of segmentation and other pattern formation genes, including those that control the dorsoventral and terminal regions (Edgar and O'Farrell, 1989).

The coupling together of the pattern formation and cell division control systems at the midblastula transition suggests that the timing of these processes is critical. The importance of timing is widely understood; the point we wish to emphasize here is that timing questions lead directly to questions about the *rate* at which gene products are synthesized. Pulling such information out of the data requires a method of analysis that follows the time development of expression patterns.

Four Ideas for Solving These Problems

The interpretation of gene expression data presents three difficult problems: isolation of effects of individual genes in a complex regulatory environment;

study of quantitative effects in gene regulation; and analysis of regulatory effects that depend on the rate of synthesis of gene products. The method we have developed for dealing with these problems is based on four main ideas. First is the choice of protein concentrations as state variables for the description of gene regulation (Turing, 1952). Second is the summary of chemical reaction kinetics by coarse-grained rate equations for protein concentrations (Glass and Kauffman, 1972; Thomas and D'Ari, 1990; Thomas, Thieffry, and Kaufman, 1995). Third is the use of a gene circuit, which is a simple method to keep *explicit* track of the activation or repression of one gene by another. Fourth is the use of least-squares fits to gene expression data to measure phenomenological parameters occurring in the gene circuit (as resistances occur in electrical circuits), so that the circuit is fully determined and usable as an analytical tool. This circle of ideas, and its range of applicability, requires very careful explanation, which is provided elsewhere (Reinitz and Sharp, 1995) and is recapitulated, briefly, in a later section of this chapter.

The problem of finding least-squares fits to gene expression data is solved by the method of simulated annealing. This is computationally intensive, and it is here that our work makes contact with the subject of computational modeling of gene regulation. A computational solution of the gene circuitry problem would not have been feasible until the advent, over the past decade, of sufficiently fast computers and adequate supporting software. Even so, computational requirements place limits on the size and complexity of the problem that can be analyzed by these methods. Adaptation of the simulated annealing method for use on parallel computers is under way as a means to push back these limits.

The gene circuit method has been used to analyze several regulatory circuits in *Drosophila*. What has been understood with this method that was not understood before? We have identified the precise interplay of regulatory actions that control the timing of *eve* stripe formation and result in the establishment of the borders of *eve* stripes 2 through 5. As an example of this work, we summarize later the analysis demonstrating that one gap domain controls each of the borders of stripes 2 through 5 and identifies the controlling gap gene for each of the eight borders occurring in these four stripes.

We have also used gene circuits to understand an important point about positional information. Cells in an embryo organize themselves into appropriate structures at correct locations by interpreting positional information. In the *Drosophila* embryo, positional information is supplied by maternal genes and interpreted by zygotic genes. The gene circuit method has been used to show that positional information in the presumptive middle body of *Drosophila* is cooperatively determined by gradients of maternal products of the *bicoid* and *hunchback* genes. This work correctly explains the anomalous displacements observed in expression patterns (Driever and Nusslein-Volhard, 1988b) and also makes predictions about the expression of gap genes under differing doses of *bcd* in embryos lacking hb^{mat}. The use of gene circuits was essential in obtaining these results, because it enabled one to unravel the effects of gradients of *bcd* and *hb^{mat}* when acting simultaneously. The analysis is reviewed in the section "Simultaneous Specification of Positional Information by Two Gradients."

HOW GENE CIRCUITS WORK

The gene circuit method is based on a few simple ideas. We will outline these ideas here in a brief and informal way so that one can understand how gene circuits work.

State Variables

The first idea is that gene interactions can be represented by their effect on the synthesis rate of gene products. Concentrations of gene products are therefore the fundamental state variables in the gene circuit method. These state variables are observable in the problems we study because the active genes have been cloned and their expression patterns can be monitored with specific reagents such as antibodies. A consequence of this first idea is that coarse-grained information about gene regulation is substantively reflected in the way in which gene products are distributed in space and time.

The second idea is that it is possible to write a fairly simple set of equations that describe how concentrations of products of a given gene change in time and space and how these changes are influenced by the activating or repressing effects of other genes. Let us see how this can be done.

The only regulatory molecules represented in our current work on the *Drosophila* blastoderm are proteins synthesized by certain gap and pair-rule genes. RNA is not included because there is no evidence of which we are aware for a direct role of RNA in the regulation of zygotic segmentation genes. The rate of protein synthesis is controlled by chemical kinetics. The very complex details of these kinetic processes are summarized here by coarse-grained rate equations for protein concentrations.

Chemical Kinetics

The change in time of concentrations of proteins is governed by three basic processes: direct regulation of protein synthesis from a given gene by the protein products of other genes (including autoregulation as a special case); transport of molecules between cell nuclei; and decay of protein concentrations. Each of these processes *must* be represented in any method that has a chance of being correct, and the equations we shall write down are the simplest and most direct, or minimal, set of equations that do this.

The form of the term accounting for the direct regulation of protein synthesis is suggested by general experience with regulated enzymatic reactions. Such reactions are characterized by a maximum reaction rate and the property that they can vary from zero up to this maximum rate. We thus introduce a scale factor setting the maximum rate of synthesis for each gene as well as a term that describes the regulatory effect of uniformly distributed transcription factors. We next introduce a regulation-expression function, which makes the synthesis rate of a given gene a monotonic, saturating (sigmoidal) function of the regulating gene products. This choice of regulation-expression function is the simplest way to interpolate between the state in which a gene is completely turned off and its state of maximum activity. Savageau points out elsewhere (chapter 6) that sigmoidal kinetics are to be expected for reactions that are confined to regions of less than three dimensions, such as chromatin.

It is an important and necessary aspect of our approach that its main results do not depend sensitively on the exact choice of this function. This feature of networks with sigmoidal interactions has been pointed out by Kaufman and Thomas here (chapter 8) and elsewhere (Kaufman and Thomas, 1987; Thomas and D'Ari, 1990).

The dependence of the regulation-expression function on protein concentrations must allow for the regulatory effects of various gene products on the expression of a particular gene. We believe that the precise binding states of regulator proteins are not required for a useful coarse-grained description of the regulative state of a eukaryotic gene. In certain cases where the relation between binding and expression is understood, it is clear that functional binding sites are redundant in the sense that regulatory specificity is conferred by groups of sites, so that mutation of any individual binding site does not eliminate function (Small, Blair, and Levine, 1992). A similar redundancy exists in the regulatory molecules themselves: It is widely accepted that individual polypeptide molecules are interchangeable and so their effect can be summarized by a concentration. The interchangeability of individual binding sites implies that their effects can also be summarized by a suitable average, in this case over the whole promoter.

These considerations, which are a restatement of our coarse-graining assumptions, lead to the important consequence that the direct regulatory action of one gene on another can be described simply in terms of its effect on concentrations. We further suppose that these interactions can be characterized by a *single* real number (for each *pair* of genes). In a specific problem, we generally will need to allow for an interaction between any pair of genes represented in the gene circuit, so that we must consider a collection of regulatory coefficients T^{ab}. This collection of numbers is conveniently represented as a matrix, $\mathbf{T}$. The matrix of regulatory coefficients defines a gene circuit. It is thus the fundamental theoretical object in the gene circuit method. In the future, it may be possible to relate the coefficients T^{ab} to quantities having direct biochemical significance. We emphasize that at the present stage of work these coefficients have the status of phenomenological parameters that must be determined by experimental data. The way in which this is done will be described shortly.

The gene circuit method includes transport of gene products between cell nuclei and decay of gene products. Exchange of gene products is modeled as a classical diffusion process (with discretized space derivatives [cf. Turing, 1952; Glass and Kauffman, 1972]). This way of modeling molecular transport is not likely to be correct at a fine level of description, but it is our judgment that classical diffusion is sufficiently accurate for our present purposes. Decay of gene products is modeled as a simple exponential, with a rate constant that must be supplied from experiment.

To summarize the discussion thus far our model for the change in concentration of a gene product leads to an ordinary differential equation having the following schematic form:

$$\left(\begin{array}{c} \text{Time rate of change} \\ \text{of protein concentration} \end{array} \right) = \text{Regulation} + \text{diffusion} + \text{decay} \qquad (1)$$

Nuclear Divisions

A differential equation of this kind holds during interphase while gene products are synthesized. It is essential to represent mitosis in the model. We do not model in detail the process of mitosis, which is quite complicated. Instead, we introduce a further coarse-graining assumption and model mitosis as an elementary event in which just three things occur: The synthesis of gene products is suspended; the number of cell nuclei is doubled; and the gene products present in a cell nucleus at the onset of mitosis are distributed among the progeny. At the mathematical level, this is accomplished by adjoining a rule to the differential equation that suspends the equation and reinitializes the state variables according to the schedule of mitosis. In general, it is possible to arrange for this rule to be triggered when the dynamics bring a given cell to a particular state (so that, for example, a specific gene product exceeds a concentration threshold). In applying the equation to the *Drosophila* blastoderm, however, we will take advantage of the simplifying fact that the timing of cell division is under maternal control to implement a predetermined schedule of mitosis. In either case, the resulting mathematical framework is an example of a hybrid dynamical system, consisting of linked continuous and discrete time evolution (Mjolsness, Sharp, and Reinitz, 1991).

Dynamical Equations for Interphase

To apply the gene circuit method, further information must be used that is specific to the model system under study. This information consists mainly of identifying the genes that are active and some general properties of their interactions. Our work has been concerned with the *Drosophila* blastoderm. We analyze this blastoderm during cleavage cycles 11 through 14 and focus on the region of the blastoderm that generates the thoracic and abdominal segments. As explained (Reinitz and Sharp, 1995), the four gap genes *Kruppel* (*Kr*), *knirps* (*kni*), *giant* (*gt*), and *hunchback* (*hb*), all under the control of *bcd*, and the pair-rule gene *eve* form an approximately isolated set of mutually regulating genes in this part of the blastoderm, during the interval of time under consideration. Furthermore, in the middle region of the embryo, the level of expression of these genes is approximately a function only of position along the anteroposterior axis. This implies that one can consider a one-dimensional system consisting of a line of nuclei along this axis.

We now can write the schematic equation 1 in explicit form. Let the position of a cell nucleus along the anteroposterior axis be indexed by i, such that nucleus $i + 1$ is immediately posterior to nucleus i. Each cell nucleus contains a copy of a regulatory circuit composed of N genes that is determined by an $N \times N$ matrix $\mathbf{T}$. The concentration of the ath gene product in nucleus i is a function of time, denoted by $v_i^a(t)$.

$$\frac{dv_i^a}{dt} = R_a g_a \left(\sum_{b=1}^{N} T^{ab} v_i^b + m^a v_i^{bcd} + h^a \right) + D^a(n)[(v_{i-1}^a - v_i^a) + (v_{i+1}^a - v_i^a)]$$

$$- \lambda_a v_i^a \tag{2}$$

where N is the number of zygotic genes included in the circuit (five in the present application). The first term on the right-hand side of the equation describes gene regulation and protein synthesis, the second describes exchange of gene products between neighboring cell nuclei, and the third represents the decay of gene products.

In equation 2, T^{ab} is the previously discussed matrix of genetic regulatory coefficients whose elements characterize the regulatory effect of gene b on gene a. This matrix does not depend on i, the nucleus index, a reflection of the fundamental fact that the cell nuclei of a multicellular organism contain identical genetic material. The bcd input is given by $m^a v_i^{bcd}$, where v_i^{bcd} is the concentration of bcd protein in nucleus i and m^a is the regulatory coefficient of bcd acting on zygotic gene a. The regulation-expression function is g_a, which we assume takes the form $g_a(u^a) = (1/2)[(u/\sqrt{u^2 + 1}) + 1]$ for all a, where $u^a = \sum_{b=1}^{N} T^{ab} v_i^b + m^a v_i^{bcd} + h^a$. R_a is the maximum rate of synthesis from gene a, and h^a summarizes the effect of general transcription factors on gene a. The diffusion parameter $D^a(n)$ depends on the number n of cell divisions that have taken place and varies inversely with the square of the distance between nuclei. We assume that the distance between adjacent nuclei is halved after a nuclear division. The decay rate of the product of gene a is denoted by λ_a.

We comment briefly on the motivations behind our choice of the thresholding function $g(u)$. We did not choose to use Hill functions as thresholding functions (Glass and Kauffman, 1972, 1973; Kaufman and Thomas, 1987; Snoussi and Thomas, 1993) for two reasons. First, the original justification for Hill functions was an elementary model of cooperative binding (Hill, 1985). In the absence of binding data, Hill functions were a reasonable first approach to the chemistry, but current data rule out this simple model without suggesting a unique replacement. Second, choosing the Hill function for thresholding leads to extra free parameters: N^2 thresholds for N genes rather than the N h^as in equation 2.

The protein concentration of hb^{mat} is incorporated into the circuit as an initial value (at cleavage cycle 11) for the concentration of hb product. This is a mathematical expression of the fact that the observed concentration of hb protein consists of both maternal and zygotic components. Initial values of the other, purely zygotic, gene products are taken to be zero at cleavage cycle 11. Finally, in implementing the rule for cell division, we incorporate the facts that in the *Drosophila* blastoderm mitosis lasts approximately 4 minutes and gene products appear to be equally distributed between the two daughter nuclei.

The formulation of the gene circuit method is not complete until one has specified a procedure that takes gene expression data as input and produces a gene circuit T^{ab} as output. Our procedure has two steps. The first is to formulate the problem of finding T^{ab} as that of obtaining a least-squares fit to gene expression data; the second step is to solve this problem using the method of simulated annealing. We refer the reader to Reinitz and Sharp (1995) for an explanation of these numerical procedures, in the context of fits to gene expression data.

EVE STRIPE FORMATION

Long-standing questions about development in general and *eve* stripe formation in particular center around two issues: How do genes interact to make stripes, and what is the relative role of genetic and nongenetic processes in making stripes? These are broad questions, and we will focus here on two specific points. First, we shall show how gene circuits enable one to identify the gap genes that control the borders of *eve* stripes 2 and 5. A striking feature of the analysis is that *eve* stripes do not form unless Eve protein has an extremely low diffusivity. This fact has an important bearing on the observed apical localization of pair-rule message, as we explain at the end of this section.

The Dataset

The data used for the gap expression domains specify the concentrations of gap gene products in early, middle, and late cleavage cycle 14. These data, together with data specifying maternal gradients of *bcd* and *hb*, are extracted from the data set given in Reinitz, Mjolsness, and Sharp (1995). Here we use the central 32 nuclei (i.e., nucleus 17 to nucleus 48) of this dataset. Published observations (Frasch, Hoey, Rushlow, Doyle, and Levine, 1987) were used to generate *eve* data for early and late cycle 13 as well as early cycle 14. At each of these times, the *eve* data are spatially uniform in the region considered. Finally, we included *eve* data from late cycle 14 (Levine, unpublished data; Stanojevic, Hoey, and Levine, 1989), at which time the stripes have clearly formed (figure 13.1); the combined gap gene and *eve* data for late cleavage cycle 14 are given elsewhere (Reinitz and Sharp, 1995). Because we did not

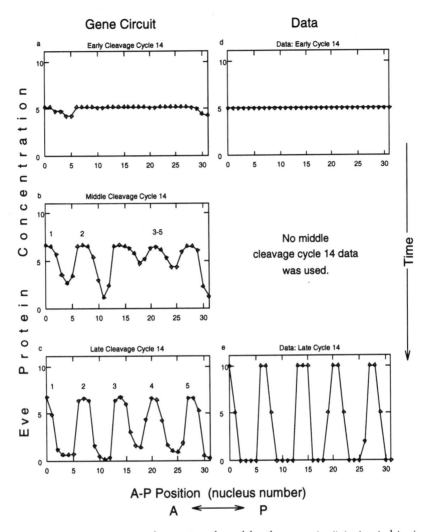

Figure 13.1 Comparison of *eve* stripes formed by the gene circuit to input data. (a–c) Behavior of the circuit at early, middle, and late cleavage cycle 14, respectively. In b, stripes 1 and 2 are distinct and are labeled to indicate this fact, whereas stripes 3–5 have just begun to form. In c, all five stripes have formed and are labeled. (d, e) The *eve* data from early and late cycle 14 that were used for the fit; no data for middle cycle 14 were used. Each graph shows the concentration of Eve protein on the vertical axis. The horizontal axis shows position along the anteroposterior axis, with anterior to the left, in terms of the number of nuclei posterior to the middle of *eve* stripe 1. One nucleus is approximately 1% egg length. Each small diamond indicates the *eve* concentration in a single nucleus. The left-hand column shows the behavior of the circuit and the right-hand column the *eve* expression data used for the fit. Additional spatially uniform *eve* expression data for early and late cleavage cycle 13 are not shown. The *eve* expression level at early cycle 13 was set to 4.0, and at late cycle 13 to 4.5. Time is measured from the completion of the tenth nuclear division. As measured from this marker, early cycle 13 occurs at 30.66 minutes, late cycle 13 at 48 minutes, early cycle 14 at 57.33 minutes, middle cycle 14 at 73.66 minutes, and late cycle 14 (the onset of gastrulation) at 89.66 minutes (Foe and Alberts, 1983). In the figure, time increases downward as indicated.

use *eve* data for midcycle 14, we included *eve* data for two different times during cycle 13. This ensured that the same total number of data points was available for each gene, so that the data fit was not biased in favor of any particular gene's expression pattern.

Our dataset also includes expression data for *eve*— for the same times at which we have wild-type data. Although, in general, the mutant expression patterns would be very different, in this particular case they are the same because gap expression is unaltered in *eve* mutants and *eve* expression is unaltered in *eve*— until after gastrulation. We incorporate this information into the method as follow: For each evaluation of the cost function, the equations are integrated and compared with data twice—first to the wild-type data and then to the mutant data. When integrating the equation for the purpose of comparing with mutant data, the **T**-matrix elements that describe the regulatory effect of *eve* on all the other genes are set to zero. This describes a mutant that is functionally null but nevertheless synthesizes inactive protein. This procedure incorporates the fact that gap gene expression patterns are unchanged in mutants for *eve*.

A Stripe-Forming Circuit

Application of the simulated annealing procedure determined a set of parameters for equation 2 that gave the closest possible fit to the data. Several simulated annealing runs gave scores that agreed to within 1 percent and nearly identical values for the parameters, so that the results are reproducible and can be reliably taken to represent the global minimum.

Solving the equation using these parameters gave the results shown in figure 13.1. This figure shows the *eve* expression patterns at early, middle, and late cycle 14, as well as some of the experimental data used.

If the only result of applying the gene circuit method were to produce the fit to the data shown in figure 13.1, this exercise would hold little interest. What is important is that the experimental data have been used to determine the parameters defining an entire circuit, which allows one to draw extremely important conclusions about the regulatory effects of the system of genes under investigation.

As explained previously, a gene circuit is characterized by a **T** matrix. The elements of this matrix, as determined by the data, are given in table 13.1. We call attention to five notable features of these matrix elements:

1. The diagonal (autoregulatory) terms of the matrix are positive.

2. The off-diagonal (cross-regulatory) terms are negative. Specifically, all the gap genes repress *eve*.

3. The input from *bcd* is positive for all genes with the exception of *kni*.

4. All *eve* outputs are zero; *eve* does not regulate gap genes.

5. An exception to items 1 and 2 is that a few of the **T**-matrix elements are zero.

Table 13.1 **T** matrix determined by the experimental data in figures 13.2 and 13.3

| Target | Regulator | | | | | |
	Kr	hb	gt	kni	eve	bcd
Kr	+0.34	−0.86	−0.44	−1.6	+0.005	+1.8
hb	+0.013	+0.11	−0.90	−0.23	+0.002	+7.6
gt	−2.5	−0.28	+0.15	−0.24	+0.001	+1.1
kni	−0.33	−4.31	−1.6	−0.076	+0.003	−1.8
eve	−2.3	−3.0	−1.9	−1.4	+0.039	+14.

Note: The *ab*th element of **T** specifies the regulatory effect of *b* on *a*. The last column, with *bcd* as regulator, corresponds to m^a in the equations. These terms may be thought of as an extra column of T^{ab}. Each entry in the table has dimensions of concentration^{-1}.

We can infer directly from this a picture in which there is generalized activation of *eve* expression by *bcd* and general transcription factors. This activation then is modulated by local repression to form stripes.

Control of Stripe Borders by One Gap Domain

The explicit representation of this process in equation 2 means that we can refine this picture further. To do so, one must take a *detailed* look at the spatial variation of gap gene expression and its regulatory effect on *eve*. In analyzing stripe formation, we need to consider only the genes that are expressed in the region where the stripe is formed. The effect of spatially varying regulation of gene *a* on *eve* is given by $T^{eve \leftarrow a} v_i^a$. These contributions are summed at each spatial location, but different gap genes will be making contributions to the sum at the various locations along the anteroposterior axis at which different stripes will form.

The key step is to consider the repressive effects of pairs of gap genes on *eve* expression. The four panels in figure 13.2 show the expression of *eve* at late cycle 14 as given by the gene circuit model. Each individual panel also shows the expression of a particular pair of gap genes, together with their net repressive effect on *eve*. Because repression takes the form of negative contributions to u^{eve} (see equation 2), the repression curves are all on the negative portions of the *y*-axis, and so a peak in a repression curve represents a minimum of repression.

The existence of a minimum of pairwise repression coinciding with stripe 2 (figure 13.2A; repression by *Kr* and *gt*) and stripe 5 (figure 13.2D; repression by *gt* and *kni*) suggests that pairwise repression is sufficient to form a stripe. For stripes 3 (figure 13.2B; repression by *hb* and *Kr*) and 4 (figure 13.2C; repression by *Kr* and *kni*), the situation is more ambiguous: A clear feature in the pairwise repression curve corresponds to the peak of each stripe but, in this case, the feature is a shoulder in the curve rather than a minimum. Because each side of these minima or shoulders of repression is formed by a

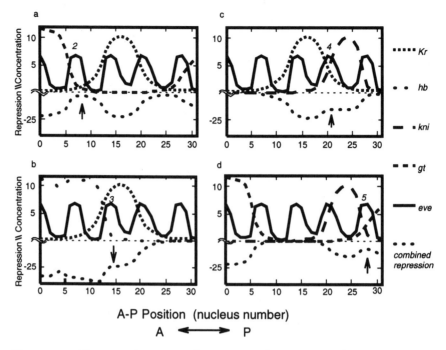

Figure 13.2 Placement of *eve* stripes relative to pairs of repressing gap domains as given by the gene circuit. The *x*-axis of each graph shows anteroposterior position as described for figure 13.1. Positive values along the *y*-axis represent relative concentrations as in figure 13.1, whereas the negative values indicate the net repressive effects of the pair of gap genes shown. Different protein species and net pairwise repression are plotted as shown in the key at the right of the figure. (a) Placement of *eve* stripe 2 (labeled) relative to *gt* and *Kr* expression domains. The combined repression curve shows $T^{eve \leftarrow Kr} v^{Kr} + T^{eve \leftarrow gt} v^{gt}$, the combined repressive effect of *gt* and *Kr* on *eve*. The arrow indicates a gap in repression corresponding to stripe 2. (b) Placement of *eve* stripe 3 (labeled) relative to domains of *hb* and *Kr* expression. The combined repression curve shows $T^{eve \leftarrow Kr} v^{Kr} + T^{eve \leftarrow hb} v^{hb}$, the combined repressive effect of *Kr* and *hb* on *eve*. The arrow indicates a shoulder in repression corresponding to stripe 3. (c) Placement of *eve* stripe 4 (labeled) relative to domains of *kni* and *Kr* expression. The combined repression curve shows $T^{eve \leftarrow Kr} v^{Kr} + T^{eve \leftarrow kni} v^{kni}$, the combined repressive effect of *kni* and *Kr* on *eve*. The arrow indicates a shoulder in repression corresponding to stripe 4. (d) Placement of *eve* stripe 5 (labeled) relative to domains of *gt* and *kni* expression. The combined repression curve shows $T^{eve \leftarrow kni} v^{kni} + T^{eve \leftarrow gt} v^{gt}$, the combined repressive effect of *gt* and *kni* on *eve*. The arrow indicates a gap in repression corresponding to stripe 5.

different gap domain, it is reasonable to suppose that each stripe border is under the control of a particular gap domain. A more exhaustive analysis given elsewhere (Reinitz and Sharp, 1995) shows that this is indeed the case. The resulting picture of the control of stripe borders is summarized in figure 13.3.

The Role of Diffusion

Table 13.2 shows that *eve* product has an extremely low diffusivity. This is an extremely robust feature of the fits whenever each gene product is allowed to

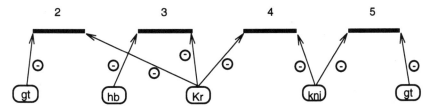

Figure 13.3 Summary of gap gene control of the eight borders of *eve* stripes 2 through 5. Circles containing dashes indicate repression.

Table 13.2 Additional parameters of equation 2 as determined from experimental data

| Parameter | Symbol | Gene | | | | |
		Kr	*hb*	*gt*	*kni*	*eve*
Threshold (effect of general transcription factors)	h^a	-1.5	$-18.$	$+0.59$	-4.5	$+4.4$
Maximum synthesis rate of promoter (minutes^{-1})	R^a	0.49	0.96	1.1	1.1	0.80
Protein half-life (minutes)	$\ln 2/\lambda^a$	19.	8.2	7.6	8.7	6.0
Diffusion operator (minutes^{-1})	D^a	0.18	3.7×10^{-4}	0.070	0.030	7.6×10^{-6}

have its own diffusivity, as opposed to fits in which all proteins were constrained to have the same diffusivity. Our analysis showed that correct formation of *eve* stripes, in conjunction with correctly formed gap domains, *requires* that *eve* have extremely low diffusivity while the gap gene proteins, notably Kruppel, have comparatively high diffusivities. This is consistent with the observation that pair-rule proteins commonly colocalize with transcripts (Riddihough and Ish-Horowicz, 1991), whereas gap gene proteins are more widely expressed than their transcripts. This could reflect immobilization of the protein by cytoskeletal elements. It has been shown that *ftz* RNA is immobilized (Edgar, Odell, and Schubiger, 1987); perhaps the same is true for *eve* protein.

We believe the evidence better supports a different explanation, which has been previously proposed (Edgar et al., 1987; Davis and Ish-Horowicz, 1991). The mRNA of pair-rule genes, including *eve* and *ftz*, is confined to cytoplasm on the apical side of each blastoderm nucleus that expresses it, whereas gap gene message is distributed throughout the cortex of the egg. Hence, the sources of gap and pair-rule proteins are distributed differently in each energid (cytoskeletal unit in the syncytium that will become a cell). Because cell membranes invaginate in an apical to basal direction, this growing geometrical obstacle will affect apically synthesized pair-rule proteins more than gap proteins synthesized both apically and basally. This geometrical effect of subcellular localization in the embryo would map to a sharp difference in diffusivity of gap and pair-rule proteins in our one-dimensional circuit, as illustrated in figure 13.4.

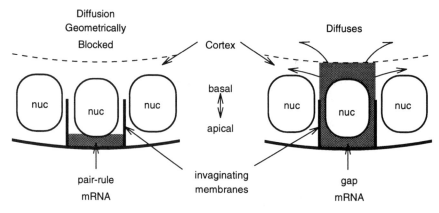

Figure 13.4   Illustration of proposed mechanism linking apical localization of pair-rule message to the observed low diffusivity of Eve protein.

This result provides an interesting example of the contribution that computational methods can make to developmental biology. It has been demonstrated that the mechanism described in figure 13.4 does in fact work: The sequences controlling localization were mapped and used to make constructs in which the mRNA for the bacterial β-galactosidase gene was apically localized, basally localized, or unlocalized (Davis and Ish-Horowicz, 1991). β-Galactosidase protein from apically localized message diffused much less than protein from unlocalized or basally localized message, and it was conjectured that apical localization was required for stripe formation because it prevented pair-rule protein from diffusing. The causal link could not be determined experimentally, however, because the sequences that control localization also control message lifetime. Our result that *eve* cannot form stripes unless it cannot diffuse provided that causal link. This is an important example of a problem in developmental biology that can best be solved by numerical methods in conjunction with experiment.

SIMULTANEOUS SPECIFICATION OF POSITIONAL INFORMATION BY TWO GRADIENTS

The idea that cells in an embryo arrange themselves into appropriate structures at correct locations by interpreting positional information has been widely discussed as an organizing principle in development. The utility of this idea depends on finding a scientifically precise way to implement the notion of interpretation of positional information, especially in the complex setting actually encountered in development.

An early proposal for doing this was put forward by Wolpert (1969) and is based on two ideas. The first is that cell fate is determined by the concentration level of morphogens and that the same cell fate always results from the same morphogen concentration. The second idea is that a gradient in concentration of morphogens is established by diffusion or other mechanisms, so that the fate of a cell depends on its position with respect to the gradient.

The first problem raised by this proposal is to demonstrate that there are chemical substances that function as morphogens, in the sense just alluded to. Early work showed that the concentrations of certain gene products indeed formed gradients but, to show that a chemical gradient is a morphogen gradient, one must perturb the chemical gradient experimentally and see whether cell fate is altered in the perturbed system.

This was done in a classical *Drosophila* experiment (Driever and Nusslein-Volhard, 1988b). The authors took advantage of the fact that levels of protein are proportional to gene dosage for one to three copies of a gene. Specifically, published photometric results (Driever and Nusslein-Volhard, 1988a) indicate that the *bcd* gradient remains close to exponential for one, two, or three doses of *bcd* and that the absolute scale of the gradient is roughly a linear multiple of the copy number, so that changing the *bcd* dose by a factor of 1 or 3 changes the *bcd* gradient by a factor of 0.5 or 1.5, respectively. Driever and Nusslein-Volhard (1988b) used the cephalic furrow and the first *eve* stripe as markers of cell fate. They observed that these markers changed position with changes in *bcd* dosage, as predicted by Wolpert's model, but that the markers did not remain at the same Bicoid concentrations (figure 13.5, curve d). If anteroposterior positional information were autonomously specified by the concentration of Bicoid, one would expect that a given landmark used to assay positional information would always be found at the same concentration of *bcd* product. Thus, the results obtained by Driever and Nusslein-Volhard (1988b) indicate that *bcd* does not autonomously specify positional information. These authors attributed this anomaly to gap gene cross-regulation.

Gene Circuit Analysis

It is a relatively simple matter to model the experiments of Driever and Nusslein-Volhard (1988b) using the gene circuit method. Doing so, we correctly predict the behavior seen in these experiments, and we are led also to an alternative explanation of the experimental results.

To carry out the investigation, we use circuit parameters obtained with fits to wild-type data only, with the normal dosage of two copies of *bcd*. We model the effect of alterations in the *bcd* dose without redetermining the circuit parameters simply by multiplying the Bicoid gradient used by an appropriate scale factor. Because neither *eve* stripe 1 nor gastrulation currently is included in the circuit, we have used the anterior edge of the *Kr* domain at the end of cleavage cycle 14 as a positional information marker. This is a suitable marker because it is invariant with respect to the fate map for the range of genotypes considered: The anterior margin of the *Kr* domain demarcates the posterior boundary of the second *eve* stripe (Stanojevic et al., 1989). Figure 13.5 shows that the model predicts the same relationship between

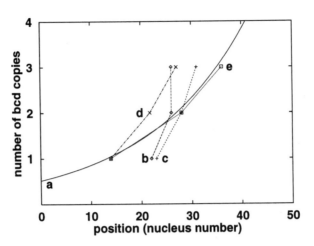

Figure 13.5 The *bcd* dose response. Figure shows that the model correctly predicts the observed *bcd* dose-response curve and also gives the predicted dose-response curve in the absence of *hb^mat*. Gene dosage is plotted on the vertical axis in terms of the number of copies of the wild-type *bcd* gene that are present. The horizontal axis is in the same units as figure 13.1, with the exception that nucleus zero is at the anterior pole rather than at *eve* stripe 1. Five curves are shown. Solid curve a shows the location of a fixed concentration of Bicoid protein at different doses of the *bcd* gene, as given by an amplitude scaling of the exponential Bicoid gradient. This curve is to be compared to curve e (hollow squares), which shows the variation in position of the anterior margin of the *Kr* domain as a function of *bcd* dosage when *hb^mat* was not included in the model. There was no significant variation in this curve among all the fits examined. Curves b (diamonds) and c (plus symbols) show the variation in position of the anterior margin of the *Kr* domain as a function of *bcd* dosage when both *bcd* and *hb^mat* are included in our model. Curves b and c are to be compared to curve d (x symbols), which represents the published experimental data (Driever and Nusslein-Volhard, 1988b) on the dose response. Curves b and c show some variation in dose response but have generally the same shape and orientation as curve d. Note that curve d uses the first *eve* stripe as a landmark, so that curves b and c are expected to be horizontally displaced from d. Curve e was obtained by fitting to data in which *hb^mat* was set equal to zero.

Bicoid concentration and positional information as described by the slightly more anterior markers used in Driever and Nusslein-Volhard (1988b).

This result is interesting, but even more interesting is the next step, which shows how the gene circuit method can be used to learn new biology. To do this, we determined the gene circuit parameters from a dataset that did not include *hb^mat* but with data that was otherwise identical to that used previously. We again varied the *bcd* dosage. We see that in the absence of *hb^mat* (see figure 13.5), the same fate always is found at the same concentration of Bicoid. This result has a straightforward interpretation: Without *hb^mat*, the same fate is found at the same Bicoid concentration. When *hb^mat* is present that is not the case, so *hb^mat* must be the agent responsible for deviations from the Wolpert model. In other words, positional information in the presumptive thoracic and anterior abdominal region of the *Drosophila* blastoderm is determined cooperatively by *bcd* and *hb^mat*.

Cooperative determination of cell fate by these two gradients at first seems surprising: hb^{mat} can be eliminated and embryos survive, but embryos from *nanos* mutant mothers in which hb^{mat} at is expressed uniformly instead of as a gradient are missing abdominal structures. The progeny of mothers mutant for *bcd* are missing head structures. Because the two areas of disrupted body structures are separate, it seemed that these gradients controlled completely different parts of the body plan. Our theoretical prediction (Reinitz, Mjolsness, and Sharp, 1992, 1995) that *bcd* and hb^{mat} cooperate in assigning cell fate along the anteroposterior axis has now been confirmed experimentally (Simpson-Brose, Treisman, and Desplan, 1994). A more quantitative prediction is that in the absence of hb^{mat}, cell fate markers will always be found at the same Bicoid concentration as the *bcd* dosage is varied. This prediction remains to be tested.

CONCLUSIONS

In conclusion, we reiterate two ways in which the work reviewed here has exhibited the utility of gene circuits. First, gene circuits enable us to isolate the effect of a specific gene in a mutually interacting set, as in the work on *eve* stripe formation, or the detailed effects of each of two concentration gradients when these are operating simultaneously, as in the work on *bcd* dosage studies. Second, gene circuits enable us to predict the results of experiments that have not been done or that are quite difficult or time-consuming to carry out. Experiments on embryos lacking hb^{mat} would be an example of the former, whereas biochemical experiments to show that low diffusivity of Eve protein is necessary for proper *eve* stripe formation would be an example of the latter.

There are potentially useful points of contact between the gene circuit method and the approach of Thomas and his coworkers as discussed in chapter 8. Their approach provides a way to decompose regulatory networks into feedback loops and to deduce the qualitative properties of these networks directly from their feedback structure (Thomas and D'Ari, 1990; Snoussi and Thomas, 1993; Thomas et al., 1995). In addition, it affords a way to go back and forth between Boolean and continuous descriptions while preserving the fixed points of the system (Thieffry et al., 1993). Thomas's method requires a mathematical description of the network as a starting point. The gene circuit approach is able to deduce the regulatory circuit directly from data but, as yet, our methods for analyzing the properties of a network are quite problem-specific. Synthesizing the two methodologies into a general-purpose tool is a worthwhile task for the future.

ACKNOWLEDGMENTS

We thank Hewlett-Packard for a grant of equipment to the Yale Center for Computational Ecology and Sun Microsystems for a grant of equipment to the Yale Center for Medical Informatics. This work was supported by grants LM 07056 and RR 07801 from the National Institutes of Health. This chapter also appears as publication no. 171 from the Brookdale Center for Molecular Biology.

SUGGESTED READING

Gilbert, S. (1994). *Developmental biology*. Sunderland, MA: Sinauer Associates. This excellent textbook provides a good introduction to the fly blastoderm and many other important developmental biological problems.

Kirkpatrick, S., Gelatt, C. D., and Vecchi, M. P. (1983). Optimization by simulated annealing. *Science, 220*, 671–680. This classic article serves as an excellent introduction to simulated annealing.

Lawrence, P. A. (1992). *The making of a fly*. Oxford: Blackwell Scientific. Written by a leader in the field.

Reinitz, J., and Sharp, D. H. (1995). Mechanism of *eve* stripe formation. *Mechanisms of Development, 49*, 133–158. The most complete written explanation of the gene circuit approach, including an appendix that describes the simulated annealing method.

14 Fallback Positions and Fossils in Gene Regulation

Boris Magasanik

As shown in the preceding chapters of this book, many of the systems responsible for the regulation of gene expression in both prokaryotes and eukaryotes require the precise interaction of many proteins. Thus, the proper response to changes in the availability of nitrogen by enteric bacteria requires the interaction of six proteins to regulate the expression of the complex *glnALG* operon (table 14.1, figures 14.1 and 14.2). This operon contains *glnA*, the structural gene for glutamine synthetase (GS), as well as *glnG* and *glnL*, the structural genes for nitrogen regulators NR_I and NR_{II}, respectively (Magasanik, 1988).

The expression of the genes of the *glnALG* operons can be initiated at three promoters, *glnAp1*, *glnAp2*, and *glnLp*. In cells growing in a nitrogen-rich medium, the transcription of *glnA* is initiated at *glnAp1*, and that of *glnLG* at *glnLp*. It is regulated negatively by NR_I, the product of *glnG*, which binds to one site overlapping *glnLp* and to two sites overlapping *glnAp1*. In this manner, NR_I maintains its intracellular concentration at approximately five molecules per cell, sufficient to reduce the expression of *glnA* to maintain a level of GS adequate for providing enough glutamine for incorporation into protein but insufficient for its role as sole agent of ammonia assimilation in cells grown in a nitrogen-poor medium. A transcriptional terminator located between the end of *glnA* and the start of *glnL* allows only one of four transcripts initiated at *glnAp1* to transcribe the *glnLG* portion of this operon. Consequently, in cells grown with nitrogen excess, due to the inhibition exerted by NR_I on transcription initiation at *glnAp1*, *glnA* and *glnLG* function as separate units of transcription. In addition, the high intracellular concentration of glutamine causes partial inactivation of GS, due to its adenylylation catalyzed by the enzyme adenylyl transferase (ATase) combined with protein II (P_{II}). P_{II} also combines with NR_{II} and prevents this protein from exerting its ability to phosphorylate NR_I (see figures 14.2A, B).

The replacement of the nitrogen rich medium by one deficient in nitrogen results in a drop in the intracellular concentration of glutamine which, as shown in figure 14.2, enables the enzyme uridylyl transferase (UTase) to uridylylate P_{II}. The resulting P_{II}-UMP combines with ATase to cause deadenylylation and consequent activation of GS, and releases NR_{II} to

Tabie 14.1 Regulators of the activity and synthesis of glutamine synthetase (GS)

Gene	Protein	Function
glnA	GS	Sensor
glnE	ATase	Modulator
glnG	NR$_I$	Response regulator
glnL	NR$_{II}$	Modulator
glnB	P$_{II}$	Signal transducer
glnD	UTase	Signal transducer

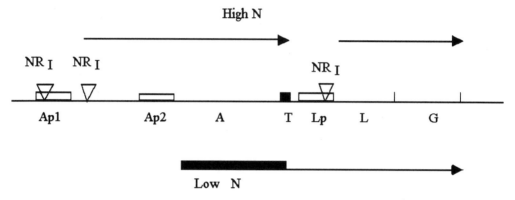

Figure 14.1 Transcription of the *glnALG* operon in cells grown with an excess of nitrogen (High N) or limiting nitrogen (Low N). White boxes indicate promoters; white triangles indicate NR$_I$-binding sites; black boxes indicate terminators. (Reprinted from Magasanik, 1988.)

phosphorylate NR$_I$. NR$_I$-phosphate in turn activates the initiation of transcription at the σ^{54}-RNA polymerase–dependent promoter *glnAp2* from its binding sites overlapping *glnAp1*, which are located 100 and 140 base pairs upstream from the start of transcription at *glnAp2*.

It is possible to move these binding sites as far as 1,000 bp upstream or downstream from the transcriptional start site at *glnAp2* without affecting the ability of NR$_I$-phosphate bound to these sites to activate transcription. Therefore, these sites are prokaryotic enhancers. The phosphorylation enables the dimeric NR$_I$ to form a tetramer that possesses the adenosine triphosphatase (ATPase) activity required for activation of transcription at σ^{54}-dependent promoters (Weiss, Claverie-Martin, and Magasanik, 1992). The strong activation of transcription raises the level of GS to that required to serve as sole agent of ammonia assimilation and raises the level of NR$_I$ from 5 molecules to approximately 70 molecules per cell. Under these conditions, the initiation of transcription at *glnAp1* and *glnLp* is blocked totally, so that *glnALG* functions as an operon, with transcription initiated exclusively at *glnAp2*.

The rise in the intracellular concentration of NR$_I$-phosphate initiates transcription at other σ^{54}-dependent, nitrogen-regulated promoters, also associated with two binding sites for NR$_I$, though these have less affinity for NR$_I$

than those associated with $glnAp2$. Among these promoters are those for operons whose products are uptake systems for glutamine and histidine, as well as enzymes and permeases required for the utilization of nitrate as a source of ammonia. Additional promoters of this type are those for the operon containing the gene for the activator for the σ^{54}-dependent promoters of the nif operons whose products are responsible for the ability of *Klebsiella pneumoniae* to use atmospheric dinitrogen, and for the gene for the activator for certain σ^{70}-dependent promoters for operons whose products are responsible for the degradation of histidine, proline, and urea (Magasanik, 1993).

An increase in the extracellular concentration of ammonia throws the whole process into reverse. The resulting increase in the intracellular concentration of glutamine causes UTase to convert P_{II}-UMP to P_{II}; this in turn stimulates ATase to adenylylate and thus inactivate GS. P_{II} also combines with NR_{II} and causes it to dephosphorylate NR_I-phosphate, halting transcription initiation at $glnAp2$. Growth in the nitrogen-replete medium reduces the intracellular concentrations of GS and NR_I, leading eventually to the partial reactivation of GS and to the initiation of transcription at $glnAp1$ and $glnLp$.

It is clear that every one of the proteins listed in table 14.1 plays an important role in regulating the nitrogen control system. The nitrogen status of the environment is sensed by GS, which generates the intracellular signal, a low or high level of glutamine. This signal is transduced by UTase to P_{II}, which then informs ATase to adenylylate GS or to deadenylylate GS-AMP and informs NR_{II} to phosphorylate NR_I or to dephosphorylate NR_I-phosphate. The final result is regulation of the activity of GS and of the σ^{54}-dependent RNA polymerase bound to $glnAp2$. Consequently, it may be expected that elimination of any one of the six components should result in a crash of the system. It is therefore surprising that most of the components can be eliminated without drastic effects on the growth of the cells.

Elimination of GS by mutations in $glnA$ has the most serious consequences. In contrast to some other bacterial species, the enteric organisms possess only one GS and loss of it incites a need for glutamine. The uptake of glutamine is inefficient and so mutants that lack GS cannot maintain during growth, with ammonia as a major source of nitrogen, an intracellular level of glutamine rivaling those organisms with functional GS. Consequently, in missense $glnA$ mutants growing on glucose, glutamine, and ammonia, NR_I is phosphorylated and raises its own intracellular level to that required for the activation of transcription of nitrogen-regulated operons; in contrast, in nonsense $glnA$ mutants, transcription initiated at $glnAp2$ is aborted and the low level of NR_I-phosphate precludes the expression of nitrogen-regulated operons (Magasanik, 1982).

The effect of the loss of NR_I is not as serious as may have been expected: In cells of the wild type growing on glucose, with ammonia as the source of nitrogen, transcription of $glnA$ is almost exclusively initiated at $glnAp2$, and

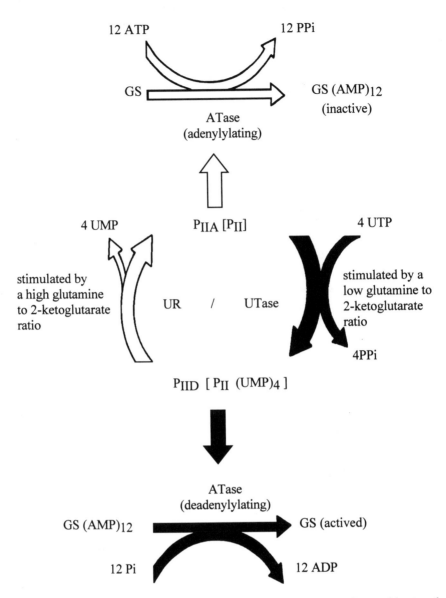

Figure 14.2 (A) Covalent modification of glutamine synthetase. (B) Covalent modification of NR$_I$. (Reprinted from Magasanik, 1988.)

so the lack of NR$_I$ could have resulted in insufficient GS and therefore a requirement for glutamine. However, the loss of NR$_I$ relieves the inhibition exerted by NR$_I$ on the initiation of transcription at *glnAp1*, allowing sufficient GS to be produced for it to be incorporated into protein and to serve as a nitrogen donor in the synthesis of purine and pyrimidine nucleotides and certain amino acids. In that way, *glnAp1* serves as a backup system for *glnAp2*. Nevertheless, the loss of NR$_I$ precludes the utilization of poorer nitrogen sources, such as dinitrogen, nitrate, and histidine.

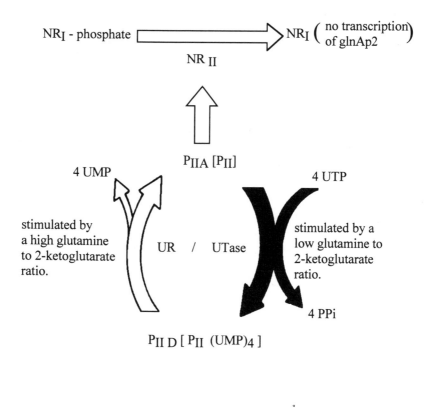

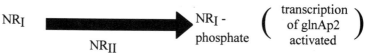

Figure 14.2 (continued)

The loss of NR_{II}, because of its role in the phosphorylation of NR_I, should have affected the activation of transcription initiation at *glnAp2* in the same manner as the loss of NR_I. However, a deletion of *glnL*, the structural gene for NR_{II}, has surprisingly little effect on the regulation of nitrogen-responsive genes and operons. The only effect of this loss is a very slow activation of transcription at *glnAp2* in response to nitrogen deprivation and an equally slow termination of this initiation when the culture is supplied with a good nitrogen source. The fallback position in this case is the phosphorylation of NR_I by acetylphosphate and the autogenous dephosphorylation of NR_I-phosphate (Magasanik, 1993).

The fact that the incubation of NR_I with acetylphosphate results in the correct phosphorylation of NR_I at the aspartate in position 54 shows that NR_I itself is the catalyst for its phosphorylation. It can accept a phosphate group from acetylphosphate but not from ATP. The role of NR_{II} is to catalyze the transfer of the γ-phosphate of ATP to its histidine residue in position 139, from which it can be transferred to aspartate 54 of NR_I by NR_I. The other important role of NR_{II} is to catalyze, in association with P_{II}, the rapid removal of the phosphate group from NR_I-phosphate. However, this function is not essential because, even in the absence of NR_{II}, NR_I-phosphate catalyzes its own dephosphorylation. Nonetheless, the presence of NR_{II} and P_{II} shortens the half-life of NR_I-phosphate from 5 to 10 minutes to 1 to 2 minutes (Liu and Magasanik, 1995).

The increased phosphorylation of NR_I in cells lacking NR_{II} in response to nitrogen depletion reflects an increase in the intracellular concentration of acetylphosphate. Such an increase is caused by the continuing catabolism of the major energy source, glucose, in the face of slowed use of the catabolites for synthesizing macromolecules. The intracellular concentration of acetyl-phosphate and, consequently, the concentration of NR_I-phosphate by its autocatalytic dephosphorylation decline in the face of restoration of a good nitrogen source. However, when NR_{II} is present, acetylphosphate plays no role in the phosphorylation of NR_I because NR_{II}, in conjunction with P_{II} and UTase, controls both the phosphorylation of NR_I and the dephosphorylation of NR_I-phosphate in response to the intracellular concentration of glutamine (Feng et al., 1992).

An important reason for controlling the activity of GS is the danger that highly active GS would deprive the cell of essential glutamate by converting it to glutamine more rapidly than it can be synthesized from α-ketoglutarate and ammonia. The controls are of two types: regulation of the activation of *glnA* expression by NR_I-phosphate and regulation of GS activity by ATase. As shown in figure 14.2, P_{II} plays an important role in both cases: It stimulates the dephosphorylation of NR_I-phosphate, which results in the arrest of transcription initiation at *glnAp2*, and it stimulates the adenylylation of GS by ATase, which results in the inactivation of GS. Nevertheless, a mutation in *glnB* leading to the loss of P_{II} does not have a deleterious effect on the growth of the affected cells. Although these cells have a high level of GS during growth in a medium with an excess of nitrogen, GS is largely adenyly-lated and therefore inactive. The effect of the lack of P_{II} on the adenylylation of GS is apparent only when one compares the rate of adenylylation of GS in these cells with that in wild-type cells following the addition of ammonia to such cells growing on a poor source of nitrogen: The rate is faster in cells containing P_{II}. This effect may be explained by the fact that glutamine stimu-lates the adenylylation of GS by ATase even in the absence of P_{II}. Although the deadenylylation of GS-AMP by ATase requires P_{II}-UMP, the lack of P_{II}-UMP in the *glnB* mutant does not prevent but merely slows the dead-enylylation of GS-AMP when the cells are deprived of ammonia. Apparently, there are nonspecific diesterases present in the cell that are able to catalyze this reaction (Foor, Reuveny, and Magasanik, 1980).

The lack of UTase resulting from a mutation in *glnD* makes it more difficult for the cell to increase the level of GS in response to nitrogen deprivation but does not result in a complete inability to respond with such an increase. One possible explanation for this is the fact that, in the absence of UTase, glutamate is required for P_{II} combined with NR_{II} to bring about the dephosphorylation of NR_I-phosphate. In nitrogen-deficient cells, the low intracellular concentration of glutamate may prevent the interference of P_{II} with the ability of NR_{II} to phosphorylate NR_I (Liu and Magasanik, 1995).

Finally, the lack of ATase resulting from a mutation in its structural gene *glnE* has almost no effect on the growth of the cells. In this case, the regulation of transcription initiation at *glnAp2* serves as a fallback position, preventing excessive GS activity. The effect of the lack of ATase becomes apparent only when ammonia is added to a culture of the mutant grown on a poor nitrogen source and therefore containing a high level of GS. In the case of the wild-type cell, this addition results in the immediate inactivation of GS and in an almost uninterrupted continuation of growth; in the case of the mutant, the loss of glutamate resulting from its conversion to glutamine by the active GS causes a long period of slow growth in order to reduce the level of GS sufficiently to allow the resumption of rapid growth (Kustu, Hirschman, Burton, Jelesko, and Meeks, 1984).

The deliberate design of complex regulatory systems by engineers generally will include the provision of backup mechanisms to compensate for the failure of a component of the system, but how can one account for the existence of biological backup mechanisms? It is my contention that the backup mechanisms discussed in the preceding pages are fossils of earlier stages in the evolution of this complex and elegant system for controlling the response to nitrogen availability. Possibly the activity of GS was regulated initially through adenylylation by ATase in response to glutamine and that of NR_I by its autophosphorylation in response to acetylphosphate. The next stage in the evolution of the system may have been the appearance of NR_{II} and P_{II} regulating the phosphorylation of NR_I and the dephosphorylation of NR_I-phosphate in response to the intracellular concentration of glutamate. P_{II} may also have interacted with ATase, perhaps in endowing the ATase with the ability to deadenylylate GS-AMP in response to a low intracellular level of glutamate. The final stage in the evolution of the system would have been the appearance of UTase, capable of acting on P_{II} in response to the intracellular concentration of glutamine. It will be of interest to discover whether fallback positions in other complex regulatory systems provide any clues to the evolution of these systems.

SUGGESTED READING

Hoch, J. A., and Silhavy, T. J. (Eds.). (1995). *Two component signal transduction*. Washington, DC: ASM Press. Contains three chapters providing a more detailed description of nitrogen regulation: "Historical Perspective," by B. Magasanik; "Control of assimilation by the NR(I)-NR(II) two-component system of enteric bacteria," by A. Ninfa; and "Mechanisms of transcriptional activation by NtrC," by S. C. Porter et al.

15 The Language of the Genes

Robert C. Berwick

IS DNA A LANGUAGE?

Both DNA and what people speak are commonly referred to as *languages*. The analogy holds, at least in the formal sense. Both DNA and human languages encode and transmit information. Both, like beads on a string, form concatenative symbol-systems. Murkier by far is how much further down the scientific road this analogy can carry us. That is the question this chapter tries to answer: Is there indeed a "language of the genes"? Can linguistic science repair what Collado-Vides (chapter 9) correctly pinpoints as the weak link in current molecular biology—namely, the relative poverty of *explanatory* molecular biology, as opposed to *descriptive* molecular biology? Modern molecular biology's reductionism comes at a steep price, leaving us chock full of complex visibles but largely bereft of corresponding simple invisibles. Why do the bacterial sigma 70 and 54 promoters look this way rather than some other way? To be sure, evolution and physical science ultimately fix these answers. Even so, explanation-seeking scientists rightly posit intermediate, theoretical selections to account for such things as a quark's spin, an electron's valence or, more to the point, a person's genes.

Our initial question about genetic language—regarding whether linguistics can shed light on molecular biology—must have two simple answers: Yes, molecular biology can benefit from linguistic science (as Collado-Vides notes) simply by providing the right general scientific scaffolding, including modern linguistic theory's modularity principles and parameterized abstraction. Specifically, as we describe later, the mechanics of both molecular biology and natural languages are grounded on the notion of *adjacency* as a fundamental principle; there is no "action at a distance" (figure 15.1). Just as the intron and extron machinery pastes together previously disconnected pieces of genetic code into an adjacent working whole, grammatical relations such as the agreement between sentence subject and verb (in the figure, between the plural *s* ending on *the guys* and the *non*existence of an *s* on *like*) are defined only under strict adjacency, and almost the whole point of Chomsky's transformational grammar is to paste together previously disconnected sentence elements.

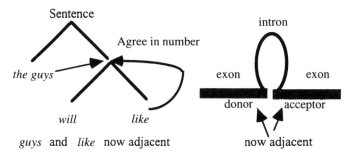

Figure 15.1 Both natural languages and the genetic language contain machinery that converts superficially "long-distance" relationships into adjacent ones.

However, no is an equally correct answer to our fundamental question because, as we shall see, natural languages form a much simpler computational system than the genetic code and transcription machinery. In a nutshell, whereas transcription exploits the three-dimensional twists and turns of biochemistry and resembles a general programming language (as noted in chapter 3), in contrast our current understanding is that natural language exploits *only* adjacency as its programming "trick." Adjacency is enough to derive (hence, explain) most of what we see in natural languages. No such corresponding explanation of why the genetic "programming language" looks the way it does has been forthcoming. The conclusion, then, is that the language of the genes is not like a natural language but more like a general programming language, the details of which we still do not fully understand. It is akin to looking at the input and output of a spreadsheet and, from that, trying to figure out not only the specific programming language instructions used but also which programming language was used—whether C, Fortran, or Pascal. As Lewontin notes in chapter 1, to understand this is probably the most difficult task of reverse engineering that anyone has ever undertaken. If this insight is accurate, it suggests that molecular biologists might do better to study the methods used by "clean-room" programmers to reverse-engineer spreadsheet programs than to try to figure out whether DNA or its transcription mechanisms generate certain kinds of non-context-free languages. More specifically, if we search through the space of context-free or non-context-free languages, we are simply searching through the *wrong* space. For natural languages, this is a space of restricted adjacency relations (described later). For the language of the genes, the appropriate representation is not yet clear, but it may be that something like the space of genetic "circuits" is more fitting (see chapters 6 and 13; McAdams and Shapiro, 1995).

The remainder of this chapter expands these points. First, we review the possible connection between the genetic code and formal language theory, showing that formal language theory serves as a poor proxy for studying programming languages and natural languages and hence is an unlikely candidate for investigating either one. The argument carries over to attempts to detect various patterns in the genetic code via different kinds of pattern-matching languages. Here (as discussed in chapter 3) many popular algorithms are based on linear string matching, including so-called hidden Markov models. Though such linear models *have* been successful in mirroring some aspects of human language, it is crucial to observe that these linear models have largely been successful in modeling speech—that is, exactly that area of human language that is strictly linear and left-to-right. Second, we turn to the differences between genetic transcription and natural languages, demonstrating how much simpler natural languages are than DNA transcription. We also demonstrate that by using a more appropriate representation—defined over four natural configurations such as *subject*—one can build better search routines for natural language patterns. Finally, we argue that the language of the genes might best be expressible as a programming language or some such constraint system, perhaps like the genetic circuits discussed elsewhere in this book. This is an area for future research.

FORMAL, NATURAL, AND BIOLOGICAL LANGUAGES

Because DNA *is* a formal language, there is a natural temptation to wheel out the armamentarium of formal language theory, but can formal language theory help us understand DNA? To answer this question, one must first understand why formal language theory was invented. Elsewhere the argument has been made (Berwick, 1989) that it is rash to expect a complex biological system such as human language to abide by elegant mathematical rules such as those that define the Chomsky hierarchy of finite-state, context-free, context-sensitive, and Turing complete (arbitrary programming) languages. The Chomsky hierarchy itself is the wrong way to size up natural languages: Languages simply don't fall neatly into one of these classes. If that is so for human languages, then it is doubly so for DNA: Indeed, as we discuss later, although there is at least some new support for an elegant algebraic description for the "core" of natural language syntax, in this regard at least, DNA seems *more* complex than natural languages.

Formal Description of Transcription

There have been some efforts (see Searls, 1993, for a particularly illuminating and insightful study) to determine whether DNA, tRNA, their various substructures, or transcription machinery itself falls into one or another of the well-known formal language theory classes. To understand the results of such studies, we it would do well to recall both what formal language theory classes define and what role formal language theory played in aiding linguistic theory and in programming languages. Formal language theory was used in the 1960s to study both formal linguistics and the complexity of programming languages, but it has not been used much since then, because computer science has developed much keener methods for analyzing computational complexity.

finite-state: linear
concatenation

context-free: hierarchical

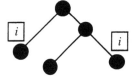
context-sensitive:
hierarchical with labels

Turing complete:
any computation

Figure 15.2   The Chomsky hierarchy: from linear to arbitrary (Turing complete) languages.

Formal Language Theory: A Brief History

Broadly speaking, formal language theory and especially the Chomsky hierarchy served as a rough *proxy* for particular complexity analyses and structural properties of both programming languages and linguistic relations. The hierarchy's relationship to computation itself is indirect. The hierarchy consists of four increasingly complex structural relations that define strict subsets of string classes (languages), as shown in figure 15.2: purely *linear concatenative* relations, or *finite-state languages*; purely nested or hierarchical, treelike relations, or *context-free languages*; tree structures augmented with labeling tags that can refer to each other across arbitrary parts of the tree, or *context-sensitive languages*; and completely arbitrary relations or arbitrary programming languages, so-called *Turing complete languages*, that can compute anything that a general programming language (such as Fortran or C) can compute.

Used diagnostically, these classes are a blunt knife because human languages do not fall neatly into any one of these classes; for example, it is by no means clear that human languages need even be computable, in the strict sense, although presumably this is so. Natural languages certainly contain recursive, hierarchical structures or *phrases*, such as "the different types of RNA polymerases," in which the group of words *of RNA polymerases* is clearly a substructure that modifies *different types*—so natural languages are at least describable as context-free languages. Beyond this, however, this blunt taxonomy has yielded very few concrete results for linguistics. Chomsky (1956) more or less established that human languages cannot be contained in the class of finite-state languages. Similarly, Searls (1993, p. 73) shows that nucleic acids are more complex than simple linear finite-state languages: They encode palindromes, embeddings, and the like (with one technical caveat that we elucidate later). Though this is an interesting discovery about nucleic acids, and though it does suggest that pattern-matching techniques for analyzing sequences will have to do more than just look at linear models, again it is important to ask whether we gain by this any new insight. Searls (1993) himself notes that it does not gain us much. The real question is whether formal language theory could ever hope to tell us much.

The answer to this last question for linguistic theory has been plain. Beyond Chomsky's original discovery (1956), formal language theory has not contributed substantially to our understanding of human language structure. Chomsky showed that linear analysis does not suffice to model human language; we need at least some notion of hierarchy. (In a later section, we show just what kind of hierarchy is required). The problem with going beyond this is that the formal language theory classes do not correspond to human languages. An infinite number of context-free (strictly hierarchical) languages are not natural languages, and these include sequences found in nucleic acids. For instance, consider the example that Searls (1993) uses to show that nucleic acid sequences are not purely linear, or finite-state: palindromes, or mirror-image, nucleic acid sequences. Such sequences are very easily generated by simple first-in, first-out push-down stacks—like placing a pile of dinner plates one on top of another and then removing the last one put on first—so one might expect to find such patterns, in the form $w_1 w_2 w_3 w_3 w_2 w_1$ or a *nested dependency*, in human languages. Instead, we find that a pattern more commonly found in natural languages, as in German or Dutch, is the *opposite* of push-down stack order—that is, the pattern $w_1 w_2 w_3 w_1 w_2 w_3$ or an *intersected dependency*. Evidently, this intersected pattern can also be found in some of the substructures of gene regulation. In this sense, nucleic acids patterns are *not* like natural languages—they contain more than do natural languages. Similarly, computer programming languages such as Fortran are syntactically context-free yet are certainly not natural languages; unlike natural languages, they require explicit instruction to learn, as any beginning programmer could tell you.

Of course, there also is no reason to believe that natural languages are some subset of the context-free languages. In hindsight, formal language theory turned out to be eminently helpful in describing programming languages but not natural languages. One might then wonder why formal language theory was wheeled out at all to attack the problem of natural languages. This appears to be simply an instance of the "lamplight fallacy"—looking where the mathematical light shines brightest—as is discussed elsewhere (Berwick and Weinberg, 1979): Researchers turned to formal language theory because it had clean mathematical properties and none of the unruly tangles of human language.

This aesthetic urge still surfaces even in the recent formal demonstrations about nucleic acids mentioned earlier. Most commonly, the argument runs this way: (1) We isolate some subset pattern in, say, English; (2) we show that this subset pattern has property P; and (3) we conclude therefore that English has property P. For example, for nucleic acid sequences, we might point out, as Searls does, that they contain palindrome sequences (step 1). Because palindromes cannot be generated by any finite-state or purely linear automaton, but can be generated as a strictly context-free language, we could, following step 2, identify *non-context-free* as the property P we want to isolate. Finally, according to step 3 of the argument, we conclude that nucleic acid sequences are not finite-state. However, this argument is flawed: Whereas this subset of the nucleic acid sequences is not finite-state, it does *not* follow that the entire system is not finite-state. To explain further, note that the language of all possible nucleotide sequences of A, T, G, C—that is, the language Σ* defined over the alphabet A, T, G, C—is certainly a regular or finite-state language but just as surely contains palindrome sequences, because it contains *all* possible sequences. To make the three-step argument apply, one must *intersect* the language studied—English or nucleic acid sequences—with some specially constructed filter designed to pick out just those sequences we know to be palindromes. Of course, this filter itself must be finite-state, and we must show that the filtering operation also is finite-state or regular (in the usual proofs, one can use set intersection as the filter because finite-state languages are closed under intersection), otherwise, we could introduce spurious non-finite-state complexity. To be sure, this point about subset properties is not easy to see. In fact, even Chomsky's original demonstration that English is not finite-state (1956) suffered from exactly this fallacy: Chomsky demonstrated that English contained patterns that were *not* finite state but did not precisely spell this proof out via intersection with an English-like "test pattern" so that the proof would apply formally. At least on first glance, then, Searls's demonstration (1993) that DNA is not context-free contains the same problems. However, it is usually easy to patch such proofs, so this is not meant as a damning critique. Rather, we should remain aware that it is too easy to single out mathematical purity at the expense of biological reality: Formal language theory does not naturally correspond to the theory of human languages, and we should not expect it to.

The Case of Hidden Markov Models

Another lesson to be learned from the lamplight fallacy relates to currently popular methods such as neural networks and hidden Markov models for discovering structure in sequence data. Here too one must be extremely careful in considering the assumptions about sequence or linguistic structure that these models make; otherwise, one will get back only what these models are able to find.

Consider the case of hidden Markov models (HMMs). These are a subcase of the finite-state languages (i.e., a linear sequence of states) but with the addition of *probabilities* on state transitions (which are hidden from our explicit view; hence the term) and associated probabilities on the actual output letters (e.g., the base alphabet) that are observed. The rough idea behind the HMM method is an update "learning" loop based on Bayes's rule: Start with some prior estimate of the hidden transition probabilities between states (say, a uniform one that assigns equal probabilities to all transitions) and then update those probabilities based on counting the sequences that are actually found, as opposed to those that are not found. (The exact method uses the Dempster-Shafer expectation maximization, or EM, algorithm.) After some initial set of sequence data has been processed in this way, we arrive at some "final" estimate of the hidden state transitions., which then can presumably be used as a more accurate reflection of the "true" model underlying the sequence generation.

To understand what HMMs can buy us in both the linguistic and the molecular biology worlds, we must understand their limiting assumptions. First, HMMs make strict assumptions about the generative processes creating the observed nucleotide sequence—namely, that it is a linear, memoryless process. Clearly, this does not encompass the long-distance cutting and pasting of intron and extron machinery, let alone more complex transcription programs. Thus, HMMs can discover linear patterns or classifications, but we cannot expect them to discover the transcription "program" because HMMs cannot even represent such sophisticated properties. Further, the EM search method has its own limitations: It is a local-gradient-ascent, or hill-climbing system; the probability estimation algorithm tries locally to improve its current estimate based on where it currently is in a search space. Such an algorithm is guaranteed to find a maximum, or best estimate, but only a local maximum. If the search space contains sharp ridges or peaks, then the algorithm can get stuck there (one reason why heuristics and parallel search methods such as those described in chapter 3 often are appealed to).

Not surprisingly, then, for natural languages, HMMs have been most successfully used for precisely those representations that are linear—namely, sound sequences. They are used for speech recognition because, for the most part, one single articulated sound depends on just the one or two sounds preceding it. For more complex linguistic descriptions that go beyond local linear descriptions, HMMs perform much worse. For example, consider a sentence such as, "How many guys do you think were arrested?" (in which *guys* and *were* must both be plural, we say that they agree in number). Note the problem with a sentence that violates this constraint, such as "How many guys do you think was arrested?" Here, there is a relation between *guys* (the subject) and *were* or *was* (the verb) separated by a long distance. The whole point of modern transformational grammar (indeed, all modern grammatical theories) is to propose descriptive levels where these two elements are brought into adjacency (so that their features can be checked for agreement). In this case, Chomsky's modern transformational theory posits an unpronounced element (seen in the representation, but not heard) that serves as the object of *arrested* and is linked to *guys*: "How many guys$_i$ do you think were arrested [empty]$_i$" (where the index i indicates the link). Using this representation the verb form *were* and the word *guys* are adjacent to one another. However, the operation that puts them together—a transformation—is not linear or local: The single transformational operation in current linguistic theory says that one can move a phrase such as *many guys* anywhere. This is beyond the descriptive power of HMMs, because HMMs, by definition, describe *memoryless* processes and, in an example such as this, one has in effect "remembered" the position of the object of *arrested* so as to link it arbitrarily far away from the position where *many guys* actually is spelled out. Thus, we would not expect HMMs to provide a good discovery procedure for such linguistic relations.

One way to shore up the weaknesses of linear HMMs is to add some notion of hierarchical patterns. This has been affected in the basic EM algorithm and is used also in hierarchical pattern-matching algorithms; for natural languages, the analog is to use stochastic context-free grammars. However, here too one can show that these methods work mostly to the extent that the right hierarchical structure is prebuilt into them. The "topology" of the relations—what variable is linked to what other variable—must be understood in advance; otherwise, the search algorithm will not find the correct representation for us.

For instance, it can be shown that the EM algorithm simply will not find the right structure for a simple phrase such as *walking on air* (the true structure being a verb phrase in the form verb–prepositional phrase, with the prepositional phrase subtree consisting of the preposition *on* followed by the noun *air*). Instead, it will converge to a local minimum, wherein the verb is clustered erroneously with the preposition as a unit, apart from the noun *air*. If one examines more closely just why this is so, it turns out that the context-free rule space is not searched completely by the HMM algorithm; instead, there are two "peaks" or local maxima that force the system to cluster either the verb with the preposition first (the wrong result) or the preposition with the noun (the right result), and most of the space leads one to the first, erroneous conclusion. This search space is simply the wrong one to look at. Put another way, context-free rules seem to be the wrong representation to describe the linguistic relations in this case, and therefore no amount of clever searching can repair the representational defect. The right move is to use the correct representation from the start, to say that phrases consist of a particular grammatical relation—the function-argument relation (the relation between the verb *walking* and its object, such as the whole unit *on air*; or the relation between a preposition such as *on* and its object, such as *air*. As usual in artificial intelligence, finding the correct representation is 90 percent or more of the battle; it is the cornerstone of building theories. The search engine is secondary.

Turning now to the biological world, the same morals carry over. HMMs can find only linear patterns. Stochastic context-free grammars are far too broad a class of hierarchical patterns, so it is likely that search engines grounded on these will miss important transcription programs. What we need to understand first is the vocabulary of the transcription programs before we go looking for the programs themselves. It is unlikely that these insights will come from general inductive inference methods, except in an exploratory sense.

Neural Network Models

What about neural network (NN) approaches? Here too it is important to understand what work NNs can do. They cannot work magic. Today it is widely known that what they provide is function approximation: Given a set of data, NNs fit that data to a particular curve. For example, in the simplest case, it is known that a single-layer NN (one intermediate layer, one set of inputs, and one output) actually is carrying out classical principal components analysis. Conceptually, the picture is this: Given some cloud of data in, say, x, y, z space, where z is the dependent variable to be explained in terms of the variables x and y (we can think of z as the nucleotide sequence and x and y as factors that account for the observed sequence), then the network learning algorithm finds two things. First, it finds two *axes*—the principal components x' and y'—that optimally account for the dependent variable z. These components correspond to the NN "units" or "cells." Second, the system finds the optimal *weights* to assign to each component to give the best fit to the data z. These correspond to the weights assigned to each NN cell or unit.

In this sense, NNs are doing statistical curve fitting. As statisticians know, one cannot build a good statistical model out of thin air: One has to know something about which variables might be related to which other variables. If one starts with a poor set of hypothesized variables x and y to explain z, then the NN search method cannot save us. For instance, if these are (obviously) poor descriptors such as say, the number of stop sequences, then no amount of NN learning can inform us adequately. In this sense, like HMMs, NNs can greatly help us explore a space of possible theories and can be extraordinarily efficient search engines for finding patterns in sequence data. Nonetheless, in the area of natural languages, NNs have not proved to be very useful except in the same places HMMs have been—for instance, in building systems that learn how to map text to speech. This is true, as it is for HMMs, because the topology of simple NNs best reflects the literally linear properties of a spoken sound sequence. Though there have been attempts to capture some of the hierarchical structure of human language via such networks (using recurrent [i.e., recursive or reentrant] nets), such attempts have been generally unsuccessful. If it is true that genetic transcription is far more sophisticated than natural language—as we show in the next section—then this result means that NNs will never give us the correct answers about transcription. What is needed is a new theory about the space of transcription programming language constructs.

In sum, NN learning algorithms can be efficient search engines for *existing* theories about linear language or nucleotide sequences and transcription, but their value for higher-order natural language or sequence constructs is more dubious. NNs can suggest possibly valuable new combinations of proposed theoretical variables, just as does principal components analysis does. However, NNs cannot invent new theoretical variables out of whole cloth. Once again, starting with the correct representations, the right search spaces, is *the* most important factor.

THE SIMPLICITY OF NATURAL LANGUAGES AND THE COMPLEXITY OF GENETIC LANGUAGE

If natural languages and the language of the genes are not formal languages, then what are they? We have mentioned several times now that both nucleotide sequences and the transcription machinery itself seem more akin to a programming language than to natural languages, and that natural languages may be much simpler than the language of the genes. In this section, we show exactly how simple natural languages may be—specifically, that natural language syntax might be grounded on just a single, simple, computational combinatorial operation. Further, this operation, which seems central to all grammatical relations, does not seem to be directly reflected in the language of the genes.

Natural Grammatical Relations

Let us begin by defining what we mean by natural grammatical relations, the relationships that natural language syntax does seem to use. Surprisingly, there seem to be relatively few central relations (perhaps only four), defined over a local domain of binary branching tree structures. This constraint is interesting because, of course, given arbitrary tree structures—such as those available if we posited arbitrary hierarchical relationships—there could just as well be an *infinite* number of distinct grammatical relations. Yet most of these are not ever used in natural languages. For instance, we could well imagine that there is a relation between, say, the subject and the object of a sentence. Indeed, if we adopted an HMM or a context-free grammar model, there is nothing at all to block such a relationship. It is in this sense that HMMs and context-free grammars are too general and therefore cannot explain why natural languages are the way they are rather than some other way. Still worse, from the point of view of discovery procedures, is the fact that search algorithms that use only the space of possibilities defined by HMMs or context-free grammars use the wrong space.

What kind of space is right then? Here we can follow recent work of Epstein (1995). The basic natural grammatical relations are perhaps best exemplified by a simple picture, where X and Y denote *nodes* or entire subtrees or phrases, such as sentences, noun phrases, or prepositional phrases. We first note that the configurations are all binary branching (not a necessary property of tree structures generally).

Reviewing the configurations in figure 15.3, the first relation is essentially that of verb-object, or preposition-object (e.g., *ate ice cream* or *on the table*); more generally, this is the function-argument relation. The second relation is almost that of tree dominance, which is essential for hierarchical description,

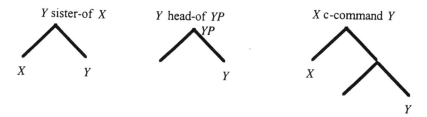

Figure 15.3 Three basic grammatical relations in natural language syntax.

as in a phrase such as *ate ice cream*, where the entire phrase is a verb phrase, denoted by the tree node *YP*, and *Y* is a subpart of the tree—in this case, *ice cream*. This is actually the notion "head-of": Note that perhaps the most prominent property of a phrase—its type—depends on the feature propagated or inherited from the word that heads it up. For example, a verb phrase such as *ate ice cream* is built around a scaffolding that consists of first a *verb*: That is, we may think of the lexical property of the verb as being propagated up from *Y* to *YP* (= a verb phrase, or VP) in the figure 15.3. In this sense, this second relation defines the *kinds of phrases* that one can find in a language. The third relation may be more unfamiliar to nonlinguistic readers but is, in fact, one of the most important in natural language syntax: It is dubbed *constituent command*, or *c-command*: A node *X* c-commands a node (phrase) *Y* just in case the first branching node that dominates *X* also dominates *Y*. In our picture, *X* does c-command *Y* (because if we go up to the first branching node that dominates *X*, we find that this dominates *Y*), but *Y* does *not* c-command *X* (so the relation is not symmetrical). Intuitively, c-command is the notion of scope in natural language, similar to the notion of scope in logical calculi or programming languages: C-command defines the domain over which a variable can be bound. In natural languages, this corresponds to sentences such as "Whom did John think that Mary saw?" which can be rendered roughly as, "For which z, z a person, did John think that Mary saw z?" where the variable z is linked to *whom*. Note that if one drew out the syntactical structure for this sentence, we would have something akin to figure 15.4, wherein the variable z is c-commanded by *whom*. Because this kind of linking shows up again and again in modern linguistic theory as the foundation of what used to be called *transformations*, one can see that this configuration is an important one. These basic relations—function-argument, head-of, and c-command—seem to be the primitive building blocks for all other linguistic relationships.

Explaining the "Natural" Grammatical Relationships

We next show, following Epstein (1995), that these basic relationships all are accounted for by a single elementary computational operation based on the adjacent concatenation of tree structures. (The syntactical reflex of this idea was first proposed by Epstein [1995].) Note that this is a "natural" result in

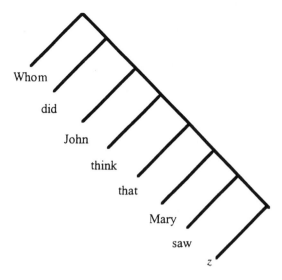

Figure 15.4 The c-command relation between *whom* and the (unpronounced) variable z, the object of *see*, is like the relationship between a quantifier and the variable it binds.

the sense that we know, on independent grounds, that human language syntactical structure is treelike (rather than purely linear, like beads on a string). Note also that if this result is correct, it automatically explains why HMM models based on linear concatenation do not do a very good job of accounting for human syntax. The central idea here is that hierarchical concatenation is the chief operation we need to derive natural language sentences. To show how this works, let us see how a sentence such as "John likes the ice cream" might be derived.

Hierarchical Concatenation
Let us first describe the concatenation operation itself. It is simply a *bottom-up tree composition*: We take two subtrees, X and Y, and "glue" them together, forming a new larger tree in a special way: Either the features of X or the features of Y are propagated to the new larger tree, forming a node of either type XP or YP. For example, suppose we have a verb *eat* (actually a subtree), and a subtree corresponding to *the ice cream* (a noun phrase, or NP). We combine these to form a VP. This abstract combination of X and Y as well as the specific example combining a verb and a noun phrase are shown in figure 15.5. The left half of the figure shows the two initial subtrees, drawn as triangles. The first triangle consists of the verb; the second consists of the noun phrase. The right half of the figure shows the result of the combinatory operation that glues these two triangles into a single larger one: (1) The features of the verb subtree are propagated up to a new node, the verb phrase (VP) node, that is the root or top of the new larger triangle, labeled Z; (2) we represent this top most point via a special set notation that marks *likes* as the "head" of this phrase; (3) the noun phrase subtree is pasted in place below.

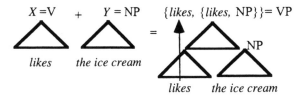

Figure 15.5 The basic operation of hierarchical (tree) concatenation. V, verb; NP, noun phrase; VP, verb phrase.

We dub this operation *bottom-up* because it pastes two smaller adjacent trees into a single larger one; it is *computational* in that we take this to be the operation of a parser proceeding from left to right through the sentence. In fact, this operation corresponds to one of the most common ways of parsing programming languages, so-called *LR parsing*, in which we paste together larger trees out of smaller ones, as shown in figure 15.5.

Deriving a Full Sentence

In this view, then, the derivation of a sentence proceeds by a sequence of hierarchical concatenation sets (what were called *derivation lines* in the original theory of Chomsky [1956]). In this case, the derivation steps are as follows, where by *form* we mean "construct a hierarchical structure like the triangle dominating *likes*":

1. Form the hierarchical (triangle, "tree") representation for *John*

2. Form the hierarchical representation for *the ice cream* (= Y in figure 15.5)

3. Form the hierarchical representation for *likes* (= X in figure 15.5)

4. Concatenate X and Y, forming an extended verb phrase, Z, corresponding to *likes the ice cream*

5. Concatenate Z with the triangle representation for *John*, forming a complete sentence, "John likes the ice cream"

In summary, note that thus far the whole sentence is derived by a sequence of hierarchical concatenation steps, and only these.

Deriving Grammatical Relations

The important point now is to show that if we assume this operation of tree concatenation to be the basic primitive of syntax, then it follows that the only grammatical relations we see will be precisely those described earlier. In this sense, we may say that natural language syntax uses only a single operation of hierarchical, adjacent tree concatenation.

Let us see why *these* basic relations follow and no others. The central insight is that two elements may be *related* in the grammar if and only if they are adjacent or visible to each other at the time of tree concatenation: that is, at the derivation step that glues the two trees together. What *visible* means is this: Let us say that a tree such as the noun phrase *the guys*, ordinarily

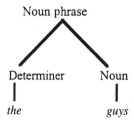

Figure 15.6   A conventional tree representation for the noun phrase *the guys*.

represented as in figure 15.6, is represented by the following set of *terms* (following Epstein, 1995):

Noun phrase = {Determiner–*the*, {Determiner–*the*, Noun–*guys*}}

Figure 15.6 shows that the tree or noun phrase corresponding to *the guys* was built out of the composition operation that pasted together *the* and *guys*, forming a new tree with a new root (topmost) node. Initially, *the* was simply the set (subtree) {Determiner–*the*}, where we have tacked on the syntactical category *Determiner* for ease of reading. Similarly, *guys* was the set (subtree) {Noun–*guys*}. The concatenation operation is as described previously, and we select one of the two combines as the name of the new root tree. We now propose simply that *two syntactical elements (i.e., hierarchical structures, trees) can enter into a grammatical relation if and only if there is some point in a derivation (sequence of concatenations) at which both trees at which both trees are made connected terms (members) of the same common subtree (via the concatenation operation).*

For example, in our figure, note that *the* and *guys* can enter into a common grammatical relation because they are both terms of some other set at the time of their concatenation, and they are directly related via concatenate— viz., the set that represents the noun phrase. This relation is, of course, simply the *sister-of* relation described in figure 15.3 (and also described as the *function-argument* relation).

The same property holds for the other two basic grammatical relations mentioned. For the head-of relation, note that in the mother tree {{*the*, {Determiner–*the*, Noun–*guys*}}, the first term in the set, *the*, can, by definition, be related to either of the other two terms: In other words, the root node can be related to either of its immediate daughters. However, this is just the head-of relation, as in figure 15.3. For c-command, note that X and Y are hierarchically concatenated; then X c-commands all the elements (terms) of Y. For instance, if $X =$ the noun phrase *tree* corresponding to *the guys*, then $X = \{the, \{\text{Determiner}–the, \text{Noun}–guys\}\}$. If $Y =$ the verb phrase tree corresponding to *like the ice cream*, then $Y =$ the five-term set {*like* {Determiner– *the*, {*the*, Noun–*ice cream*}}}, and X c-commands every one of those terms (and not vice versa, crucially). In contrast, *ice cream* cannot c-command *guys* because, at the time the noun phrase corresponding to *the ice-cream* was built (concatenated out of two parts), *guys* was not part of that set of terms. In this way, the asymmetrical nature of c-command is derived.

In contrast, certain relations can *never* obtain: For example, because *the ice-cream* and *the guys* are never concatenated together directly but only after trees above them have been built, there can be no grammatical relation that holds between subjects and objects, as we find to be the case in natural languages. Note that there is no *logical* reason this should be so otherwise, unless we assume some fundamental constraint such as the concatenation operation. In other words, in a general context-free system, we can easily write a grammar that relates subjects and objects. Why we do not find any such relations remains a mystery, unless there is a more fundamental constraint that underlies natural languages. As we have sketched, this law seems to be a simple one: Natural languages are formed by a single algebraic operation of hierarchical concatenation. This, then, is *natural language*.

CONCLUSION: NATURAL AND GENETIC GRAMMARS

Plainly, the concatenation operation is simple. It is *adjacency* as extended from strings to trees. Just as plainly, the proposed "grammars" for genetic transcription are vastly more complicated. There appears to be nothing in natural languages corresponding to splicing followed by distinct reading frames. To take another example, Searls (1993) uses the logic programming language Prolog to describe exons and translated regions, which is fine except that with Prolog we also can describe a connection between subjects and objects in natural languages that we do not see. Of course, all this says is that the constraint lies in the particulars of the program that the scientist writes rather than in the constraints of the programming language itself. If this is so, it is left to the programmer or scientist to discover the constraints; the space of possible theoretical descriptions given by the representation language—in this case, Prolog—is vast. If this is so, then we still do not have any strong insight into what constitutes the language of the genes. We know what it is not: It is *not* a finite-state language or a context-free language, but neither is it a context-sensitive language. If anything, the language of the genes is much more like a programming language whose constraints we do not know (or whose programs we do not know). Now, the problem of identifying a program's details from observations of its input and output behavior is very, very difficult; even in the case of finite-state programs, the problem is unsolvable unless we assume that we know other constraints (such as the number of states in the program). Yet this is the task that molecular biologists seemingly have set for themselves. Given that we currently have no general means of carrying out such inductions, it would seem best to work out particular case studies of reverse engineering—looking at transcription in the way that Collado-Vides (chapter 9) has done and then determining what kind of computer program is best suited for describing the engineered constraints we do observe. Our knowledge here seems just at the starting point, so much so that we must amass many more case studies before we can come up with the generalizations that will tell us what the genetic constraints are. By way of comparison, it has taken more than 40 years to determine that, in the end, syntactical relationships in natural languages are, in fact, derivable from a single, simple algebraic operation. We can discover what language the genes are speaking to us only by more years of careful listening.

Glossary

Algorithmic complexity (of a sequence)   The minimal length of a program generating the sequence (using an appropriate scale). Given a sequence, there is no general principle that permits one to calculate its algorithmic complexity. Therefore, the algorithmic complexity of a sequence is always smaller than or equal to that of the known program of minimal length able to generate the sequence. By definition, a sequence is random if its algorithmic complexity is equal to the length of the sequence. A repetitive sequence has a low algorithmic complexity. Consequently, it can be understood that algorithmic complexity measures an aspect of the information carried by a sequence. *See also* Logical depth

Allometric morphogenesis   When two different aspects of a developing system are plotted against each other in log-log coordinates and the result is a straight line. Also called *allometry*.

Anterior-posterior axis   The head-tail body axis.

Apical   That side of an epithelium furthest away from the basement membrane. The apical side of the *Drosophila* blastoderm is toward the outside of the egg.

Biochemical systems theory   Theory of integrated biochemical systems represented within the power-law formalism, which provides the canonical nonlinear representation. The theory consists of definitions, concepts, methods of analysis, and general principles that have been elucidated for various classes of biochemical systems.

Blastoderm   Stage in embryonic development during which the embryo consists of a hollow shell of cells or nuclei.

Boolean   The formal language used in digital computers, named for the mathematician and logician Boole. Boolean variables and functions can take only two values: *true* and *false* or *0* and *1*. Boolean operations also are defined in what is called *Boolean algebra*: *or, not, and, nand,* and so on.

Bottleneck   Section of a pathway or network that is limiting the feasibility or performance of the whole system.

Canonical nonlinear representation   Generic form of nonlinear expression within which all particular expressions within a broad class are contained. One can then represent any particular case simply by selecting appropriate values for the parameters in the general expression.

Capacity for gene regulation   The ratio of maximal level of gene expression to basal level of gene expression as a particular stimulus varies over its full range. The capacity may differ for different stimuli and in different contexts.

Chaos   In modern mathematical usage, those processes that take different trajectories, and so produce very different outcomes, from indistinguishably different beginnings.

Chromatin   The substance that makes up eukaryotic chromosomes. It contains DNA, histones, and other proteins.

Cleavage cycle   The period between the completion of division $n - 1$ and division n in the series of 13 synchronous nuclear divisions that occur in the *Drosophila* embryo prior to gastrulation.

Coactivators   Proteins that are thought to form a bridge between transcriptional activators and general transcription factors, to mediate the activating properties of transcriptional activators to the basal transcription machinery.

Conditions for stability   Well-known conditions that must be met for a system to remain stable in the face of small perturbations. For a system of n dependent variables, there are n so-called eigen values that must have negative real parts. Alternatively, there are n Routh-Hurwitz criteria that must have the same sign. Given a particular system with well-defined parameter values, stability can be determined by straightforward methods according to these conditions.

Contig   A set of overlapping clones that span a contiguous region of the genome.

Derivation   In generative grammar, the set of processes used in the grammar, usually involving rewriting rules, to produce a particular sentence.

Design principles   Rules governing the design for a class of biological systems. These rules may be deduced from first principles or empirically observed correlations, but they ultimately are based on physical laws and the organizational constraints imposed by system dynamics.

Determination   The process by which a cell becomes committed to a particular path of differentiation. Not accompanied by morphological changes.

Deterministic chaos   Behavior that has the appearance of being random but is, in fact, generated by a deterministic system that is exquisitely sensitive in initial conditions.

Differentiation   The process by which a cell specializes to become a particular type (nerve cell, liver cell, skin cell, etc.). Usually accompanied by major morphological changes.

Distribution of influence   Relative to a particular dependent variable, the ratio of each independent influence to the total influence. The influence of any independent variable on a particular dependent variable in a system is defined quantitatively as the magnitude of the corresponding logarithmic gain. One can determine the magnitude of the logarithmic gain for each of the independent variables; the sum of these influences specifies the total influence.

Dorsoventral Axis   The back-stomach body axis.

Diffusion limit (encounter limit)   The upper bound on the rate of an enzymatic reaction, as dictated by diffusion (or encounter rates) of substrates and enzyme. This limit may also be made subject to thermodynamic restrictions.

Dynamic diseases   Diseases that result from altered system dynamics or aberrations in timing rather than the absence of a particular gene product.

Dynamic programming   An iterative method to find the global maximum or minimum of a function (the distance between sequences, for example) where the global solution can be obtained from a set of previously obtained global solutions for a reduced function (i.e., a set of sequences one symbol shorter).

ECO2DBASE   An electronic database containing the locations on two-dimensional gels of proteins of *Escherichia coli*.

Eukaryote   An organism that possesses cells containing a nucleus and having certain other features. All multicellular and many single-celled organisms are eukaryotes; bacteria are not.

Feedback circuit (feedback loop)   Circuit formed by elements $a, b, c, \ldots, n$ when $a, b, c, \ldots, n$ are elements of a system and a influences the rate of production of b, b of c, and c of a. One can have 1-, 2-, $\ldots$, n-element circuits. In the language of graph theory, these are oriented and signed (because each interaction can be positive or negative). Feedback circuits, or feedback loops, play an essential role in biology because they are "the wheels of regulation." Any feedback circuit is positive or negative, according to whether there is an even or an odd number of negative interactions. To these two types of circuits correspond radically different properties.

Fractal kinetics   Kinetics observed under dimensionally restricted conditions, including reactions that occur on fractal surfaces, which give rise to elementary rate laws with noninteger kinetic orders.

Gap gene   A member of a class of segmentation genes that usually are expressed in two large domains (10–15 nuclei) in the *Drosophila* blastoderm. When mutated, there is a large gap in the body pattern.

Gastrulation   Common feature of the embryogenesis of all animals (except sponges) in which the three germ layers of endoderm, mesoderm, and ectoderm form.

Gene circuitry   The topology of interactions by which the products of various transcriptional units affect one another's expression.

GeneMark   Program predicting probable coding regions that is based on higher-order Markov chain models, looking at the probability of certain bases following others at particular intervals and comparing it with a training set of known genes from the organism in question.

General (basal) transcription factors   Ancillary proteins that are required for transcription carried out by RNA polymerases.

Generalized mass action   Generalization of traditional mass action in which the kinetic orders need not be positive integers.

Generative grammar   Within formal language theory, an ordered fourtuple (Vn, Vt, S, P), where Vn and Vt are finite alphabets corresponding to the nonterminal and terminal symbols, respectively. S is a distinguished (initial) symbol of Vn, and P is a finite set of ordered production (rewriting) rules (Hopcroft and Ullman, 1979). Within the study of natural language, a *grammar* is defined as a theory that gives an explicit account of the form and meaning of sentences (Chomsky, 1986, p. 3). The term *generative* comes from mathematics (i.e., a system **G** generates a vectorial space if any vector of the space is a linear combination of the vectors in **G**).

Genome expression map   Image of a reference two-dimensional polyacrylamide gel that displays the location of every protein encoded in the genome of an organism.

Gibbs energy   Quantitative thermodynamic measure of the feasibility of a bioreaction or transformation. The Gibbs energy should be evaluated at the actual conditions of the system (including the actual metabolite concentrations), whereas the standard Gibbs energy is evaluated at a specific standard state different from the actual state of the biochemical system.

Heteroduplex mapping   A way to look at the similarities and differences between two related pieces of DNA. The two DNAs to be compared (such as the genomes of two related viruses) are mixed together, denatured to separate the strands, and then allowed to renature. Double strands will form between the two different molecules in regions of high homology, with loops or bubbles of single-stranded DNA marking significant differences.

Hill function   A sigmoid function of the form (increasing function): $y = x^n/\theta^n + x^n$, where x is a variable, θ is a threshold associated to the variable, and n is positive. The exponent n is called the *Hill coefficient*. As n increases, the steepness of the function curve increases, to the point where if $n \rightarrow$ infinity, a step function similar to a Boolean function is obtained. Hill functions with integral coefficient n are binding isotherms for a ligand with concentration x binding cooperatively to n sites.

Homeostasis   Property that consists in tending to maintain variables at or near intermediate levels (i.e., between low and high boundaries of the system). This takes place with or without oscillation.

Invagination   A folding-in of a membrane or epithelium.

Kinetic order   Characterizes the influence of a variable on the rate of a reaction. Defined as the partial derivative of the rate law with respect to one of its concentration variables multiplied by the ratio of the concentration to the rate [i.e., $(\partial v_i/\partial X_j)(X_j/v_i)$]. It also is equal to the exponent of the variable in the rate law represented in the power-law formalism. In general, the kinetic order of a rate law will vary with the operating conditions. However, in the mass-action representation, kinetic order with respect to a given metabolite has a fixed integer value equal to the number of molecules of that metabolite entering into the reaction.

Kirchhoff's node equations   One of the two dual forms in which the equations describing network behavior can be expressed. In chemical networks, they express the conservation of mass: The difference between the aggregate rate of appearance and the aggregate rate of disappearance must equal the rate of accumulation of material at the node in question.

Local representation   A representation that can be made as accurate as desired, within some local neighborhood about a nominal operating value.

Logarithmic gain   The gain or amplification of a biological signal as it is propagated from an independent variable (stimulus) to a dependent variable (response). If the logarithm of the dependent variable is plotted against the logarithm of the independent variable, then the slope at the nominal operating point is equal to the corresponding logarithmic gain.

Logical depth   The length of time necessary to generate the actual development of a sequence (e.g., the nth digit of π), as proposed by Charles Bennett. The difference with repetitive sequences, which generate an apparent complicated form, occurs because they need a long computation time to be produced. Logical depth provides another evaluation of the information of a sequence. *See* Algorithmic complexity.

Logical description   System that endows variables and functions with a limited number of values, only two (0 and 1) in simple cases. In view of the complexity of biological and other regulatory systems, it often is advisable to have a qualitative though rigorous view such as that provided by logical descriptions.

Mass-action kinetics   Kinetics of elementary chemical reactions in which the rate of the reaction is proportional to the product of the concentrations of the interacting species. This is founded on probability arguments that assume homogeneous spatial distributions of the reactants.

Mathematically controlled comparisons   The mathematical equivalent of a well-controlled experiment in which the experimental system and the control or reference system differ with respect to a change (mutation) in a single process but are otherwise identical. The changes also are carefully selected so as to minimize extraneous systemic consequences, and in this way attention is focused on the inherent or irreducible differences between the systems.

Metabolic resource   Any component or set of components with the potential to limit rates of the metabolism. There is usually a physiological upper bound applicable to any resource. Enzymes, amino acids, or the total soluble pool are examples of metabolic resources.

Michaelis-Menten formalism   A mathematical language or representation with a structure consisting of ordinary nonlinear differential equations whose elements are rate laws in the form of rational functions. Each rate law characterizes an enzyme-catalyzed reaction, described by mass-action kinetics, in which the different forms of enzyme are assumed to be in quasi-steady state and constrained to equal the total amount of enzyme. This is the underlying structure for essentially all contemporary enzyme kinetics and is based on a set of key assumptions shared with the original development of Michaelis-Menten kinetics.

Mode of gene regulation   Molecular mechanisms controlling gene expression can be manifested in one of two alternative modes—positive or negative.

Morphogen   A chemical substance that controls determination.

Multistationarity   Property of a system that has more than one steady state. In particular, there exist systems that can persist in either of two or more states of regime, which may be steady states or more complex regimes.

Negative feedback circuit (negative feedback loop)   A feedback circuit in which each element exerts a negative indirect effect on itself. Characterized by an odd number of negative interactions, they are responsible for homeostasis.

Neur(on)al networks   A mathematical and computational approach grounded on a general metaphor representing living neurons as connected points, linked to one another in a vectorial way by "synapses." Time-dependent signals pass through the network as a function of an input signal. In the most general model, a fraction of the signals go through the synapse itself but also through other synapses on the same neurons. Many different designs of neuronal networks have been proposed to simulate learning processes.

Organizational complexity   A form of complexity possessed by systems with a large number of elements that are very different in kind and maintained in highly organized states. This is in contrast to other forms of complexity possessed by systems with a large number of elements of similar kind and simple rigid organization.

Pair-rule Gene   A member of a class of segmentation genes that usually are expressed in seven stripes during the blastoderm stage. When mutated, there are gaps in the segmental pattern, with a spatial frequency of two segments.

Positive feedback circuit (positive feedback loop)   A feedback circuit in which each element exerts a positive indirect effect on itself. Characterized by an even number of negative interactions, they are responsible for multistationarity and its biological modality, differentiation.

Power-law formalism   A mathematical language or representation with a structure consisting of ordinary nonlinear differential equations whose elements are products of power-law functions. It is this formalism that underlies many of the newer approaches to the modeling and analysis of integrated biochemical systems.

Preinitiation complex   General transcription factors assembled at the TATA-box (TATA being a four-letter DNA sequence) prior to the start of transcription.

Pseudogene   Damaged descendant of a gene or product of reverse transcription that still retains some of the information content of the gene, though it is not expressed owing to lack of control signals or other problems.

Quasi-steady-state assumption   The assumption that a variable changes so rapidly in relation to the other variables of a system that the fast variable can be treated as if it has gone through its dynamics and come to a true steady state. The differential equation describing the fast variable then is set equal to zero, thereby reducing it to an algebraic equation that relates the value of the fast variable to the other variables of the system.

Regulation (regulatory systems)   The means by which biological and other complex systems are somehow informed about the value of some crucial variables and can take this information into account for modulating the future level of these variables.

Response time   The time that it takes a system to move from a predisturbance steady state to within $\pm 5\%$ of the new steady state that is established following the imposition of a sustained change in the system's environment.

Robustness   The quality of a system where by the system's normal behavior is changed very little in response to perturbations in the values of the various parameters that define the system. Robust systems have selective value, whereas those whose normal behavior is drastically altered by small perturbations tend to be selected against.

Sigma factors   Bacterial proteins that interact with RNA polymerase and dictate promoter specificity.

Steady-state expression characteristic   A log-log plot depicting level of gene expression versus intensity of the associated stimulus. All other independent variables are held constant, and the system is allowed to reach a steady state for each value of the stimulus.

Syncytial   Not divided into cells.

Systematically structured formalisms   A canonical formalism that, by virtue of its regular structure, makes mathematical analysis tractable.

Systems strategy   A framework that allows for systematic development and evaluation of our understanding. The basic approach involves definition of a system for study, separation and characterization of the underlying mechanisms, and reconstruction and verification of integrated behavior. A more articulated form of the basic scientific method.

Thermodynamic bottleneck   Section of a pathway that is thermodynamically infeasible or nearly infeasible. If this bottleneck is altered, then the pathway as a whole becomes feasible.

Transcriptional activators   Proteins that bind to specific DNA sequences to modulate the expression of nearby promoters.

UspA   A protein induced under a wide variety of stress conditions in *Escherichia coli*.

References

Abarbanel, R. M. (1984). *Protein structural knowledge engineering*. Unpublished doctoral thesis, University of California, San Francisco.

Abarbanel, R. M., Wieneke, P. R., Mansfield, E., Jaffe, D. A., and Brutlag, D. L. (1984). Rapid searches for complex patterns in biological molecules. *Nucleic Acids Research, 12,* 263−280.

Abel, Y., and Cedergren, R. (1990). *The normalized gene designation database* [data file]. Available electronically via ftp from ncbi.nlm.nih.gov.

Aho, A. V., Hopcroft, J. E., and Ullman, J. D. (1983). *Data structures and algorithms*. Reading, MA: Addison-Wesley.

Akam, M. (1987). The molecular basis for metameric pattern in the *Drosophila* embryo. *Development, 101,* 1−22.

Albery, W. J., and Knowles, J. R. (1976). Evolution of enzyme function and the development of catalytic efficiency. *Biochemistry, 15,* 5631−5640.

Alexandrov, N. N. (1992). Local multiple alignment by consensus matrix. *Computer Applications in the Biosciences, 8,* 339−345.

Alexandrov, N. N., and Mironov, A. A. (1987). Recognition of *Escherichia coli* promoters given the DNA primary structure. *Molekuliarnaia Biologiia* (Moskva), *21,* 242−249.

Alff-Steinberger, C. (1987). Codon usage in *Homo sapiens*: Evidence for a coding pattern on the non-coding strand and evolutionary implications of dinucleotide discrimination. *Journal of Theoretical Biology, 124,* 89−95.

Almagor, H. (1985). Nucleotide distribution and the recognition of coding regions in DNA sequences: An information theory approach. *Journal of Theoretical Biology, 117,* 127−136.

Altschul, S. F. (1989). Gap costs for multiple sequence alignment. *Journal of Theoretical Biology, 138,* 297−309.

Altschul, S. F. (1991). Amino acid substitution matrices from an information theoretic perspective. *Journal of Molecular Biology, 219,* 555−565.

Altschul, S. F., and Erickson, B. W. (1985). Significance of nucleotide sequence alignments: A method for random sequence permutation that preserves dinucleotide and codon usage. *Molecular Biology and Evolution, 2,* 526−538.

Altschul, S. F., Gish, W., Miller, W., Myers, E. W., and Lipman, D. J. (1990). Basic local alignment search tool. *Journal of Molecular Biology, 215,* 403−410.

Anfinsen, C. B. (1973). Principles that govern the folding of protein chains. *Science, 181,* 233−240.

Bachar, O., Fischer, D., Nussinov, R., and Wolfson, H. (1993). A computer vision–based technique for 3-D sequence-independent structural comparison of proteins. *Protein Engineering*, *6*, 279–288.

Banerji, J., Rusconi, S., and Schaffner, W. (1981). Expression of a beta-globin gene is enhanced by remote SV40 DNA sequences. *Cell*, *27*, 299–308.

Barton, G. J. (1992). Computer speed and sequence comparison [letter]. *Science*, *257*, 1609–1610.

Barton, G., and Sternberg, M. J. E. (1987a). Evaluation and improvements in the automatic alignment of protein sequences. *Protein Engineering*, *1*(2), 89–94.

Barton, G. J., and Sternberg, M. J. E. (1987b). A strategy for the rapid multiple alignment of protein sequences. *Journal of Molecular Biology*, *198*, 327–337.

Barton, G. J., and Sternberg, M. J. E. (1990). Flexible protein sequence patterns a sensitive method to detect weak structural similarities. *Journal of Molecular Biology*, *212*, 389–402.

Bazett-Jones, D. P., Leblanc, B., Herfort, M., and Moss, T. (1994). Short-range DNA looping by the *Xenopus* HMG-box transcription factor, xUBF. *Science*, *264*, 1134–1137.

Beckwith, J. (1987). The operon: An historical account. In F. C. Neidhardt, J. L. Ingraham, K. Brooks Low, B. Magasanik, M. Schaechter, and H. E. Umbarger (Eds.), Cellular and molecular biology: Escherichia coli *and* Salmonella typhimurium (Vol. 2). Washington, DC: American Society for Microbiology.

Bengio, Y., and Pouliot, Y. (1990). Efficient recognition of immunoglobulin domains from amino acid sequences using a neural network. *Computer Applications in the Biosciences*, *6*(4), 319–324.

Bennett, C. H. (1988). Logical depth and physical complexity. In *The universal Turing machine: A half-century survey*. Oxford, UK: Oxford University Press.

Bennetzen, J. L., and Hall, B. D. (1982). Codon selection in yeast. *Journal of Biological Chemistry*, *257*, 3026–3031.

Benoist, C., and Chambon, P. (1981). *In vivo* sequence requirements of the SV40 early promoter region. *Nature*, *290*, 304–310.

Benzécri, J. P. (Ed.). (1984). L'analyse des données. In *La taxinomie*. Paris: Dunod.

Benzer, S. (1956). The elementary units of heredity. In W. D. McElroy and B. Glass (Eds.). *A symposium on the chemical basis of heredity*. Baltimore: Johns Hopkins University Press.

Bergey, D. H. (1984). *Bergey's manual of systematic bacteriology*. Baltimore/London: Williams & Wilkins.

Bernstein, H., and Bernstein, C. (1989). Bacteriophage T4 genetic homologies with bacteria and eucaryotes. *Journal of Bacteriology*, *171*, 2265–2270.

Bertalanffy, L. von. (1960). Principles and theory of growth. In W. W. Nowinski (Ed.), *Fundamental aspects of normal and malignant growth*. New York: Elsevier, pp. 137–259.

Berwick, R. (1989). Natural language, computational complexity and generative capacity. *Computers and Artificial Intelligence*, *8*(5), 423–441.

Berwick, R., and Weinberg, A. (1979). Parsing efficiency, computational complexity, the evaluation of grammatical theories. *Linguistic Inquiry*, *13*, 265–191.

Blaisdell, B. E. (1991). Average values of a dissimilarity measure not requiring sequence alignment are twice the averages of conventional mismatch counts requiring sequence alignment for a variety of computer-generated model systems. *Journal of Molecular Evolution*, *32*, 521–528.

Blake, R. D., and Hinds, P. W. (1984). Analysis of the codon bias in *E. coli* sequences. *Journal of Biomolecular Structure and Dynamics, 2,* 593–606.

Bloch, P. L., Phillips, T. A., and Neidhardt, F. C. (1980). Protein identifications of O'Farrell two-dimensional gels: Locations of 81 *Escherichia coli* proteins. *Journal Bacteriology, 141,* 1409–1420.

Blundell, T. L., Sibanda, B. L., Sternberg, M. J. E., and Thornton, J. M. (1986). Knowledge-based prediction of protein structures and the design of novel molecules. *Nature, 326,* 347–352.

Bock, A., Forchhammer, K., Heider, J., Leinfelder, W., Sawers, G., Veprek, B., and Zinoni, F. (1991). Selenocysteine: The 21st amino acid. *Molecular Microbiology, 5,* 515–520.

Bode, H. W. (1945). *Network analysis and feedback amplifier design.* Princeton: Van Nostrand.

Bohr, H., Bohr, J., Brunak, S., Cotterill, R. M. J., Lautrup, B., Norskov, L., Olsen, O. H., and Petersen, S. B. (1988). Protein secondary structure and homology by neural networks. *FEBS Letters, 241,* 223–228.

Bohr, H., Brunak, S., Cotterill, R., Fredholm, H., Lautrop, B., and Petersen, S. (1990). A novel approach to prediction of the 3-dimensional structures of protein backbones by neural networks. *FEBS Letters, 261,* 43–46.

Bork, P., and Grunwald, C. (1990). Recognition of different nucleotide-binding sites in primary structures using a property-pattern approach. *European Journal of Biochemistry, 191,* 347–358.

Borodovsky, M., and McIninch, J. (1992). Prediction of gene locations using DNA Markov chain models. In H. Lim, J. Fickett, C. Cantor, and R. Robbins (Eds.), *Proceedings of the Second International Conference on Bioinformatics, Supercomputing and Complex Genome Analysis.* Singapore: World Scientific.

Borodovsky, M., Rudd, K. E., and Koonin, E. V. (1994). Intrinsic and extrinsic approaches for detecting genes in a bacterial genome. *Nucleic Acids Research, 22,* 4756–4767.

Boswell, D. R. (1988). A program for template matching of protein sequences. *CABIOS, 4,* 345–350.

Botstein, D. (1980). A theory of modular evolution for bacteriophages. *Annals of the New York Academy of Sciences, 354,* 484–491.

Bowie, J. U., Lüthy, R., and Eisenberg, D. (1991). A method to identify protein sequences that fold into a known three-dimensional structure. *Science, 253,* 164–170.

Bradley, M. K., Smith, T. F., Lathrop, R. H., Livingston, D. M., and Webster, T. A. (1987). Consensus topography in the ATP binding site of the simian virus 40 and polyomavirus large tumor antigens. *Proceedings of the National Academy of Sciences of the United States of America, 84,* 4026–4030.

Bramson, M., and Lebowitz, J. L. (1988). Asymptotic behavior of densities in diffusion dominated annihilation reactions. *Physical Review Letters, 61,* 2397–2400.

Breese, K., Friedrich, T., Andersen, T. T., Smith, T. F., and Figge, J. (1991). Structural characterization of a 14-residue peptide ligand of the retinoblastoma protein: Comparison with a nonbinding analog. *Peptide Research, 4,* 220–226.

Brent, R., and Ptashne, M. (1981). Mechanism of action of the *lexA* gene product. *Proceedings of the National Academy of Sciences of the United States of America, 78,* 4202–4208.

Brent, R., and Ptashne, M. (1985). A eukaryotic transcriptional activator bearing the DNA specificity of a prokaryotic repressor. *Cell, 43,* 729–736.

Bright, M. W., Hurson, A. R., and Pakzad, S. H. (1992). A taxonomy and current issues in multidatabase systems. *IEEE Computer , 25*(3), 50.

Brouillet, S., Risler, J.-L., Henaut, A., and Slonimski, P. P. (1992). Evolutionary divergence plots of homologous proteins. *Biochimie, 74*, 571–580.

Bulmer, M. (1990). The effect of context on synonymous codon usage in genes with low codon usage bias. *Nucleic Acids Research, 18*, 2869–2873.

Buratowski, S., Hahn, S., Guarente, L., and Sharp, P. A. (1989). Five intermediate complexes in transcription initiation by RNA polymerase II. *Cell, 56*, 549–561.

Burns, S. A., and Locascio, A. (1991). A monomial-based method for solving systems of non-linear algebraic equations. *International Journal of Numerical Methods in Engineering, 31*, 1295–1318.

Cairns, J., Stent, G. S., and Watson, J. D. (1966). *Phage and the origins of molecular biology*. Cold Spring Harbor, NY: Cold Spring Harbor Laboratory.

Campbell, A., and Botstein, D. (1983). Evolution of the lambdoid phages. In R. Hendrix, J. Roberts, F. Stahl, and R. Weisberg (Eds.), *Lambda II*. Cold Spring Harbor, NY: Cold Spring Harbor Laboratory.

Casjens, S., Hatfull, G., and Hendrix R. (1992). Evolution of dsDNA tailed-bacteriophage genomes. *Seminars in Virology, 3*, 383–397.

Caspersson, T. (1936). Über den chemischen Aufbau der Strukturen des Zellkernes. *Acta Medica Scandinavica, 73* (Suppl. 8), 1–151.

Chamberlin, M. J. (1995). New models for the mechanism of transcription elongation and its regulation. In *The Harvey lecture series*. (Vol. 33). New York: Wiley-Liss.

Cherkassky, V., and Vassilas, N. (1989). *Associative database retrieval using back propagation networks*. Presented at the International Symposium on Computational Intelligence, Milan, Italy.

Chomsky, N. (1956). Three models for the description of languages. *IRE Transactions on Information Theory, 2*(3), 113–124.

Chomsky, N. (1959). On certain formal properties of grammars. *Information and Control, 2*(2), 137–167.

Chomsky, N. (1986). *Knowledge of language*. New York: Praeger.

Chomsky, N., and Halle, M. (1968). *The sound pattern of English*. Cambridge, MA: MIT Press.

Chou, P. Y., and Fasman, G. D. (1978). Empirical predictions of protein conformation. *Annual Review of Biochemistry, 47*, 251–276.

Cinkosky, M. J., Fickett, J. W., Gilna, P., and Burks, C. (1991). Electronic data publishing and GenBank. *Science, 252*, 1273–1277.

Claverie, J.-M., and Sauvaget, I. (1985). Assessing the biological significance of primary consensus patterns using sequence databanks. *Computer Applications in the Biosciences, 1*, 95–104.

Claverie-Martin, F., and Magasanik, B. (1992). Positive and negative effects of DNA bending on activation of transcription from a distant site. *Journal of Molecular Biology, 227*, 996–1008.

Clegg, J.S. (1984). Properties and metabolism of the aqueous cytoplasm and its boundaries. *American Journal of Physiology, 246*, R133–R151.

Cohen, B. I., Presnell, S. R., and Cohen, F. E. (1991). Pattern based approaches to protein structure prediction. *Methods in Enzymology, 202*, 252–268.

Cohen, C., and Parry, D. A. D. (1986). Alpha-helical coiled coils—a widespread motif in proteins. *Trends in the Biochemical Sciences, 11*, 245.

Cohen, F. E., Abarbanel, R. M., Kuntz, I. D., and Fletterick, R. J. (1983). Secondary structure assignment for alpha/beta proteins by a combinatorial approach. *Biochemistry*, 22, 4894–4904.

Cohen, F. E., Abarbanel, R. M., Kuntz, I. D., and Fletterick, R. J. (1986). Turn prediction in proteins using a complex pattern matching approach. *Biochemistry*, 25, 266–275.

Cohen, J. (1967) Feathers and patterns. *Advances in Morphogenesis*, 5, 1–38.

Cohen, J. (1969). Dermis, epidermis and dermal papilla interacting. In W. Montagna (Ed.), *Advances in the biology of skin: Hair growth* (Vol. 9). Oxford: Pergamon.

Cohen, J. (1992). The case for and against sperm selection. In B. Baccetti (Ed.), *Comparative spermatology—twenty years after* [Ares-Serono Symposia 75]. New York: Raven.

Cohen, J., and Adeghe, J.-H. A. (1987). The other spermatozoa; fate and functions. In H. Mohri (Ed.), *New horizons in spermatozoal research*. The Japanese Science Societies Press.

Cohen, J., and Stewart, I. N. (1994). *The collapse of chaos*. New York: Viking.

Cohn, M., and Horibata, K. (1959). Inhibition by glucose of the induced synthesis of the β-galactoside-enzyme system of *Escherichia coli*. Analysis of maintenance. *Journal of Bacteriology*, 78, 601–612.

Collado-Vides, J. (1991a). The search for a grammatical theory of gene regulation is formally justified by showing the inadequacy of context-free grammars. *Computer Applications in the Biosciences*, 7, 321–326.

Collado-Vides, J. (1991b). A syntactic representation of units of genetic information. *Journal of Theoretical Biology*, 148, 401–429.

Collado-Vides, J. (1992). Grammatical model of the regulation of gene expression. *Proceedings of the National Academy of Sciences of the United States of America*, 89, 9405–9409.

Collado-Vides, J. (1993a). A linguistic representation of the range of transcription initiation of sigma 70 promoters: I. An ordered array of complex symbols with distinctive features. *Biosystems*, 29, 87–104.

Collado-Vides, J. (1993b). A linguistic representation of the range of transcription initiation of sigma 70 promoters: II. Distinctive features of promoters and their regulatory binding sites. *Biosystems*, 29, 105–128.

Collado-Vides, J. (1995). Some ideas towards a grammatical model of the sigma 54 bacterial promoters. In *Proceedings of the First International Symposium on Intelligence in Neural and Biological Systems*. Los Alamitos, CA: IEEE Computer Society Press.

Collado-Vides J. (in press). Towards a unified grammatical model of the σ^{70} and the σ^{54} types of bacterial promoters. *Biochimie*.

Collado-Vides, J., Magasanik, B., and Gralla, J. D. (1991). Control site location and transcriptional regulation in *Escherichia coli*. *Microbiology Reviews*, 55, 371–394.

Cornette, J. L., Cease, K. B., Margalit, H., Spouge, J. L., Berzofsky, J. A., and DeLisi, C. (1987). Hydrophobicity scales and computational techniques for detecting amphipathic structures in proteins. *Journal of Molecular Biology*, 195, 659–685.

Cote, J., Quinn, J., Workman, J. L., and Peterson, C. L. (1994). Stimulation of GAL4 derivative binding to nucleosomal DNA by the yeast SWI/SNF complex. *Science*, 265, 53–60.

Cowan, J., d'Acci, K., Guttman, B., and Kutter, E. (1994). Gel analysis of T4 prereplicative proteins. In J. D. Karam (Ed.), *Molecular biology of bacteriophage T4*. Washington, DC: American Society for Microbiology.

Cox, E. C., and Yanofsky, C. (1967). Altered base ratios in the DNA of an *Escherichia coli* mutator strain. *Proceedings of the National, 58*, 1895−1902.

Dalma-Weiszhaus, D., and Brenowitz, M. (1992). Interactions between DNA-bound transcriptional regulators of the *Escherichia coli gal* operon. *Biochemistry, 31*, 6980−6989.

Danchin, A. (1979). The generation of immune specificity: A general selective model. *Molecular Immunology, 16*, 515−526.

Danchin, A. (1986). Foreword. In Erwin Schrödinger, *Qu'est-ce que la vie?* Paris: Christian Bourgois.

Danchin, A. (1990). *Une aurore de pierres. Aux origines de la vie.* Paris: Le Seuil.

Danchin, A. (in press). La profondeur critique: Une conséquence de la métaphore alphabétique de l'hérédité. *Revue d'Histoire des Sciences.*

Dardel, F., and Bensoussan, P. (1988). DNAid: A Macintosh full screen editor featuring a built-in regular expression interpreter for the search of specific patterns in biological sequences using finite state automata. *Computer Applications in the Biosciences, 4*, 483−486.

d'Aubenton Carafa, Y., Brody, E., and Thermes, C. (1990). Prediction of rho-independent *Escherichia coli* transcription terminators. A statistical analysis of their RNA stem-loop structures. *Journal of Molecular Biology, 216*, 835−858.

Davis, I., and Ish-Horowicz, D. (1991). Apical localization of pair-rule transcripts requires 3′ sequences and limits protein diffusion in the *Drosophila* blastoderm embryo. *Cell, 67*, 927−940.

Davis, R., and Lenat, D. (1982). *Knowledge-based systems in artificial intelligence.* New York: McGraw-Hill.

Day, W. H. E., and McMorris, F. R. (1993). *Alignment, comparison and consensus of molecular sequences: A bibliography.* (Available on the Internet: http://www.pitt.edu/~hirtle/day.html.)

Dayhoff, M. O., Schwartz, R. M., and Orcutt, B. C. (1978). A model of evolutionary change in proteins. In M. O. Dayhoff (Ed.), *Atlas of protein sequence and structure.* Washington, DC: National Biomedical Research Foundation.

de Candolle, A. (1991). Laws of botanical nomenclature. Quoted in D. H. Nicolson (Ed.), A history of botanical nomenclature. *Annals of the Missouri Botanical Garden, 78*, 33−56. (Original work published 1867.)

Delbrück, M. (1949). Discussion in: *Unités biologiques douées de continuité génétique. Colloques Internationaux du Centre National de la Recherche Scientifique, 8*, 33−35.

Delorme, M.-O., and Hénaut, A. (1988). Merging of distance matrices and classification by dynamic clustering. *Computer Applications in the Biosciences, 4*, 453−458.

Dickerson, R. E. (1971). Sequence and structure homologies in bacterial and mammalian-type cytochromes. *Journal of Molecular Biology, 57*, 1−15.

Dodd, I. B., and Egan, J. B. (1987). Systematic method for the detection of potential lambda cro-like DNA-binding regions in proteins. *Journal of Molecular Biology, 194*, 557−564.

Donoghue, M. J. (1992). Homology. In: E. F. Keller and E. A. Lloyd (Eds), *Keywords in evolutionary biology.* Cambridge, MA: Harvard University Press.

Doolittle, R. F., Hunkapillar, M. W., Hood, L. E., Davare, S. C., Robbins, K. C., Aaronson, S. A., and Antoniades, H. N. (1983). Simian sarcoma virus onc gene, v-sis, is derived from the gene (or genes) encoding a platelet-derived growth factor. *Science, 221*, 275−277.

Drake, J. W., and Kreuzer, K. N. (1994). DNA transactions in T4-infected *Escherichia coli* p. 11−13., In J. D. Karam (Ed.), *Molecular biology of bacteriophage T4.* Washington, DC: American Society for Microbiology.

Dreyfus, H. L. (1992). *What computers still can't do. A critique of artificial reason.* Cambridge, MA: MIT Press.

Driever, W., and Nusslein-Volhard, C. (1988a). A gradient of *bicoid* protein in *Drosophila* embryos. *Cell, 54,* 83–93.

Driever, W., and Nusslein-Volhard, C. (1988b). The *bicoid* protein determines position in the *Drosophila* embryo in a concentration-dependent manner. *Cell, 54,* 95–104.

Dunn, J., and Studier, F. W. (1993). Complete nucleotide sequence of bacteriophage T7 DNA and the locations of T7 genetic elements. *Journal of Molecular Biology, 166,* 477–535.

Dynlacht, D. B., Hoey, T., and Tjian, R. (1991). Isolation of coactivators associated with the TATA-binding protein that mediate transcriptional activation. *Cell, 66,* 563–576.

Dyson, N., Bernards, R., Friend, S. H., Gooding, L. R., Hassell, J. A., Major, E. O., Pipas, J. M., Vandyke, T., and Harlow, E. (1990). Large T antigens of many polyomaviruses are able to form complexes with the retinoblastoma protein. *Journal of Virology, 64,* 1353–1356.

Dyson, N., Howley, P. M., Münger, K., and Harlow, E. (1989). The human papilloma virus-16 E7 oncoprotein is able to bind to the retinoblastoma gene product. *Science, 243,* 934–937.

Eddy, S., and Gold, L. (1991). The phage T4 nrdB intron: A deletion mutant of a gene found in the wild. *Genes and Development, 5,* 1032–1041.

Edgar, B. A., Odell, G. M., and Schubiger, G. (1987). Cytoarchitecture and the patterning of *fushi tarazu* expression in the *Drosophila* blastoderm. *Genes and Development, 1,* 1226–1237.

Edgar, B., and O'Farrell, P. (1989). Genetic control of cell division patterns in the *Drosophila* embryo. *Cell, 57,* 177–187.

Edgar, B., and O'Farrell, P. (1990). The three postblastoderm cell cycles of *Drosophila embryogenesis are regulated in G2 by string.* Cell, *62,* 469–480.

Eisenberg, D., Wilcox, W., and Eshita, S. (1987). Hydrophobic moments as tools for analysis of protein sequences. In J. J. L'Italien (Ed.), *Proteins: Structure and function.* New York: Plenum.

Epstein, S. D. (1995). *Unprincipled syntax and the derivation of syntactic relations.* Unpublished manuscript, Department of Linguistics, Harvard University, Cambridge, MA.

Fagin, B., Watt, J. G., and Gross, R. (1993). A special-purpose processor for gene sequence analysis. *Computer Applications in the Biosciences, 9,* 221–226.

Farber, R., Lapedes, A., and Sirotkin, K. (1992). Determination of eukaryotic protein coding regions using neural networks and information theory. *Journal of Molecular Biology, 226,* 471–479.

Feaver, W. J., Gileadi, O., Li, Y., and Kornberg, R. D. (1991). CTD kinase associated with yeast RNA polymerase II initiation factor b. *Cell, 67,* 1223–1230.

Feng, J., Atkinson, M. R., McCleary, W., Stock, J. B., Wanner, B. L., and Ninfa, A. J. (1992). Role of phosphorylated metabolic intermediates in the regulation of glutamine synthetase synthesis in *Escherichia coli. Journal of Bacteriology, 174,* 6061–6070.

Fersht, A. (1977). *Enzyme structure and mechanism.* Reading, MA: W. H. Freeman.

Fickett, J. W., Torney, D. C., and Wolf, D. R. (1992). Base compositional structure of genomes. *Genomics, 13,* 1056–1064.

Figge, J., Webster, T., Smith, T. F., and Paucha, E. (1988). Prediction of similar transforming regions in simian virus 40 large T, adenovirus E1A, and *myc* oncoproteins. *Journal of Virology, 62*(5), 1814–1818.

Fischel-Ghodsian, F., Mathiowitz, G., and Smith, T. F. (1990). Alignment of protein sequences using secondary structure: A modified dynamic programming method. *Protein Engineering, 3,* 577–581.

Fischer, D., Bachar, O., Nussinov, R., and Wolfson, H. (1992). An efficient automated computer vision–based technique for detection of three dimensional structural motifs in proteins. *Journal of Biomolecular Structure and Dynamics, 9,* 769–789.

Fishleigh, R. V., Robson, B., Garnier, J., and Finn, P. W. (1987). Studies on rationales for an expert system approach to the interpretation of protein sequence data. *FEBS Letters, 214,* 219–225.

Foe, V. A., and Alberts, B. M. (1983). Studies of nuclear and cytoplasmic behaviour during the five mitotic cycles that precede gastrulation in *Drosophila* embryogenesis. *Journal of Cell Science, 61,* 31–70.

Foor, F., Reuveny, Z., and Magasanik, B., (1980). Regulation of the synthesis of glutamine synthetase by the P_{II} protein in *Klebsiella aerogenes*. *Proceedings of the National Academy of Sciences of the United States of America, 77,* 2636–2640.

Frasch, M., Hoey, T., Rushlow, C., Doyle, H. J., and Levine, M. (1987). Characterization and localization of the even-skipped protein of *Drosophila*. *The EMBO Journal, 6,* 749–759.

Galas, D. J., Eggert, M., and Waterman, M. S. (1985). Rigorous pattern-recognition methods for DNA sequences. *Journal of Molecular Biology, 186,* 117–128.

Galfi, L., and Racz, Z. (1988). Properties of the reaction front in an A + B → C type reaction-diffusion process. *Physical Reviews [A], 38,* 3151–3154.

Garnier, J., Osguthorpe, D. J., and Robson, B. (1978). Analysis of the accuracy and implications of simple methods for predicting the secondary structure of globular proteins. *Journal of Molecular Biology, 120,* 97–120.

Gascuel, O. (1993). Inductive learning and biological sequence analysis. The PLAGE program. *Biochimie, 75,* 363–370.

Gascuel, O., and Danchin, A. (1986). Protein export in prokaryotes and eukaryotes: Indications of a difference in the mechanism of exportation. *Journal of Molecular Evolution, 24,* 130–142.

Ge, H., and Roeder, R. G. (1994). Purification, cloning and characterization of a human coactivator, PC4 that mediates transcriptional activation of class II genes. *Cell, 78,* 513–523.

Geourjon, C., and Deleage, G. (1993). Interactive and graphic coupling between multiple alignments, secondary structure predictions and motif/pattern scanning into proteins. *Computer Applications in the Biosciences, 9,* 87–91.

Glass, B. (1955). Pseudoalleles. *Science, 122,* 233.

Glass, L. (1975). Classification of biological networks by their qualitative dynamics. *Journal of Theoretical Biology, 54,* 85–107.

Glass, L., and Kauffman, S. (1972). Co-operative components, spatial localization and oscillatory cellular dynamics. *Journal of Theoretical Biology, 34,* 219–237.

Glass, L., and Kauffman, S. A. (1973). The logical analysis of continuous non-linear biochemical control networks. *Journal of Theoretical Biology, 39,* 103–129.

Glass, L., and Pasternak, J. S. (1978). Prediction of limit cycles in mathematical models of biological oscillations. *Bulletin of Mathematical Biology, 40,* 27–44.

Goodman, S. D., and Nash, H. A. (1989). Functional replacement of a protein-induced bend in a DNA recombination site. *Nature, 341,* 251–254.

Goodwin, B. C. (1963). *Temporal organisation in cells: A dynamic theory of cellular control processes.* London: Academic Press.

Gorbalenya, A. (1994). Self-splicing group I and group II introns encode homologous (putative) DNA endonucleases of a new family. *Protein Science, 3,* 1117–1120.

Gotoh, O. (1990a). Consistency of optimal sequence alignments. *Bulletin of Mathematical Biology, 52,* 509–525.

Gotoh, O. (1990b). Optimal sequence alignment allowing for long gaps. *Bulletin of Mathematical Biology, 52,* 359–373.

Gouy, M., and Gautier, C. (1982). Codon usage in bacteria: Correlation with gene expressivity. *Nucleic Acids Research, 10,* 7055–7074.

Gracy, J., Chiche, L., and Sallantin, J. (1993). Learning and alignment methods applied to protein structure prediction. *Biochimie, 75,* 353–361.

Graniero-Porati, M. I., and Porati, A. (1988). Informational parameters and randomness of mitochondrial DNA. *Journal of Molecular Evolution, 27,* 109–113.

Greenberg, G. R., He, P., Jilfinger, J., and Tseng, M.-J. (1994). Deoxyribonucleoside triphosphate synthesis and T4 DNA replication. In J. D. Karam (Ed.), *Molecular biology of bacteriophage T4.* Washington, DC: American Society for Microbiology.

Gribskov, M., Devereux, J., and Burgess, R. R. (1984). The codon preference plot: Graphic analysis of protein coding sequences and prediction of gene expression. *Nucleic Acids Research, 12,* 539–549.

Gribskov, M., McLachlan, A. D., and Eisenberg, D. (1987). Profile analysis: Detection of distantly related proteins. *Proceedings of the National Academy of Sciences of the United States of America, 84,* 4355–4358.

Grob, U., and Stuber, K. (1988). Recognition of ill-defined signals in nucleic acid sequences. *Computer Applications in the Biosciences, 4,* 79–88.

Guarente, L., Yocum, R., and Gifford, P. (1982). A GAL10-CYC1 hybrid yeast promoter identifies the GAL4 regulatory region as an upstream site. *Proceedings of the National Academy of Sciences of the United States of America, 79,* 7410–7414.

Guibas, L. J., and Odlyzko, A. M. (1980). Long repetitive patterns in random sequences. *Zeitscrift für Wahrscheinlichkeitstheorie und verwandte Gebiete, 53,* 241–262.

Gusein-Zade, S. M., and Borodovsky, M. Y. (1990). An improved distribution of codon frequencies allowing for inhomogeneity of DNA's primary-structure evolution. *Journal of Biomolecular Structure and Dynamics, 7,* 1185–1197.

Haiech, J., and Sallatin, J. (1985). Computer search of calcium binding sites in a gene data bank: Use of learning techniques to build an expert system. *Biochimie, 67,* 555–560.

Hammer, M., and McLeod, D. (1981). Database description with SDM: A semantic database model. *ACM Transactions on Database Systems, 6,* 351–386.

Hammes, G., and Schimmel, P. (1970). Rapid reactions and transient states. In P. D. Boyer (Ed.), *The enzymes.* New York: Academic Press.

Han, K., and Kim, H.-J. (1993). Prediction of common folding structures of homologous RNA's. *Nucleic Acids Research, 21,* 1251–1257.

Hanai, R., and Wada, A. (1989). Novel third-letter bias in *Escherichia coli* codons revealed by rigorous treatment of coding constraints. *Journal of Molecular Biology, 207,* 655–660.

Heinrich, R., and Rapoport, T. A. (1974). A linear steady-state treatment of enzymatic chains. *European Journal of Biochemistry, 42*, 89–95.

Helmann, J. D., and Chamberlain, M. J. (1988). Structure and function of bacterial sigma factors. *Annual Review of Biochemistry, 57*, 839–872.

Hénaut, A., Limaiem, J., and Vigier, P. (1985). The origins of the strategy of codon use. *Biochimie, 67*, 775–783.

Henikoff, S., and Henikoff, J. G. (1991). Automated assembly of protein blocks for database searching. *Nucleic Acids Research, 19*(23), 6565–6572.

Henikoff, S., Keene, M. A., Fechtel, K., and Fristrom, J. W. (1986). Gene within a gene: Nested *Drosophila* genes encode unrelated proteins on opposite strands. *Cell, 44*, 33–42.

Henneke, C. M. (1989). A multiple sequence alignment algorithm for homologous proteins using secondary structure information and optionally keying alignments to functionally important sites. *Computer Applications in the Biosciences, 5*, 141–150.

Herendeen, D. R., Kassavetis, G. A., and Geiduschek, E. P. (1992). A transcriptional enhancer whose function imposes a requirement that proteins track along DNA. *Science, 256*, 1298–1303.

Herendeen, S. L., VanBogelen, R. A., and Neidhardt, F. C. (1978). Levels of the major proteins of *Escherichia coli* during growth at different temperatures. *Journal of Bacteriology, 139*, 185–194.

Hertz, G. Z., Hartzell, G. W., and Stormo, G. D. (1990). Identification of consensus patterns in unaligned DNA sequences known to be functionally related. *Computer Applications in the Biosciences, 6*(2), 81–92.

Hill, C. M., Waight, R. D., and Bardsley, W. G. (1977). Does any enzyme follow the Michaelis-Menten equation? *Molecular and Cellular Biochemistry, 15*, 173–178.

Hill, M. O. (1974). Correspondence analysis: A neglected multivariate method. *Applied Statistics, 23*, 340–353.

Hill, T. L. (1985). *Cooperativity theory in biochemistry: Steady-state and equilibrium systems.* Berlin: Springer-Verlag.

Hiromi, K. (1979). *Kinetics of fast enzymatic reactions.* New York: Halsted Press.

Hirosawa, M., Hoshida, M., Ishikawa, M., and Toya, T. (1993). MASCOT: Multiple alignment system for protein sequences based on three-way dynamic programming. *Computer Applications in the Biosciences, 9*, 161–167.

Hirschhorn, J. N., Brown, S. A., Clark, C. D., and Winston, F. (1992). Evidence that SNF2/SW12 and SNF5 activate transcription in yeast by altering chromatin structure. *Genes and Development, 6*, 2288–2298.

Hirschman, J., Wong, P. K., Sei, K., Keener, J., and Kustu, S. (1985). Products of nitrogen regulatory genes ntrA and ntrC of enteric bacteria activate glnA transcription in vitro: Evidence that the ntrA product is a sigma factor. *Proceedings of the National Academy of Sciences of the United States of America, 82*, 7525–7529.

Hirst, J. D., and Sternberg, M. J. E. (1991). Prediction of ATP-binding motifs: A comparison of a perceptron-type neural network and a consensus sequence method. *Protein Engineering, 4*, 615–623.

Hlavacek, W. S., and Savageau, M. A. (1995). Subunit structure of regulator proteins influences the design of gene circuitry: Analysis of perfectly coupled and completely uncoupled circuits. *Journal of Molecular Biology, 248*, 739–755.

Hodgman, T. C. (1989). The elucidation of protein function by sequence motif analysis. *Computer Applications in the Biosciences, 5,* 1–14.

Höfestadter, D. (1979). *Godel, Escher, Bach: An eternal golden braid.* New York: Basic Books.

Holbrook, S., Muskal, S., and Kim, S. (1990). Predicting surface exposure of amino acids for protein sequence. *Protein Engineering, 3*(8), 659–665.

Holley, L. H., and Karplus, M. (1989). Protein structure prediction with a neural network. *Proceedings of the National Academy of Sciences of the United States of America, 86,* 152–156.

Holm, L., and Sander, C. (1993). Structural alignment of globins, phycocyanins and colicin-A. *FEBS Letters, 315,* 301–306.

Hopcroft, J. E., and Ullman, J. D. (1979). *Introduction to automata theory, languages, and computation.* Reading, MA: Addison-Wesley.

Hope, I., and Struhl, K. (1986). Functional dissection of a eukaryotic transcriptional activator protein, GCN4 of yeast. *Cell, 46,* 885–894.

Horiuchi, J., Silverman, N., Marcus, G. A., and Guarente, L. (1995). ADA3, a putative transcriptional adaptor, consists of two separable domains and interacts with ADA2 and GCN5 in a trimeric complex. *Molecular and Cellular Biology, 15,* 1203–1209.

Horton, P. B., and Kanehisa, M. (1992). An assessment of neural network and statistical approaches for prediction of *E. coli* promoter sites. *Nucleic Acids Research, 16,* 4331–4338.

Huang, W. M., Ao, S. Z., Casjens, S., Orlandi, R., and Zeikus, R. (1988). A persistent untranslated sequence within bacteriophage T4 DNA topoisomerase gene 60. *Science, 239,* 1005–1012.

Hunter, L. (1993). *Artificial intelligence and molecular biology.* Cambridge, MA: AAAI Press/The MIT Press.

Hunter, W. N., Langlois d'Estainot, B., and Kennard, O. (1989). Structural variation in d(CTCTAGAG). Implications for protein-DNA interactions. *Biochemistry, 28,* 2444–2451.

Hurson, A. R., Bright, M. W., and Pakzad, S. (1994). *Multidatabase systems: An advanced solution for global information sharing.* Los Alamitos, CA: IEEE Computer Society Press.

Ingham, P. W. (1988). The molecular genetics of embryonic pattern formation in *Drosophila. Nature, 335,* 25–34.

Ingraham, J. L., Maaløe, O., and Neidhardt, F. C. (1983). *Growth of the bacterial cell.* Sunderland, MA: Sinauer Associates.

Inostroza, J., Flores, O., and Reinberg, D. (1991). Factors involved in specific transcription by mammalian RNA polymerase II. Purification and functional analysis of general transcription factor IIE. *Journal of Biology Chemistry, 266,* 9304–9308.

Irvine, D. H. (1991). The method of controlled mathematical comparison. In E. O. Voit, (Ed.). *Canonical nonlinear modeling: S-system approach to understanding complexity.* New York: Van Nostrand Reinhold, pp. 90–109.

Irvine, D. H., and Savageau, M. A. (1990). Efficient solution of nonlinear ordinary differential equations expressed in S-system canonical form. *Society for Industrial and Applied Mathematics, Journal on Numerical Analysis, 27,* 704–735.

Jiang, Z., and Ebner, C. (1990). Simulation study of reaction fronts. *Physical Reviews [A], 42,* 7483–7486.

Johnson, M. S., Overington, J., and Blundell, T. L. (1993). Alignment and searching for common protein folds using a data bank of structural templates. *Journal of Molecular Biology, 231,* 735–752.

Jones, D. T., Orengo, C. A., Taylor, W. R., and Thornton, J. M. (1993). Progress towards recognising protein folds from amino acid sequence. Presented at the conference, Advances in Gene Technology: Protein Engineering and Beyond, Miami.

Jones, D. T., Taylor, W. R., and Thornton, J. M. (1992). The rapid generation of mutation data matrices from protein sequences. *Computer Applications in the Biosciences, 8,* 275–282.

Kabsch, W., and Sander, C. (1983). How good are predictions of protein secondary structure? *FEBS Letters, 155,* 179–182.

Kacser, H., and Burns, J. A. (1973). Control of enzyme flux, *Symposia of the Society for Experimental Biology, 27,* 65–104.

Kang, K., and Redner, S. (1984). Scaling approach for the kinetics of recombination processes. *Physical Review Letters, 52,* 955–958.

Karam, J. D. (Ed.). (1994). *Molecular biology of bacteriophage T4.* Washington, DC: American Society for Microbiology.

Karlin, S. (1990). Distribution of clusters of charged amino acid in protein sequences. In R. H. Sarma and M. H. Sarma (Eds.), *DNA protein complexes and proteins* (Vol. 2). Schenectady, NY: Adenine Press.

Karlin, S., Blaisdell, B. E., Mocarski, E. S., and Brendel, V. (1989a). A method to identify distinctive charge configurations in protein sequences, with applications to human herpesvirus polypeptides. *Journal of Molecular Biology, 205,* 165–177.

Karlin, S., and Brendel, V. (1992). Chance and statistical significance in protein and DNA sequence analysis. *Science, 257,* 39–49.

Karlin, S., and Macken, C. (1991). Some statistical problems in the assessment of inhomogeneities of DNA sequence data. *Journal of the American Statistical Association, 86*(413), 27–35.

Karlin, S., Ost, F., and Blaisdell, B. E. (1989b). Patterns in DNA and amino acid sequences and their statistical significance. In M. S. Waterman (Ed.), *Mathematical methods for DNA sequences.* Boca Raton, FL: CRC Press.

Karp, P. (1992). A knowledge base of chemical compounds of intermediary metabolism. *Computer Applications in the Biosciences (CABIOS), 8,* 347–357.

Karp, P., and Riley, M. (1994). Representation of metabolic knowledge. In L. Hunter, D. Searls, and J. Shavlik (Eds.), *Proceedings of the First International Conference on Intelligent Systems and Molecular Biology.* Washington, DC: AAI Press.

Karp, R. M., Miller, R. E., and Rosenberg, A. L. (1972). Rapid identification of repeated patterns in strings, trees and arrays. *Proceedings of the 4th Annual ACM Symposium on the Theory of Computing,* 125–136.

Kashlev, M., Nudler, E., Goldfarb, A., White, T., and Kutter, E. (1993). Bacteriophage T4 Alc protein: A transcription termination factor sensing local modification of DNA. *Cell, 75,* 147–154.

Kauffman, S. (1993). *The origins of order.* Oxford, UK: Oxford University Press.

Kauffman, S. A. (1969). Metabolic stability and epigenesis in randomly constructed genetic nets. *Journal of theoretical Biology, 22,* 437–467.

Kaufman, M., and Thomas, R. (1987). Model analysis of the bases of multistationarity in the humoral immune response. *Journal of Theoretical Biology, 129,* 141–162.

Kim, B., and Little, J. W. (1992). Dimerization of a specific DNA-binding protein on the DNA. *Science, 255,* 203–206.

Kim, J. S., and Davidson, N. (1974). Electron microscope heteroduplex study of sequence relations of T2, T4 and T6 bacteriophage DNAs. *Virology*, *57*, 93–111.

Kneller, D. G., Cohen, F. E., and Langridge, R. (1990). Improvements in protein secondary structure prediction by an enhanced neural network. *Journal of Molecular Biology*, *214*, 171–182.

Knighton, D. R., Zheng, J., Ten Eyck, L. F., Ashford, V. A., Xuong, N.-H., Taylor, S. S., and Sowadski, J. M. (1991). Crystal structure of the catalytic subunit of cyclic adenosine monophosphate-dependent protein kinase. *Science*, *253*, 407–414.

Kohara, Y., Akiyama, K., and Isono, K. (1987). The physical map of the whole *E. coli* chromosome: Application of a new strategy for rapid analysis and sorting of a large genomic library. *Cell.*, *50*, 495–508.

Koleske, A. J., and Young, R. A. (1994). An RNA polymerase II holoenzyme responsive to activators. *Nature*, *368*, 466–469.

Koo, Y.-E. L., and Kopelman, R. (1991). Space- and time-resolved diffusion-limited binary reaction kinetics in capillaries: Experimental observation of segregation, anomalous exponents, and depletion zone. *Journal of Statistical Physics*, *65*, 893–918.

Koonin, E., Bork, P., and Sander, C. (1994). Yeast chromosome III: New gene functions. *EMBO Journal*, *13*, 493–503.

Kopelman, R. (1986). Rate processes on fractals: Theory, simulations, and experiments. *Journal of Statistical Physics*, *42*, 185–200.

Kretzschmar, M., Kaiser, K., Lottspeich, F., and Meisterernst, M. (1994). A novel mediator of class II gene transcription with homology to viral immediate–early transcriptional regulators. *Cell*, *78*, 525–534.

Kruskal, J. B., and Sankoff, D. (1983). An anthology of algorithms and concepts for sequence comparison. In K. Sankoff and J. B. Kruskal (Eds.), *Time warps, string edits, and macromolecules: The theory and practice of sequence comparison*. Reading MA: Addison-Wesley.

Kustu, S., Hirschman, J., Burton, D., Jelesko, J., and Meeks, J. C. (1984). Covalent modification of bacterial glutamine synthetase: Physiological significance. *Molecular and General Genetics*, *197*, 309–317.

Kutter, E., D'Acci, K., Drivdahl, R., Gleckler, J., McKinney, J. C., Peterson, S., and Guttman, B. S. (1994a). Identification of bacteriophage T4 prereplicative proteins on two-dimensional polyacrylamide gels. *Journal of Bacteriology*, *176*, 1647–1654.

Kutter, E., Gachechiladze, K., Poglazov, A., Marusich, E., Shneider, M., Aronsson, P., Napuli, A., Porter, D., and Mesyanzhinov, V. (1996). Evolution of T4-related phages. *Virus Genes*, *11*, 213.

Kutter, E., Stidham, T., Guttman, B., Kutter, E., Batts, D., Peterson, S., Djavakhishvili, T., Arisaka, F., Mesyanzhinov, V., Rüger, W., and Mosig, G. (1994b). Genomic map of bacteriophage T4. In J. D. Karam (Ed.), *Molecular biology of bacteriophage T4*. Washington, DC: American Society for Microbiology.

Kutter, E., White, T., Kashlev, M., Uzan, M., McKinney, J., and Guttman, B. (1994c). Effects on host genome structure and expression. In J. D. Karam (Ed.), *Molecular biology of bacteriophage T4*. Washington, DC: American Society for Microbiology.

Landès, C., Hénaut, A., and Risler, J.-L. (1992). A comparison of several similarity indexes used in the classification of protein sequences: A multivariate analysis. *Nucleic Acids Research*, *20*, 3631–3637.

Landès, C., Hénaut, A., and Risler, J.-L. (1993). Dot-plot comparisons by multivariate analysis (DOCMA): A tool for classifying protein sequences. *Computer Application in the Biosciences*, *9*, 191–196.

Médigue, C., Viari, A., Hénaut, A., and Danchin, A. (1991b). *Escherichia coli* molecular genetic map (1500 kbp): Update II. *Molecular Microbiology, 5,* 2629–2640.

Médigue, C., Viari, A., Hénaut, A., and Danchin, A. (1993). Colibri: A functional data base for the *Escherichia coli* genome. *Microbiological Reviews, 57,* 623–654.

Meinhardt, H., and Gierer, A. (1974). Applications of a theory of biological pattern formation based on lateral inhibition. *Journal of Cell Science, 15,* 321–346.

Mengeritsky, G., and Smith, T. F. (1987). Recognition of characteristic patterns in sets of functionally equivalent DNA sequences. *Computer Applications in the Biosciences, 3*(3), 223–227.

Merrill, P. T., Sweeton, D., and Wieschaus, E. (1988). Requirements for autosomal gene activity during precellular stages of *Drosophila melanogaster*. *Development, 104,* 495–509.

Miller, E., Karam, J., and Spicer, E. (1994). Control of translation initiation: mRNA structure and protein repressors. In J. D. Karam (Ed.), *Molecular biology of bacteriophage T4*. Washington, DC: American Society for Microbiology.

Miller, J. H., and Reznikoff, W. S. (Eds.). (1980). *The Operon*. New York: Cold Spring Harbor Laboratory.

Minsky, M. (1968). Matter, mind, and models. In M. Minsky (Ed.), *Semantic information processing*. Cambridge, MA: MIT Press.

Minton, A. P. (1992). Confinement as a determinant of macromolecular structure and reactivity. *Biophysical Journal, 63,* 1090–1110.

Miyazawa, S., and Jernigan, R. L. (1993). A new substitution matrix for protein sequence searches based on contact frequencies in protein structures. *Protein Engineering, 6,* 267–278.

Mjolsness, E., Sharp, D. H., and Reinitz, J. (1991). A connectionist model of development. *Journal of Theoretical Biology, 152,* 429–453.

Monod, J., Wyman, J., and Changeux, J.-P. (1965). On the nature of allosteric transitions: A plausible model. *Journal of Molecular Biology, 12,* 88–118.

Moore, R. (1986). Computational techniques. In G. Bristow (Ed.), *Electronic speech recognition*. New York: McGraw-Hill.

Moreau, P., Hen, R., Wasylyk, B., Everett, R., Gaub, M. P., and, Chambon, P. (1981). The SV40 72 base repair repeat has a striking effect on gene expression both in SV40 and other chimeric recombinants. *Nucleic Acids Research, 9,* 6047–6068.

Muskal, S., Holbrook, S., and Kim, S. (1990). Prediction of the disulfide-bonding state of cysteine in proteins. *Protein Engineering, 3*(8), 667–672.

Myers, E. W., and Miller, W. (1989). Approximate matching of regular expressions. *Bulletin of Mathematical Biology, 51,* 5–37.

Myhill, J. (1952). Some philosophical implications of mathematical logic: I. Three classes of ideas. *The Review of Metaphysics, 6,* 165–198.

Nakata, K., Kanehisa, M., and Maizel, J. V., Jr. (1988). Discriminant analysis of promoters regions in *Escherichia coli* sequences. *Computer Applications in the Biosciences, 4,* 367–371.

Needleman, S. B., and Wunsch, C. D. (1970). A general method applicable to the search for similarities in the amino acid sequence of two proteins. *Journal of Molecular Biology, 48,* 443–453.

Neer, E. J., Schmidt, C. J., Nambudripad, R., and Smith, T. F. (1994). The ancient regulatory-protein family of WD-repeat proteins. *Nature, 371,* 297–300.

Neidhardt, F. C., Appleby, D. B., Sankar, P., Hutton, M. E., and Phillips, T. A. (1989). Genomically linked cellular protein databases derived from two-dimensional polyacrylamide gel electrophoresis. *Electrophoresis, 10,* 116–122.

Neidhardt, F. C., Vaughn, V., Phillips, T. A., and Bloch, P. L. (1983). Gene-protein index of *Escherichia coli* K–12. *Microbiological Reviews, 47*, 231–284.

Newhouse, J. S., and Kopelman, R. (1988). Steady-state chemical kinetics on surface clusters and islands: Segregation of reactants. *Journal of Physical Chemistry, 92*, 1538–1541.

Nierlich, D. (1992). Genetics nomenclature enters the computer age. *ASM News, 58*, 645–646.

Nivinskas, R., Vaiskunaite, R., and Raudonikiene, A. (1992). An internal AUU codon initiates a smaller peptide encoded by bacteriophage T4 baseplate gene 26. *Molecular and General Genetics, 232*, 257–261.

Nomenclature Committee of the International Union of Biochemistry. (1984). *Enzyme nomenclature*. New York: Academic Press.

Nossal, N. G. (1994). The bacteriophage T4 DNA replication fork. In J. D. Karam (Ed.), *Molecular biology of bacteriophage T4*. Washington, DC: American Society for Microbiology.

Novick, A., and Weiner, M. (1957). Enzyme induction, an all or none phenomenon. *Proceedings of the National Academy of Sciences of the United States of America, 43*, 553–566.

Nussinov, R., and Wolfson, H. J. (1991). Efficient detection of three-dimensional structural motifs in biological macromolecules by computer vision techniques. *Proceedings of the National Academy of Sciences of the United States of America, 88*, 10495–10499.

Nyström, T., and Neidhardt, F. C. (1992). Cloning, mapping and nucleotide sequencing of a gene encoding a universal stress protein in *Escherichia coli*. *Molecular Microbiology, 6*, 3187–3198.

Nyström, T., and Neidhardt, F. C. (1993). Isolation and properties of a mutant of *Escherichia coli* with an insertional inactivation of the *uspA* gene, which encodes a universal stress protein. *Journal of Bacteriology, 175*, 3949–3956.

Nyström, T., and Neidhardt, F. C. (1994). Expression and role of the universal stress protein, UspA, of *Escherichia coli* during growth arrest. *Molecular Microbiology, 11*, 537–544.

O'Farrell, P. H. (1975). High resolution two-dimensional electrophoresis of proteins. *Journal of Biological Chemistry, 250*, 4007–4021.

O'Farrell, P. Z., Goodman, H. M., and O'Farrell, P. H. (1977). High resolution two-dimensional electrophoresis of basic as well as acidic proteins. *Cell, 12*, 1133–1142.

Ohkuma, Y., Sumimoto, H., Horikoshi, M., and Roeder, R. G. (1990). Factors involved in specific transcription by mammalian RNA polymerase II: purification and characterization of general transcription factor TFIIE. *Proceedings of the National Academy of Sciences of the United States of America, 87*, 9163–9167.

Orengo, C. A., Flores, T. P., Jones, D. T., Taylor, W. T., and Thornton, J. M. (1993). Recurring structural motifs in proteins with different functions. *Current Biology, 3*, 131–139.

Orengo, C. A., and Taylor, W. R. (1993). A local alignment method for protein structure motifs. *Journal of Molecular Biology, 233*, 488–497.

Orengo, C. A., and Thornton, J. M. (1993). Alpha plus beta folds revisited: Some favoured motifs. *Current Biology, 1*, 105–120.

Ovchinnikov, A. A., and Zeldovich, Ya. B. (1978). Role of density fluctuations in bimolecular reaction kinetics. *Chemical Physics, 28*, 215–218.

Panjukov, V. V. (1993). Finding steady alignments: Similarity and distance. *Computer Applications in the Biosciences, 9*, 285–290.

Pao, Y.-H. (1989). *Adaptive pattern recognition and neural networks*. Reading, MA: Addison-Wesley.

Patel, N. H., Condron, B. G., and Zinn, K. (1994). Pair-rule expression patterns of evenskipped are found in both short- and long-germ beetles. *Nature, 367*, 429–434.

Patthy, L. (1987). Detecting homology of distantly related proteins with consensus sequences. *Journal of Molecular Biology, 198*, 567–577.

Pavletich, N. P., and Pabo, C. O. (1991). Zinc finger-DNA recognition: Crystal structure of a Zif268-DNA complex at 2.1 A. *Science, 252*, 809–817.

Pearson, W. R., and Lipman, D. J. (1988). Improved tools for biological sequence comparison. *Proceedings of the National Academy of Sciences of the United States of America, 85*, 2444–2448.

Pearson, W. R., and Miller, W. (1992). Dynamic programming algorithms for biological sequence comparison. *Methods in Enzymology, 210*, 575–601.

Pedersen, S., Bloch, P. L., Reeh, S., and Neidhardt, F. C. (1978). Patterns of protein synthesis in *E. coli*: A catalog of the amount of 140 individual proteins at different growth rates. *Cell, 14*, 179–190.

Petrilli, P. (1993). Classification of protein sequences by their dipeptide composition. *Computer Applications in the Biosciences, 9*, 205–209.

Petrosky, H. (1992). *To engineer is human*. New York: Vintage Books.

Phillips, T. A., Vaughn, V., Bloch, P. L., and Neidhardt, F. C. (1987). Gene-protein index of *Escherichia coli* K–12 (Vol. 2). In F. C. Neidhardt, J. L. Ingraham, K. Brooks Low, B. Magasanik, M. Schaecter, and H. E. Umbarger (Eds.), *Cellular and molecular biology:* Escherichia coli *and* Salmonella typhimurium. Washington, DC: American Society for Microbiology.

Ponder, J. W., and Richards, F. M. (1987). Tertiary templates for proteins: Use of packing criteria in the enumeration of allowed sequences for different structural classes. *Journal of Molecular Biology, 193*, 775–791.

Pongor, S., Skerl, V., Cserzo, M., Hatsagi, Z., Simon, G., and Bevilacqua, V. (1993). The SBASE protein domain library, release 2.0: A collection of annotated protein sequence alignments. *Nucleic Acids Research, 21*, 3111–3115.

Presnell, S. R., Cohen, B. I., and Cohen, F. E. (1992). A segment-based approach to protein secondary structure prediction. *Biochemistry, 31*, 983–993.

Presta, L. G., and Rose, G. D. (1988). Helix signals in proteins. *Science, 240*, 1632–1652.

Prestridge, D. S., and Stormo, G. (1993). SIGNAL SCAN 3.0: New database and program features. *Computer Applications in the Biosciences, 9*, 113–115.

Pribnow, D. (1979). Genetic control signals in DNA. In R. F. Goldberger (Ed.), *Biological regulation and development* (Vol. 1). New York: Plenum.

Ptashne, M. (1992). *A genetic switch* (2nd ed.). Cambridge, MA: Cell Press/Blackwell Scientific.

Qian, N., and Sejnowski, T. J. (1988). Predicting the secondary structure of globular proteins using neural network models. *Journal of Molecular Biology, 202*, 865–884.

Reddy, G. P. V., and Matthews, C. K. (1978). Functional compartmentation of DNA precursors in T4 phage-infected bacteria. *Journal of Biology Chemistry, 253*, 3461–3467.

Reid, R. C., Prausnitz, J. M., and Poling, B. E. (1987). *The properties of gases and liquids* (4th ed.). New York: McGraw-Hill.

Reinitz, J., Mjolsness, E., and Sharp, D. H. (1992). *Cooperative control of positional information in* Drosophila *by* bicoid *and maternal* hunchback (Tech. Rep. No. LAUR-92-2942). Los Alamos: Los Alamos National Laboratory. (URL file://sunsite.unc.edu/pub/academic/biology/ecology +evolution/papers/drosophila_theory/Positional_Info.ps.)

Reinitz, J., Mjolsness, E., and Sharp, D. H. (1995). Cooperative control of positional information in *Drosophila* by *bicoid* and maternal *hunchback. Journal of Experimental Zoology, 271,* 47–56.

Reinitz, J., and Sharp, D. H. (1995). Mechanism of formation of eve stripes. *Mechanisms of Development, 49,* 133–158.

Reiss, M. J. (1989). *The allometry of growth and reproduction.* New York: Cambridge University Press.

Reitzer, L. J., Bueno, R., Cheng, W. D., Abrams, S. A., Rothstein, D. M., Hunt, T. P., Tyler, B., and Magasanik, B. (1987). Mutations that create new promoters suppress the sigma 54 dependence of glnA transcription in *Escherichia coli. Journal of Bacteriology, 169,* 4279–4284.

Reitzer, L. J., and Magasanik, B. (1986). Transcription of glnA in *E. coli* is stimulated by activator bound to sites far from the promoter. *Cell, 45,* 785–792.

Repoila, F., Tetart, F., Bouet, J. Y., and Krisch, H. M. (1994). Genomic polymorphism in the T-even bacteriophages. *EMBO Journal, 13,* 4181–4192.

Rhodes, D., and Klug, A. (1986). An underlying repeat in some transcriptional control sequences corresponding to half a double helical turn of DNA. *Cell, 46,* 123–132.

Richards, F. J. (1969). The quantitative analysis of growth. In F. C. Steward (Ed.), *Plant physiology* (Vol. 5A). New York: Academic Press.

Richards, F. M., and Kundrot, C. E. (1988). Identification of structural motifs from protein coordinate data: Secondary structure and first-level supersecondary structure. *Proteins, 3,* 71–84.

Riddihough G., and Ish-Horowicz, D. (1991). Individual stripe regulatory elements in the *Drosophila* hairy promoter respond to maternal, gap, and pair-rule genes. *Genes and Development, 5,* 840–854.

Riley, M. (1993). Functions of the gene products of *Escherichia coli. Microbiological Reviews, 57,* 862–952.

Riley, M., and Krawiec, S. (1987). Genome organization. In F. C. Neidhardt, J. L. Ingraham, K. Brooks Low, B. Magasanik, M. Schaechter, and H. E. Umbarger (Eds.), Cellular and molecular biology: Escherichia coli *and* Salmonella typhimurium (Vol. 2). Washington, DC: American Society for Microbiology.

Ring, C. S., and Cohen F. E. (1993). Modeling protein structures: Construction and their applications. *FASEB J, 7,* 783–790.

Risler, J.-L., Delorme, M.-O., Delacroix, H., and Hénaut, A. (1988). Amino acid substitutions in structurally related proteins. A pattern recognition approach. *Journal of Molecular Biology, 204,* 1019–1029.

Ritter, J. K., Chen, F., Sheen, Y. Y., Tran, H. M., Kimura, S., Yeatman, M. T., and Owens, I. S. (1992). A novel complex locus *UGT1* encodes human bilirubin, phenol, and other UDP-glucuronosyltransferase isozymes with identical carboxyl termini. *Journal of Biological Chemistry, 267,* 3257–3261.

Robbins, R. J. (1992). Database and computational challenges in the human genome project. *IEEE Engineering in Medicine and Biology Magazine, 11,* 25–34.

Robbins, R. J. (1993). Genome informatics: Requirements and challenges. In H. A. Lim, J. W. Fickett, C. R. Cantor, and R. J. Robbins (Eds.), *Bioinformatics, supercomputing and complex genome analysis.* Singapore: World Scientific Publishing.

Robbins, R. J. (1994a). Biological databases: A new scientific literature. *Publishing Research Quarterly, 10*, 1–27.

Robbins, R. J. (1994b). Genome informatics I: Community databases. *Journal of Computational Biology, 1*, 173–190.

Robbins, R. J. (1994c). Representing genomic maps in a relational database. In S. Suhai (Ed.), *Computational methods in genome research*. New York: Plenum.

Robbins, R. J. (1994d). *Genome informatics: Toward a federated information infrastructure* [keynote address]. Presented at the Third International Conference on Bioinformatics and Genome Research, Tallahassee, Florida, June 1–4, 1994.

Robbins, R. J. (1995). Information infrastructure. *IEEE Engineering in Medicine and Biology Magazine. 14*, 746–759.

Roberts, L. (1989). New chip may speed genome analysis. *Science, 244*, 655–656.

Robson, B., and Greaney, P. J. (1992). Natural sequence code representations for compression and rapid searching of human-genome style databases. *Computer Applications in the Biosciences, 8*, 283–289.

Roof, D. M., and Roth, J. R. (1992). Autogenous regulation of ethanolamine utilization by a transcriptional activator of the *eut* operon in *Salmonella typhimurium. Journal of Bacteriology, 174*, 6634–6643.

Rose, J., and Eisenmenger, F. (1991). A fast unbiased comparison of protein structures by means of the Needleman and Wunsch algorithm. *Journal of Molecular Evolution, 32*, 340–354.

Rosenberg, A. H., Lade, B. N., Chui, D-S., Lin, S-W., Dunn, J. J., and Studier, F. W. (1987). Vectors for selective expression of cloned DNAs by T7 RNA polymerase. *Gene, 56*, 125–135.

Rosenblatt, F. (1962). *Principles of neurodynamics*. New York: Spartan.

Rosenblenth, A., and Weiner, N. (1945). *Philosophy of Science, 12*, 316–321.

Rossman, M. G., Moras, D., and Olsen, K. W. (1974). Chemical and biological evolution of a nucleotide-binding protein. *Nature, 250*, 194–199.

Rouxel, T., Danchin, A., and Hénaut, A. (1993). METALGEN.DB: Metabolism linked to the genome of *Escherichia coli*, a graphics-oriented database. *Computer Applications in the Biosciences, 9*, 315–324.

Rudd, K. E. (1993). Maps, genes, sequences, and computers: An *Escherichia coli* case study. *ASM News, 59*, 335–341.

Ruelle, D. (1991). *Chance and chaos*. Princeton, NJ: Princeton University Press.

Rumelhart, D. E., Hinton, G. E., and Williams, R. J. (1986). *Parallel distributed processing: Explorations in the microstructure of cognition*. Cambridge, MA: MIT Press.

Russell, R. L. (1974). Comparative genetics of T-even bacteriophages. *Genetics, 78*, 967–988.

Russell, R. L., and Huskey, R. J. (1974). Partial exclusion between T-even bacteriophages: An incipient genetic isolation mechanism. *Genetics, 78*, 989–1014.

Sanderson, K. E., and Hall, C. A. (1970). F-prime factors of *Salmonella typhimurium* and inversion between *S. typhimurium* and *Escherichia coli. Genetics, 64*, 215–228.

Sankar, P., Hutton, M. E., VanBogelen, R. A., Clark, R. L., and Neidhardt, F. C. (1993). Expression analysis of cloned chromosomal segments of *Escherichia coli. Journal of Bacteriology, 175*, 5145–5152.

Sankoff, D. (1972). Matching sequences under deletion/insertion constraints. *Proceedings of the National Academy of Sciences of the United States of America, 69*, 4–6.

Sankoff, D., and Kruskal, J. B. (1983). *Time warps, string edits, and macromolecules: The theory and practice of sequence comparison*. Reading, MA: Addison-Wesley.

Saqi, M. A. S., and Sternberg, M. J. E. (1991). A simple method to generate non-trivial alternate alignments of protein sequences. *Journal of Molecular Biology, 219*, 727–732.

Savageau, M. A. (1969). Biochemical systems analysis: II. The steady state solutions for an n-pool system using a power-law approximation. *Journal of Theoretical Biology, 25*, 370–379.

Savageau, M. A. (1971). Concepts relating the behavior of biochemical systems to their underlying molecular properties. *Archives of Biochemistry and Biophysics, 145*, 612–621.

Savageau, M. A. (1976). *Biochemical systems analysis: A study of function and design in molecular biology*. Reading, MA: Addison-Wesley.

Savageau, M. A. (1977). Design of molecular control mechanisms and the demand for gene expression. *Proceedings of the National Academy of Sciences of the United States of America, 74*, 5647–5651.

Savageau, M. A. (1983). Regulation of differentiated cell-specific functions. *Proceedings of the National Academy of Sciences of the United States of America, 80*, 1411–1415.

Savageau, M. A. (1989). Are there rules governing patterns of gene regulation? In B. C. Goodwin, and P. T. Saunders (Eds.), *Theoretical biology—epigenetic and evolutionary order*. Edinburgh: Edinburgh University Press.

Savageau, M. A. (1991). Biochemical systems theory: Operational differences among variant representations and their significance. *Journal of Theoretical Biology, 151*, 509–530.

Savageau, M. A. (1992). A critique of the enzymologist's test tube. In E. E. Bittar (Ed.), *Fundamentals of medical cell biology* (Vol. 3A). Greenwich, CT: JAI Press.

Savageau, M. A. (1993a). Finding multiple roots of nonlinear algebraic equations using S-system methodology. *Applied Mathematics and Computation, 55*, 187–199.

Savageau, M. A. (1993b). Influence of fractal kinetics on molecular recognition. *Journal of Molecular Recognition, 6*, 149–157.

Savageau, M. A. (1996). Power-law formalism: A canonical nonlinear approach to modeling and analysis. In V. Lakshmikantham (Ed.), *World Congress of Nonlinear Analysts, 92*, vol. 4, Berlin: Walter de Gruyter Publishers.

Savageau, M. A. (1995). Michaelis-Menten mechanism reconsidered: Implications of fractal kinetics. *Journal of Theoretical Biology, 176*, 115–124.

Savageau, M. A., and Jacknow, G. (1979). Feedforward inhibition in biosynthetic pathways: Inhibition of the aminoacyl-tRNA synthetase by intermediates of the pathway. *Journal of Theoretical Biology, 77*, 405–425.

Savageau, M. A., and Voit, E. O. (1987). Recasting nonlinear differential equations as S-systems: A canonical nonlinear form. *Mathematical Biosciences, 87*, 83–115.

Schaeffer, L., Moncollin, V., Roy, R., Staub, A., Mezzina, M., Sarasin, A., Weeda, G., Hoeijmakers, J. H. J., and Egly, J. M. (1994). The ERCC2/DNA repair protein is associated with the class II BTF2/TFIIH transcription factor. *EMBO Journal, 13*, 2388–2392.

Schaeffer, L., Roy, R., Humbert, S., Moncollin, V, Vermeulen, W., Hoeijmakers, J. H. H., Chambon, P., and Egly, J. M. (1993). DNA repair helicase: A component of BTF2 (TFIIH) basic transcription factor. *Science, 260*, 58–63.

Schneider, T. D., Stormo, G. D., Gold, L., and Ehrenfeucht, A. (1986). Information content of binding sites on nucleotide sequences. *Journal of Molecular Biology, 188*, 415–431.

Schroeder, M. (1991). *Fractals, chaos, power laws*. New York: Freeman.

Schuler, G. D., Altschul, S. F., and Lipman, D. J. (1991). A workbench for multiple alignment construction and analysis. *Proteins, 9*, 180–191.

Schwartz, M. (1987). The Maltose regulon. In F. C. Neidhardt, J. L. Ingraham, K. Brooks Low, B. Magasanik, M. Schaechter, and H. E. Umbarger (Eds.), *Cellular and molecular biology: Escherichia coli and Salmonella typhimurium* (Vol. 2). Washington, DC: American Society for Microbiology.

Searls, D. (1992). The linguistics of DNA. *American Scientist, 80*, 579–591.

Searls, D. (1993). *Artificial intelligence and molecular biology*. Cambridge, MA: MIT Press.

Searls, D., and Dong, S. (1993). A syntactic pattern recognition system for DNA sequences. In H. Lim, J. Fickett, C. Cantor, and R. Robbins (Eds.), *Proceedings of the Second International Conference on Bioinformatics, Supercomputing and Complex Genome Analysis*. Singapore: World Scientific Publishing.

Selick, H. E., Stormo, G. D., Dyson, R. L., and Alberts, B. M. (1993). Analysis of five presumptive protein-coding sequences clustered between the primosome genes, 41 and 61, of bacteriophages T4, T2, and T6. *Journal of Virology, 67*, 2305–2316.

Selkov, E. E. (1968). Self-oscillations in glycolysis: I. A simple kinetic model. *European Journal of Biochemistry, 4*, 79–86.

Sellers, P. H. (1974). On the theory and computation of evolutionary distances. *Society of Industrial and Applied Mathematics Journal of Applied Mathematics, 26*, 787–793.

Sentenac, A. (1985). Eukaryotic RNA polymerases. *CRC Critical Reviews in Biochemistry, 18*, 31–90.

Shannon, C. E. (Ed.). (1949). *The mathematical theory of communication*. Urbana: University of Illinois Press.

Sharma, M., Ellis, R., and Hinton, D. (1992). Identification of a family of bacteriophage T4 genes encoding proteins similar to those present in group I introns of fungi and phage. *Proceedings of the National Academy of Sciences of the United States of America, 89*, 6658–6662.

Sharp, P., and Li, W.-H. (1986). An evolutionary perspective on synonymous codon usage in unicellular organisms. *Journal of Molecular Evolution, 24*, 28–38.

Sharp, P., and Li, W.-H. (1987). The codon adaptation index—a measure of directional synonymous codon bias, and its potential applications. *Nucleic Acids Research, 15*, 1281–1295.

Sharp, P. M. (1991). Determinants of DNA sequence divergence between *Escherichia coli* and *Salmonella typhimurium*: Codon usage, map position, and concerted evolution. *Journal of Molecular Evolution, 33*, 23–33.

Shepherd, J. C. W. (1981). Method to determine the reading frame of a protein from the purine/pyrimidine genome sequence and its possible evolutionary justification. *Proceedings of the National Academy of Sciences of the United States of America, 78*, 1596–1600.

Sheth, A. P., and Larson, J. A. (1990). Federated database systems for managing distributed heterogeneous, and autonomous databases. *ACM Computing Surveys, 22*, 183–236.

Shiraishi, F., and Savageau, M. A. (1992a). The tricarboxylic acid cycle in *Dictyostelium discoideum*: I. Formulation of alternative kinetic representations. *Journal of Biological Chemistry, 267*, 22912–22918.

Shiraishi, F., and Savageau, M. A. (1992b). The tricarboxylic acid cycle in *Dictyostelium discoideum*: II. Evaluation of model consistency and robustness. *Journal of Biological Chemistry, 267*, 22919–22925.

Shiraishi, F., and Savageau, M. A. (1992c). The tricarboxylic acid cycle in *Dictyostelium discoideum*: III. Analysis of steady-state and dynamic behavior. *Journal of Biological Chemistry, 267*, 22926–22933.

Shiraishi, F., and Savageau, M. A. (1992d). The tricarboxylic acid cycle in *Dictyostelium discoideum*: IV. Resolution of discrepancies between alternative methods of analysis. *Journal of Biological Chemistry, 267*, 22934–22943.

Shiraishi, F., and Savageau, M. A. (1993). The tricarboxylic acid cycle in *Dictyostelium discoideum*: V. Systemic effects of including protein turnover in the current model. *Journal of Biological Chemistry, 268*, 16917–16928.

Shub, D., Coetzee, T., Hall, D., and Belfort, M. (1994). The self-splicing introns of bacteriophage T4. In J. D. Karam (Ed.), *Molecular biology of bacteriophage T4*. Washington, DC: American Society for Microbiology.

Sibbald, P. R., and Argos, P. (1990). Scrutineer: A computer program that flexibly seeks and describes motifs and profiles in protein sequence databases. *CABIOS, 6*(3), 279–288.

Silverman, N., Agapite, J., and Guarente, L. (1994). Yeast ADA2 protein binds to the VP16 protein activation domain and activates transcription. *Proceedings of the National Academy of Sciences of the United States of America, 91*, 11665–11668.

Simpson-Brose, M., Treisman, J., and Desplan, C. (1994). Synergy between two morphogens, bicoid and hunchback, is required for anterior patterning in *Drosophila*. *Cell, 78*, 855–865.

Singer, M., and Berg, P. (1991). *Genes and genomes*. Mill Valley, CA: University Science Books.

Singer, P. A., Levinthal, M., and Williams, L. S. (1984). Synthesis of the isoleucyl- and valyl-tRNA synthetases and the isoleucine-valine biosynthetic enzymes in a threonine deaminase regulatory mutant of *Escherichia coli* K–12. *Journal of Molecular Biology, 175*, 39–55.

Smale, S. T., and Baltimore, D. (1989). The "initiator" as a transcription control element. *Cell, 57*, 103–113.

Small, S., Blair, A., and Levine, M. (1992). Regulation of *even-skipped* stripe 2 in the *Drosophila* embryo. *The EMB0 Journal, 11*, 4047–4057.

Smith, G. R., and Magasanik, B. (1971). Nature and self-regulated synthesis of the repressor of the *hut* operons in *Salmonella typhimurium*. *Proceedings of the National Academy of Sciences of the United States of America, 68*, 1493–1497.

Smith, H. O., Annau, T. M., and Chandrasegaran, S. (1990). Finding sequence motifs in groups of functionally related proteins. *Proceedings of the National Academy of Sciences of the United States of America, 87*, 826–830.

Smith, R. (1988). A finite state machine algorithm for finding restriction sites and other pattern matching applications. *Computer Applications in the Biosciences, 4*, 459–465.

Smith, R. F., and Smith, T. F. (1989). Identification of new protein kinase–related genes in three herpesviruses: herpes simplex virus, varicella-zoster virus, and Epstein-Barr virus. *Journal of Virology, 63*(1), 450–455.

Smith, R. F., and Smith, T. F. (1990). Automatic generation of primary sequence patterns from sets of related protein sequences. *Proceedings of the National Academy of Sciences of the United States of America, 87*, 118–122.

Smith, R. F., and Smith, T. F. (1992). Pattern-induced multi-sequence alignment (PIMA) algorithm employing secondary structure-dependent gap penalties for use in comparative protein modelling. *Protein Engineering, 5*(1), 35–41.

Smith, T. F., and Waterman, M. S. (1981). Identification of common molecular subsequences. *Journal of Molecular Biology, 147,* 195–197.

Smith, T. F., Waterman, M. S., and Sadler, J. R. (1983). Statistical characterization of nucleic acid sequence functional domains. *Nucleic Acids Research, 11,* 2205–2220.

Snoussi, E. H. (1989). Qualitative dynamics of piece-linear differential equations: A discrete mapping approach. *Dynamics and Stability Systems, 4,* 189–207.

Snoussi, E. H., and Thomas, R. (1993). Logical identification of all steady states: The concept of feedback loop characteristic states. *Bulletin of Mathematical Biology, 55,* 973–991.

Snyder, E. E., and Stormo, G. D. (1993). Identification of coding regions in genomic DNA sequences: An application of dynamic programming and neural networks. *Nucleic Acids Research, 21,* 607–613.

Sober, E. (Ed.). (1984). *Conceptual issues in evolutionary biology.* Cambridge, MA: Mit Press.

Sorribas, A., and Savageau, M. A. (1989a). A comparison of variant theories of intact biochemical systems: I. Enzyme-enzyme interactions and biochemical systems theory. *Mathematical Biosciences, 94,* 161–193.

Sorribas, A., and Savageau, M. A. (1989b). A comparison of variant theories of intact biochemical systems: II. Flux-oriented and metabolic control theories. *Mathematical Biosciences, 94,* 195–238.

Spouge, J. L. (1991). Fast optimal alignment. *Computer Applications in the Biosciences, 7,* 1–7.

Srere, P., Jones, M. E., and Matthews, C. (1989). *Structural and organizational aspects of metabolic regulation.* New York: Alan R. Liss.

Staden, R. (1989). Methods for calculating the probabilities of finding patterns in sequences. *Computer Applications in the Biosciences, 5,* 89–96.

Stanojevic, D., Hoey, T., and Levine, M. (1989). Sequence-specific DNA-binding activities of the gap proteins encoded by *hunchback and Kruppel* in *Drosophila. Nature, 341,* 331–335.

States, D. J., and Botstein, D. (1991). Molecular sequence accuracy and the analysis of protein coding regions. *Proceedings of the National Academy of Sciences of the United States of America, 88,* 5518–5522.

Stormo, G. D. (1988). Computer methods for analyzing sequence recognition of nucleic acids. *Annual Review of Biophysics and Biophysical Chemistry, 17,* 241–263.

Stormo, G. D. (1990). Consensus patterns in DNA. *Methods in Enzymology, 183,* 211–221.

Stormo, G. D., and Hartzell, G. W. (1989). Identifying protein-binding sites from unaligned DNA fragments. *Proceedings of the National Academy of Sciences of the United States of America, 86,* 1183–1187.

Stormo, G. D., Schneider, T. D., and Gold, L. M. (1982a). Characterization of translational initiation sites in *E. coli. Nucleic Acids Research, 16,* 2971–2996.

Stormo, G. D., Schneider, T. D., Gold, L., and Ehrenfeucht, A. (1982b). Use of the Perceptron algorithm to distinguish translational initiation sites in *E. coli. Nucleic Acids Research, 10,* 2997–3011.

Strohman, R. (1994). Epigenesis: The missing beat in biotechnology? *Bio/Technology, 12,* 156–164.

Studier, F. W., and Moffatt, B. A. (1986). Use of bacteriophage T7 RNA polymerase of direct selective high-level expression of cloned genes. *Journal of Molecular Biology, 189*, 113–130.

Studier, F. W., Rosenberg, A. H., Dunn, J. J., and Dubendorff, J. W. (1990). Use of T7 RNA polymerase to direct expression of cloned genes. *Methods in Enzymology, 185*, 60–89.

Sturtevant, A. H. (1913). The linear arrangement of six sex-linked factors in *Drosophila* as shown by their mode of association. *Journal of Experimental Zoology, 14*, 43–59.

Sturtevant, A. H., and Beadle, G. W. (1939). *An introduction to genetics.* Philadelphia: Saunders.

Subbiah, S., Laurents, D. V., and Levitt, M. (1993). Structural similarity of DNA-binding domains of bacteriophage repressors and the globin core. *Current Biology, 3*, 141–148.

Sugita, M. (1961). Functional analysis of chemical systems in vivo using a logical circuit equivalent. *Journal of Theoretical Biology, 1*, 415–430.

Sweetser, D., Nonet, M., and Young, R. A. (1987). Prokaryotic and eukaryotic RNA polymerase have homologous core subunits. *Proceedings of the National Academy of Sciences of the United States of America, 84*, 1192–1996.

Tabor, S. (1990). Expression using the T7 RNA polymerase/promoter system. In F. A. Ausubel, R. Brent, R. E. Kingston, D. D. Moore, J. G. Seidman, J. A. Smith, and K. Struhl (Ed.), *Current protocols in molecular biology.* New York: Greene Publishing and Wiley Interscience.

Tabor, S., and Richardson, C. C. (1985). A bacteriophage T7 RNA polymerase/promoter system for controlled exclusive expression of specific genes. *Proceedings of the National Academy of Sciences of the United States of America, 82*, 1074–1078.

Tanese, N., Pugh, B. F., and Tjian, R. (1991). Coactivators for a proline-rich activator purified from the multisubunit human TFIID complex. *Genes and Development, 5*, 2212–2224

Taylor, E. W., Ramanathan, C. S., Jalluri, R. K., and Nadimpalli, R. G. (1994). A basis for new approaches to the chemotherapy of AIDS: Novel genes in HIV-1 potentially encode selenoproteins expressed by ribosomal frameshifting and termination suppression. *Journal of Medicinal Chemistry, 37*, 2637–2654.

Taylor, W. R. (1986). Identification of protein sequence homology by consensus template alignment. *Journal of Molecular Biology, 188*, 233–258.

Taylor, W. R. (1988). Pattern matching methods in protein sequence comparison and structure prediction. *Protein Engineering, 2*(2), 77–86.

Taylor, W. R., and Orengo, C. A. (1989). Protein structure alignment. *Journal of Molecular Biology, 208*, 1–22.

Thauer, R. K., Jungermann, K., and Decker, K. (1977). Energy conservation in chemotropic ancerobic bacteria. *Bacteriological Reviews, 41*, 100–180.

Thieffry, D., Colet, M., and Thomas, R. (1993). Formalisation of regulatory networks: A logical method and its automatization. *Mathematical Modelling and Scientific Computing, 2*, 144–151.

Thomas, R. (1973). Boolean formalization of genetic control circuits. *Journal of Theoretical Biology, 42*, 563–585.

Thomas, R. (Ed.). (1979). *Kinetic logic: A boolean approach to the analysis of complex regulatory systems.* Special edition of *Lecture of Notes in Biomathematics, 29*.

Thomas, R. (1983). Logical vs continuous description of systems comprising feedback loops: The relation between time delays and parameters. *Studies in Physical and Theoretical Chemistry, 28*, 307–321.

Thomas, R. (1991). Regulatory networks seen as asynchronous automata: A logical description. *Journal of Theoretical Biology, 153*, 1–23.

Thomas, R., and D'Ari, R. (1990). *Biological Feedback*. Boca Raton, FL: CRC Press.

Thomas, R., Thieffry, D., and Kaufman, M. (1995). Dynamical behavior of biological regulatory networks: I. Biological role of feedback loops and practical use of the concept of the loop-characteristic state. *Bulletin of Mathematical Biology, 57*, 247–276.

Thompson, C. M., Koleske, A. J., Chao, D. M., and Young, R. A. (1993). A multisubunit complex associated with the RNA polymerase II CTD and TATA-binding protein in yeast. *Cell, 73*, 1361–1375.

Thornton, J. M., and Gardner, S. P. (1989). Protein motifs and data-base searching. *Trends in the Biochemical Sciences, 14*, 300.

Toussaint, D., and Wilczek, F. (1983). Particle-antiparticle annihilation in diffusive motion. *Journal of Chemical Physics, 78*, 2642–2647.

Turing, A. M. (1952). The chemical basis of morphogenesis. *Transactions of the Royal Society of London, Series B, 237*, 37–72.

Tyson, J. (1975). Classification of instabilities in chemical reaction systems. *Journal of Chemical Physics, 62*, 1010–1015.

United States Department of Energy. (1990). *Understanding our genetic inheritance. The U.S. Human Genome Project: The first five years*. Bethesda, MD: National Institutes of Health.

United States National Academy of Sciences, National Research Council, Commission on Life Sciences, Board on Basic Biology, Committee on Mapping and Sequencing the Human Genome. (1988). *Mapping and sequencing the human genome*. Washington, DC: National Academy Press.

Uzan, M., Brody, E., and Favre R. (1990). Nucleotide sequence and control of transcription of the bacteriophage T4 motA regulatory gene. *Molecular Microbiology, 4*, 1487–1496.

VanBogelen, R. A., Hutton, M. E., and Neidhardt, F. C. (1990). Gene protein database of *Escherichia coli* K–12 (3rd ed.). *Electrophoresis, 11*, 1131–1166.

VanBogelen, R., and Neidhardt, F. C. (1990). Global systems approach to bacterial physiology: Protein responders to stress and starvation. *FEMS Microbiology and Ecology, 74*, 121–128.

VanBogelen, R. A., and Neidhardt, F. C. (1991). Gene-protein database of *Escherichia coli* K–12 (4th ed.). *Electrophoresis, 12*, 955–994.

VanBogelen, R. A., Sankar, P., Clark, R. L., Bogan, J. A., and Neidhardt, F. C. (1992). The gene-protein database of *Escherichia coli* K–12 (5th ed.). *Electrophoresis, 13*, 1014–1054.

Vingron, M., and Argos, P. (1989). A fast and sensitive multiple sequence alignment algorithm. *CABIOS, 5*(2), 115–121.

Vogt, G., and Argos, P. (1993). Profile sequence analysis and database searches on a transputer machine connected to a Macintosh computer. *Computer Applications in the Biosciences, 9*, 25–28.

Voit, E. O. (1991). *Canonical nonlinear modeling: S-system approach to understanding complexity*. New York: Van Nostrand Reinhold.

Voit, E. O., Irvine, D. H., and Savageau, M. A. (1990). *The user's guide to ESSYNS* (2nd ed.). Charleston: Medical University of South Carolina Press.

Voit, E. O., and Rust, P. F. (1992). Tutorial: S-system analysis of continuous univariate probability distributions. *Journal of Statistical Computation and Simulation, 42*, 187–249.

Von Heijne, G. (1983). Patterns of amino acids near signal-sequence cleavage sites. *European Journal of Biochemistry, 133,* 17–21.

Wachtershauser, G. (1988). Before enzymes and templates: Theory of surface metabolism. *Microbiolical Reviews, 52,* 452–484

Walker, J. E., Saraste, M., Runswick, M. J., and Gray, N. J. (1982). Distantly related sequences in the alpha-subunits and beta-subunits of ATP synthase, myosin, kinases and other ATP-requiring enzymes and a common nucleotide binding fold. *EMBO Journal, 1*(8), 945–951.

Wallace, J. C., and Henikoff, S. (1992). PATMAT: A searching and extraction program for sequence, pattern, and block queries and databases. *CABIOS, 8,* 249–254.

Walsh, J. M., and Walsh, A. K. (1939). *Plain English hand book.* Wichita, KS: McCormick-Mathers Publishing.

Waterman, M. S. (1983). Sequence alignments in the neighborhood of the optimum with general application to dynamic programming. *Proceedings of the National Academy of Sciences of the United States of America, 80,* 3123–3124.

Waterman, M. S., Arratia, R., and Galas, D. J. (1984). Pattern recognition in several sequences: Consensus and alignment. *Bulletin of Mathematical Biology, 46,* 515–527.

Waterman, M. S., and Byers, T. H. (1985). A dynamic programming algorithm to find all solutions in a neighborhood of the optimum. *Mathematical Biosciences, 77,* 179–188.

Waterman, M. S., and Eggert, M. (1987). A new algorithm for best subsequence alignments with application to tRNA-rRNA comparisons. *Journal of Molecular Biology, 197,* 723–728.

Waterman, M. S., and Jones, R. (1990). Consensus methods for DNA and protein sequence alignment. *Methods in Enzymology, 183,* 221–237.

Watson, J. D., Hopkins, N. H., Roberts, J. W., Steitz, J. A., and Weiner, A. M. (1992). *Molecular biology of the gene.* Menlo Park, CA: Benjamin/Cummins Publishing.

Webb, E. C. (1993). Enzyme nomenclature: A personal retrospective. *The FASEB Journal, 7,* 1192–1194.

Webster, T. A., Lathrop, R. H., and Smith, T. F. (1987). Prediction of a common structural domain in aminoacyl-tRNA synthetases through use of a new pattern-directed inference system. *Biochemistry, 26,* 6950–6957.

Weiss, V., Claverie-Martin, F., and Magasanik, B. (1992). Phosphorylation of nitrogen regulator I (NR$_I$) of *Escherichia coli* induces strong cooperative binding to DNA essential for the activation of transcription. *Proceedings of the National Academy of Sciences of the United States of America, 89,* 5088–5092.

Werbos, P. (1974). *Beyond regression: New tools for prediction and analysis in the behavioral sciences.* Unpublished doctoral thesis, Harvard University, Cambridge, MA.

White, J. V. (1988). Modeling and filtering for discretely valued time series. In J. C. Spall (Ed.), *Bayesian analysis of time series and dynamic models.* New York: Marcel Dekker.

White, J. V., Stultz, C. M., and Smith, T. F. (1994). Protein classification by stochastic modeling and optimal filtering of amino-acid sequences. *Mathematical Biosciences, 119,* 35–75.

Wierenga, R. K., Terpstra, P., and Hol, W. G. J. (1986). Prediction of the occurrence of the ADP-binding beta-alpha-beta fold in proteins, using an amino acid sequence fingerprint. *Journal of Molecular Biology, 187,* 101–107.

Wieschaus, E., and Sweeton, D. (1988). Requirements for x-linked zygotic gene activity during cellularization of early *Drosophila* embryos. *Development, 104,* 483–493.

Winston, P. H. (1992). *Artificial intelligence*. Reading, MA: Addison-Wesley.

Wolffe, A. P. (1994). Architectural transcription factors. *Science, 264,* 1100–1101.

Wolpert, L. (1969). Positional information and the spatial pattern of cellular differentiation. *Journal of Theoretical Biology, 25,* 1–47.

Wolpert, L. (1993). *The triumph of the embryo*. London: Academic Press.

Won, K. (1990). *Introduction to object-oriented databases*. Cambridge, MA: MIT Press.

Woodger, J. H. (1952). *Biology and language*. Cambridge, UK: Cambridge University Press.

Wulf, W. A., Shaw, M., Hilfinger, P. N., and Flon, L. (1981). *Fundamental structures of computer science*. Reading, MA: Addison-Wesley.

Wyman, J. (1964). Linked functions and reciprocal effects in hemoglobin: A second look. *Advances in Protein Chemistry, 19,* 223–286.

Yockey, H. P. (Ed.). (1992). *Information theory and molecular biology*. Cambridge, UK: Cambridge University Press.

Yura, T., Mori, H., Nagai, H., Nagata, T., Ishihama, A., Fujita, N., Isono, K., Mizobuchi, K., and Nakata, A. (1992). Systematic sequencing of the *Escherichia coli* genome: Analysis of the 0–2.4 min region. *Nucleic Acids Research, 20,* 3305–3308.

Zawel, L., and Reinberg, D. (1992). Advances in RNA polymerase II transcription. *Current Opinion in Cell Biology, 4,* 488–495.

Zajanckauskaite, A., Raudonikiene, A., and Nivinskas, R. (1994). Cloning and expression of genes from the genomic region between genes cd and 30 of bacteriophage T4. *Gene, 147,* 71–76.

Zhou, Q., Lieberman, P. M., Boyer, T. G., and Berk, A. J. (1992). Holo-TFIID supports transcriptional stimulation by diverse activators and from a TATA-less promoter. *Genes and Development, 6,* 1964–1974.

Zhu, Z.-Y., Sali, A., and Blundell, T. L. (1992). A variable gap penalty function and feature weights for protein 3-D structure comparison. *Protein Engineering, 5,* 43–51.

Zuker, M. (1991). Suboptimal sequence alignment in molecular biology. Alignment with error analysis. *Journal of Molecular Biology, 221,* 403–420.

Zumofen, G., Blumen, A., and Klafter, J. (1985). Concentration fluctuations in reaction kinetics. *Journal of Chemical Physics, 82,* 3198–3206.

Zurek, W. H. (1989). Thermodynamic cost of computation, algorithmic complexity and the information metric. *Nature, 341,* 119–124.

Contributors

Robert C. Berwick
Departments of Electrical Engineering and Computer Science
Massachusetts Institute of Technology
Cambridge, Massachusetts

Fred E. Cohen
Department of Medicine and Pharmaceutical Chemistry
University of California, San Francisco
San Francisco, California

Jack Cohen
Ecosystems Analysis and Management Unit
School of Biological Sciences
University of Warwick, UK

Julio Collado-Vides
Centro de Investigación sobre Fijación de Nitrógeno
Universidad Nacional Autónoma de México (UNAM)

Antoine Danchin
Institut Pasteur
Unité de Régulation de l'Expression Génétique
Département de Biochimie et Génétique Moléculaire
Paris, France

Leonard Guarente
Department of Biology
Massachusetts Institute of Technology
Cambridge, Massachusetts

Elizabeth Kutter
Evergreen State College
Olympia, Washington

Richard Lathrop
Department of Information and Computer Science
University of California
Irvine, California

Richard Lewontin
Museum of Comparative Zoology
Harvard University
Cambridge, Massachusetts

Boris Magasanik
Department of Biology
Massachusetts Institute of Technology
Cambridge, Massachusetts

Michael L. Mavrovouniotis
Department of Chemical Engineering, Technological Institute
Northwestern University
Evanston, Illinois

Frederick C. Neidhardt
Department of Microbiology and Immunology
University of Michigan
Ann Arbor, Michigan

Thomas Oehler
Department of Biology
Massachusetts Institute of Technology
Cambridge, Massachusetts

John Reinitz
Mt. Sinai School of Medicine
New York, New York

Sean H. Rice
Department of Biology
Yale University
New Haven, Connecticut

Robert J. Robbins
Fred Hutchinson Cancer Research Center
Seattle, Washington

Michael A. Savageau
Department of Microbiology and Immunology
The University of Michigan
Ann Arbor, Michigan

David H. Sharp
Los Alamos National Laboratory
Los Alamos, New Mexico

Temple F. Smith
Biomolecular Engineering Research Center
College of Engineering
Boston University
Boston, Massachusetts

René Thomas
Laboratoire de Génétique des procaryotes
Université Libre de Bruxelles
Brussels, Belgium

Index

bending proteins, 208, 209
biochemical pathways, 244–247
cells and phenotypes, 245–246
cellular role, 108
computational biology, 12
language, 281–283
mapping, 244–247
program, 245
protein interaction, 181–188
sequence analysis, 13–16, 92, 109
Shannon-Weaver model, 95–97
Turing's arguments, 108
universal model, 243–244
Dorsoventral axis, 298
Drosophila, gene circuits, 253
Dynamic programs, genomic analysis, 42, 93

EC numbers, 80
Electronic data publishing, 70. *See also*
    Databases
Engineering
  evolution as, 1–2
  genetic algorithms, 2
Entropy, 97–98. *See also* Shannon's entropy
Environment of organisms, 9–10
Enzyme kinetic formalism. *See also* Enzyme
    reactions
  basic assumptions, 115–118
  biochemical systems, 129–133
  dynamics, 121–122
  fractal kinetics, 124–125
  gene circuits, 133–140
  integrative biology, 115
  new types, 125–129
  physical context, 122–123
  steady state, 121
  systems strategy, 117
Enzyme nomenclature, 85, 86
Enzyme reactions. *See also* Metabolic
    pathways
  fractal kinetics, 124–125
  global rates, 229–231
  maximal rates, 228–229
  Michaelis-Menten paradigm, 120
  ordered mechanism, 228
  rate-limit interpretation, 226
  resource allocation, 232–234
  simulation and analysis, 212–213
Enzymes
  acetate metabolism, 162–163
  allosteric, 144n
  bioreaction stoichiometry, 227
  commission numbers, 79–81

dimerization, 123
nomenclature, 85–86
protein machines, 19
time plots, 85, 86
Epigenetic regulatory networks, 167–168
Escherichia coli
  chromosomal segment library, 151–153
  CRP, 183
  DNA sequence analysis, 13, 28
  genome expression, 147, 155–158
  global regulation, 147
  growth rate, 148, 149
  nitrogen regulation, 273
  protein spots, 151–155
  sigma 70 promoters, 182
  transcription regulation (*see* Bacterial
    transcription)
  2-D gels (*see* Protein spots on 2-D gels)
  2-D gel electrophoresis, 148–150
  Uspa, 302 (*see also* Universal stress protein)
  UspA induction, 160
Eukaryotes, 298
  algorithm complexity, 102–105
  building, 245–246
  definition problems, 74–75
  gene circuit complexity, 255
  promoters, 206, 207
Eukaryotic transcription, 205–209
  activators and coactivators, 206–209
  basal factors, 206
  enhancers, 206
Evolution
  Bennett's logical depth, 106–107
  as engineering, 1–2
  fallback positions (*see* Gene regulation)
  fitness surfaces, 5, 7–8
  gene regulation, 279
  genetic algorithms, 2
  of problems, 8–10
  protein, 106–107
  T-even phage, 20
  theory, 2
Evolutionary biology, 179
Exons and cistrons, 75
Expectation maximization, 286, 288
Experiments, controlling, 250

Factorial correspondence analysis (FCA),
    101–102
Fallback positions, gene regulation, 273–279
Federation concept, databases, 71, 82
Feedback loops, 168, 298. *See also* Regu-
    latory networks